Contents

		Page
Preface		v
Chapter 1	**Study skills and examination preparation**	6
Chapter 2	**Data I: Acquisition and organization**	11
	2.1 Introduction	11
	2.2 Sources of data	11
	2.3 Primary data	13
	2.4 Principles of sampling	14
	2.5 Methods of collecting data	18
	2.6 Questionnaires	19
	2.7 Data: accuracy, error and bias	21
Chapter 3	**Data II: Presentation**	26
	3.1 Introduction	26
	3.2 Classification of data	26
	3.3 Presentation of data: tables	27
	3.4 Pictorial representation of data	29
	3.5 Presentation of data: graphs	36
	3.6 Graphs showing data changes over time	39
Chapter 4	**Descriptive statistics**	48
	4.1 Introduction	48
	4.2 Measures of centre	49
	4.3 Measures of dispersion	55
	4.4 Measures of skewness	61
	4.5 Index numbers	62
	4.6 Practical index numbers and their uses	69
Chapter 5	**Time series analysis**	77
	5.1 Introduction	77
	5.2 Patterns or components of a time series	78
	5.3 Calculating the trend	80
	5.4 Calculating the seasonal variation	84
	5.5 Forecasting by extrapolation	86
	5.6 Time series: multiplicative model	88
	5.7 Forecasting: caution needed!	89
	5.8 Deseasonalizing data	90
Chapter 6	**Linear regression and correlation**	96
	6.1 Introduction	96
	6.2 The scatter diagram	96
	6.3 Independent and dependent variables	96
	6.4 The line of best fit I	98
	6.5 Interpolation and extrapolation	99
	6.6 Correlation	100
	6.7 The line of best fit II: regression	105
	6.8 Correlation and regression: summary	107
Chapter 7	**Probability: theory and practical applications**	112
	7.1 Introduction	112
	7.2 Probability: the basic rules	113
	7.3 Basic decision criteria	117
	7.4 Expected values	118

7.5 Tree diagrams 119
7.6 Prior and posterior probabilities 120
7.7 Multi-stage decision analysis 121
7.8 Permutations and combinations 123
7.9 The binomial distribution 124
7.10 The Poisson distribution 126
7.11 The normal distribution 128

Chapter 8 **Sampling theory** 140
8.1 Introduction 140
8.2 The Central Limit Theorem 140
8.3 Point estimates 142
8.4 Confidence intervals 142
8.5 Sample size 143
8.6 Estimating population variance, σ^2: Bessel's correction factor 144
8.7 Hypothesis testing: testing a sample mean 145
8.8 Sampling error 147
8.9 Small samples: *t*-distributions 149

Chapter 9 **Introducing business mathematics** 154
9.1 Introduction 154
9.2 Number patterns: sequences, series and progressions 154
9.3 Linear equations 157
9.4 Simultaneous equations 157
9.5 Quadratic equations 158
9.6 Matrix arithmetic and algebra 160
9.7 Calculus in a business context: measure of change 165
9.8 Rounding of data 172

Chapter 10 **The time value of money and financial appraisal** 181
10.1 Introduction 181
10.2 Simple interest 181
10.3 Compound interest 182
10.4 Increasing/decreasing the sum invested 183
10.5 Depreciation including the sinking fund formula 183
10.6 Present values and discounting 185
10.7 Investment appraisal 187

Chapter 11 **Linear programming** 198
11.1 Introduction 198
11.2 Formulation of the linear program 199
11.3 Graphical solution of linear programs 200
11.4 Graphic solution: summary 203
11.5 Non-standard problems and sensitivity analysis 205

Chapter 12 **Stock control** 215
12.1 Introduction 215
12.2 Simple stock control: assumptions 215
12.3 The EOQ or EBQ stock control model 215
12.4 Stock control with discounts: fixed holding costs 218
12.5 Stock control with discounts: variable holding costs 219
12.6 Continuous replenishment 220
12.7 Sensitivity analysis 222

Appendices
1 Random numbers 227
2 Cumulative binomial probabilities 228
3 Cumulative Poisson probabilities 232
4 Areas in the tail of the normal distribution 236
5 Percentage points of the normal distribution 237
6 Percentage points of the *t* distribution 237
7 Present value factors 238

Glossary 240
Index 245

Preface

The aim of this book is to provide a base in business mathematics and statistics, for students in their first year at university or college studying a business, accounting or economics related course. It is also suitable for students taking the various professional examinations and contains numerous examples of typical examination questions. BTEC students will find that the material gives more than adequate coverage of the quantitative aspects of their course, and should find the book a useful source of reference.

It is intended that the book should provide easy access to each topic area. Students who are new to the subject areas covered in quantitative analysis, and those who have some familiarity, should find the emphasis on problem solving a useful and rewarding way of learning about the sometimes complex areas of mathematics and statistics.

The book is not an examination crammer, nor a collection of examination questions. Students should not be led into believing that they can succeed in mathematical subject areas by merely practising a few questions. It is necessary, also, to understand the underlying concepts in each topic area. This book balances the explanation of relevant concepts with providing a facility for practising the type of questions likely to be encountered in an examination situation.

I am grateful for the permission received for the reproduction of past examination questions from the following:

The Chartered Association of Certified Accountants (ACCA)
The Chartered Institute of Management Accountants (CIMA)
The Institute of Chartered Secretaries and Administrators (CIS)
The Chartered Institute of Public Finance and Accountancy (CIPFA)

Each question used is cross-referenced to the appropriate Institute or Association.

This book is dedicated to Elaine, Emily and Sam.

1 Study skills and examination preparation

It is rare for study skills to be a formal part of a course, so you must take on the responsibility for this aspect of your study yourself.

The development of study skills will enable you to learn faster and for the learning to be more efficient. Central to good study skills is the desire to study; you must have the correct attitude and want to study. If you have this, you have half the battle won.

1.1 Preliminaries

1.1.1 The syllabus

Make sure you know what your syllabus is. If you have not already done so, get a copy. Do not rely on a copy acquired from a friend who studied the same paper five years ago or the copy borrowed from your local library, unless you can be confident that it is up to date. Examining bodies are constantly revising syllabuses.

Make sure you have an up-to-date syllabus. Once acquired, it should be placed at the beginning of your file. This will provide you with a readily accessible checklist for your progress.

As your study progresses, cross-check between your notes and the syllabus to monitor the coverage of the necessary topics. Syllabuses often provide an idea of the depth of the study required. It could help you to avoid doing too much as well as ensuring you cover the necessary material to the required depth.

1.1.2 A calculator

Make sure you get a calculator and make sure it will be acceptable. Many examining boards will not allow you to use programmable calculators in examinations.

A calculator is not intended to work things out for you, so make sure you know how to use it and, when using it, check that the results you are obtaining are what you expect to get.

1.2 Plan your study time

You won't want to work all the time, but it is a good idea if you plan when you intend to work.

If you are studying at college, then your plan will be an extension of your existing timetable. If you have to fit in study around full-time employment, never an easy thing to do, it is even more important to plan your study periods carefully. Spread your work throughout the week and, just as important, give yourself time which is free from work. It is of little use to sit down to study and find you are so tired that all you can do is sit and stare at your notes. It is important to find time to do the things you like, and perhaps your studies will also become something you like doing.

1.2.1 Early bird or late bird?

It is a matter of personal preference when you decide to study. Some people find early in the morning is the best time, some prefer the quiet and peace of the night. Whichever you choose, make sure it feels natural. Working when you are tired is not efficient. You will receive no reward for sitting asleep at your desk or clocking up hundreds of hours work if at the end of the period you have been unable to learn anything.

Most people find that a period of about three hours with short breaks every hour is the most efficient way to work. During your breaks, leave your books to make a cup of tea or coffee.

1.2.2 Where to study

The choices open to you will depend on your circumstances. If it is possible, you should arrange to have a room or part of a room set aside specifically for your study. This will enable you to keep your books and notes in one place. Also, having a location which is fixed in your mind (and in that of others!) as your work area will make it easier to study.

It should be free from distractions; it is impossible to work well and watch television at the same time.

It is also important that you like the work space you choose for yourself and that it is comfortable, well-ventilated, etc. If you feel uncomfortable, too hot or too cold, you will not be able to work well, so take time to make the room feel right.

You will need a table, a place for your pens, books, etc and a reading lamp.

Having your books, notes, etc readily accessible will save time and start you off on each work session in the right frame of mind.

1.3 How to study

If you have any experience at all of studying mathematical subjects, you will be aware that it is impossible to learn mathematics by simply reading books. It is necessary to do it. Your study periods should be spent reviewing particular topics, perhaps re-writing your lecture notes or parts of them, and attempting associated problems. This book is designed to be used in just that way.

1.4 Making notes

Most students under-use their notes as a study resource. Their importance to your study cannot be too highly stressed. If well-recorded, notes provide an invaluable aid to learning and, in particular, to revision for examinations. It should be recognized that the notes you take are for your benefit only. If they are merely copies of someone else's notes, whether the lecturer's or a friend's, they are not serving the purpose for which they are intended.

To make sure your notes are in a format that you understand would be a good way to start each study session.

When referring to this and other books, you should get into the habit of making notes and working through the problems that you encounter as you study.

It is worth repeating that mathematics will not be learnt by reading alone. You must do mathematics.

1.5 Variety of study methods

Keeping notes in different formats is worth considering. For example, any formulae you feel it is necessary to remember could be recorded on index cards; summaries of background information can be recorded in a small notebook which could then be referred to while travelling to college or work.

Reference to other textbooks is an invaluable method of seeing a topic presented from a different viewpoint.

Record your difficulties as well as your successes. The sooner you do this the sooner you can refer to someone for help, and consequently bring your notes up to date. It should be clear that your notes form the focal point for all your study and should be looked upon as your most important reference and not as a collection of disjointed writings, to be looked at from time to time.

1.6 Do you want to be alone?

While there will be times when you have to spend time working alone, you should not waste one of the most valuable learning resources you have available, your colleagues.

Because mathematics is an intellectually demanding discipline, the opportunity to discuss problems and share approaches to problem solving, another central factor of learning mathematics, should prove very rewarding. It could widen your social contacts as well.

As long as you exercise reasonable discipline, group study sessions with two people or more can prove to be a very useful way of working. If organized properly, the ambiance created by a number of people tackling the same problem is very productive.

The way in which you, as a group, can best benefit is to agree certain topic areas and arrange to bring a supply of problems and/or textbooks for reference.

In practice, there is no limit to how you can organize group sessions. You might find it convenient to restrict group work sessions to your college library. A variety of study approaches should be tried. You may find they are all suitable in different ways, or that the only way for you is to work alone.

1.7 Revision techniques

Having spent the whole year (or years) working hard, keeping your notes up to date, following up problem areas at the earliest opportunity, attempting and handing in assignments in plenty of time, etc, etc, you must face the particular problems of how to organize your examination revision.

Strictly speaking, revision should be taking place all the time. Do not be misled into thinking that revision is something you do at the last moment. Mathematics is a complex and demanding subject and it requires your learning to be gradual and for you to spend time practising the necessary skills and consolidating what you have learnt as you proceed. There is too much to learn for it to be successfully done in a few weeks.

It has been shown that, for recall to remain efficient, frequent revision is necessary.

1.7.1 Final revision: three months to go

However, as examinations approach your revision will, of necessity, become more concentrated. Your final revision programme should start about three months before the examinations. Your study timetable should be replaced by a revision timetable.

As you planned your study time, you must now plan your revision time.

At this stage you should be reviewing work. Your notes, if they have been properly compiled, will prove an invaluable source of reference.

1.7.2 Examination format

Familiarize yourself with the format of your examination paper: does it involve multiple choice questions? If so, find out how it is marked. Not all multiple choice papers work on a simple '1 mark for correct', '0 marks for incorrect' principle.

Practise questions you have done before, without reference to the solutions.

Practise past examination questions. Check that they are still relevant. If the syllabus has changed, the type of examination paper and/or examination questions may have changed.

Make notes on pocket-sized 'revision cards' for use when travelling. Make charts of important factors to pin to your study wall. Test yourself frequently. Arrange with a colleague to test each other. Both aspects of testing are very useful.

While there will be topics you feel more confident about than others, do not rely too heavily on certain questions 'coming up'. That is, do not rely on 'tips'. Try not to leave significant gaps in your revision.

1.7.3 Time-constrained tests

Examinations are periods of pressure and one aspect of them which is particularly pressurizing is that they are time-constrained. Unless you have taken many examinations you may not be used to the particular demands examinations make of you.

To prepare yourself, arrange to give yourself trial examinations and/or trial examination questions, under full examination conditions. This means starting at a pre-arranged time and, without reference to notes, attempting the required number and type of questions. It would be ideal to use a past paper that you had saved for such an occasion. This not only prepares you for the types of pressure you will encounter, it also enables you to build up examination speed. Many candidates fail because they 'run out of time'!

1.8 The examination: nearly there

It is not unusual to suffer from examination nerves or to feel everything has gone out of your head. Nothing written here will stop those feelings, but it may help you to deal with them more successfully.

The most important factor is to know you have done everything possible in preparing for the day.

1.8.1 Basic preparation

Make sure you have all the necessary basic information which, though it is not concerned with the academic aspect of your study, is important to ensure you are properly prepared for the day.

1 Make sure you know the date, day, time and location of each paper.

2 Keep a note of your examination number, the examination centre number and the telephone number of the examination centre or of anyone you would need to contact in an emergency.

3 If the examination hall location is unfamiliar to you, practise getting there, or at least be aware of how to get there in plenty of time.

Make sure that any equipment you may want to use is suitable and in good condition.

1 Do not change your pens at the last moment. Use ones that you are used to and that are comfortable to use.

2 Pencils (already sharpened), a sharpener and rubber.

3 Drawing instruments: ruler, set square, compasses, etc.

4 Check that your calculator is in good order and provide yourself with replacement batteries, which you should practise replacing. You will, of course, have already checked that your calculator is one that is allowed!

5 A watch or clock.

1.8.2 The day before

Relax! If you have missed something out, it's too late now. What you require is rest.

Check your equipment is ready and that you know where you are going and how to get there.

Don't be one of those looking bleary-eyed on the morning of the exam. Try to get to bed early.

1.8.3 The big day

Get an early start; there's nothing worse than arriving late and knowing that you've lost time before you even start and that you have probably missed a vital announcement from the invigilators.

Assuming that you have arrived in plenty of time and that you have not forgotten anything vital, such as your calculator, sit quietly and listen to what the invigilators have to say. It could be very important, an amendment to the paper, for instance. Fill in your candidate number, centre number, etc, when instructed to do so by the invigilators.

1.8.3.1 Read the instructions

When instructed to start, take some time to check a number of things.

1 Is it the correct paper?

2 How many sections are there to the paper?

3 Are there compulsory questions?

4 Is there any choice of questions?

5 How many questions do you need to do?

6 Are there any special instructions regarding the presentation of questions. For instance, BTEC questions tend to demand a particular format for solutions. Some questions may demand sketches, while others may want detailed graphs. All of this should be familiar to you before the examination. Nevertheless, you should check.

7 Read through the whole paper. Decide which questions you are going to do. Check that your choice does not run counter to the instructions. Cross out the ones you do not intend to attempt. Decide in which order you are going to attempt them.

8 Check how much time can be spent on each question. This is not always straightforward. Some questions are awarded more marks than others. You should

come prepared with this information, but do not rely on it. Use the information on the examination paper in front of you.

1.8.3.2 Answering the questions

Read the question through again to check you understand what is required. Note down any formulae, or particular points you want to remember and that you think you might need. You are ready to start.

1 Make sure your writing is legible.

2 If a question asks for a diagram or a graph, make sure it is properly and clearly labelled. If units are required don't forget to include them.

3 Set your answer out in a logical manner, numbering the steps if appropriate.

4 If using a calculator, make a rough calculation to check that your calculator result is what you would expect it to be and note, where possible, the steps of the calculation. If you make a mistake you must give the examiner some idea of whether your methodology was correct. Marks are often given for the workings, not only for the correct answer.

5 Only provide the information asked for. If proofs are not asked for do not waste time doing them.

6 Use your time appropriately. If part (a) of a question carries only 5 marks and part (b) carries 12 marks, make sure your answer reflects the time spent on those parts.

7 Make sure you answer the question as it is asked, not as you would like to have it asked.

8 If you come unstuck with a question, read through it again to check that you have not missed some information or a clue.

9 When you have completed a question, check through it to see that nothing has been missed out.

10 Check to see that you are keeping to your schedule.

11 Be careful about answering too many questions. Some examiners will mark all the questions and ignore the worst, others ignore the last question.

1.8.3.3 When you have finished

1 Allow some time at the end of the examination for checking your paper.

2 Check that the title page is correctly completed. Do not forget to include your name and/or examination number.

3 Make sure any loose pages are attached to the main answer book and are identifiable with your name and/or examination number, however instructed.

4 Check that you have numbered your answers correctly.

To summarize:

1 Read the paper carefully.

2 Think about the questions you have to do and/or you are choosing.

3 Allocate time appropriately; don't be tempted to overrun. You can always return to a question if you have time.

4 Don't panic and don't expect to get 100 per cent. The aim is to pass.

If you've worked consistently throughout the year and you've planned your study time and your revision time well, then you will have done all you can. Good luck in your examinations. You probably won't need it.

Recommended reading

De Leeuw, M., and De Leeuw, E., *Read Better, Read Faster* (Penguin Books, 1986).

Good, M., and South, C., *In the Know, 8 Keys to Successful Learning* (BBC Books, 1988).

Chart Foulkes Lynch, *100 Questions & Answers: Statistics and Quantitative Methods* (Holt, Rinehart & Winston, 1984).

2 Data I: Acquisition and organization

2.1 Introduction

There are many ways in which statistics can be applied in the business world. A common usage is the collection of information or data concerning a business, in order to analyse the present position and/or to make decisions concerning the future of the business. This chapter is concerned with the stages involved in acquiring and organizing the information or data (the analysis of data will be considered in later chapters). The stages range from identification of objectives, before data is collected, to presentation of the collected data, after data has been collected.

2.1.1 Examination questions

This area of the syllabus commonly prompts questions of a discursive nature. The questions are easy as long as you are completely familiar with the terminology. The question may be general, requiring knowledge of a range of topics within this particular area, or it may relate to a particular topic, requiring you to apply your knowledge of a number of aspects of that topic.

2.1.2 Areas to be studied

We will consider the following stages:

1 **Definition of objectives:** We need to be aware of precisely what is required from the investigation; mistakes in identifying objectives could prove costly and/or time-consuming.

2 **Definition of units:** Production data may be required in terms of number of items produced or the amount of raw material used.

3 **Sources of data:** It is necessary to identify where information can be obtained. This involves identifying a population (see later) and establishing whether the data we require already exists, as the result of some other investigation (this, from our point of view, would be termed secondary data; see section 2.2.1), or does not exist in a satisfactory state, making it necessary for us to carry out our own research (consequently acquiring primary data; see section 2.3).

4 **Accuracy of data:** A high degree of accuracy may not be required, for example, it may be necessary only to know that sales have increased by approximately 200 000 and not the actual increase of 219 456 items. On the other hand it is important to be aware of the potential for inaccuracy in data collected, because of random error or some inherent bias in the collecting process.

5 **Processing of data:** Once collected, the data needs to be classified or grouped together in some coherent manner, to facilitate analysis.

6 **Presentation of data:** It is often worth presenting data in some diagrammatic form to provide an overview or to draw attention to certain aspects of the information, for example skewness; see section 4.4.

In the rest of this chapter, consideration will be given to the stages involved in a statistical investigation up to the point of presenting the data in the form of tables, charts, diagrams, etc.

2.2 Sources of data

Once our objectives have been established, one of the first steps is to check whether statistics exist which satisfy these objectives. As stated above, data can be either primary data (data used for the specific purpose for which it was collected) or secondary data (data used by someone other than the compilers for a purpose for which it was not originally intended). Secondary data is a relative concept. To the original compilers the data was primary.

2.2.1 Secondary data

In referring to a secondary source of data, care must be taken to ensure that it provides the information which is being sought. Suppose we wish to investigate the types of occupation which have wage rates similar to the wage rates of middle managers employed in the chemical industry. If a publication uses the term 'middle management' we must be sure it is used in the same sense as we are using it in our investigation. If a publication gives statistics on earnings we must be sure it includes all occupations we are interested in. Secondary data, whether published by the government or some other organization, can save time and give valuable information; but we must be sure that it is appropriate.

Secondary data exists in many forms. In the example above we may be able to use existing data from sources within the chemical industry or from published sources such as *Economic Trends*, published monthly by the Central Statistical Office (CSO), or the *Department of Employment Gazette*, published monthly by the Department of Employment. Both publications contain statistics on earnings. There is a wide range of published economic, business and accounting data, and to expect to be familiar with it all is impracticable. Nevertheless, it is advisable to know something about some of the more important sources of published statistics and their content.

The Central Statistical Office (CSO) coordinates the United Kingdom's statistical service, while the European Community has the Statistical Office of the European Community (SOEC) which plays a similar role for member countries of the EC. Statistics on the world economy are published in, for example, the *United Nations: Statistical Yearbook*. If you contact these organizations it is often possible to obtain information about their publications and useful summaries of data they provide; some of these are free.

2.2.2 Publications

It is not possible to list the full range of publications, but some of the more important ones are as follows:

1 *Annual Abstract of Statistics:* This is a useful general reference book, published by the CSO. It contains most government statistics of economic and business interest. It contains mainly annual data and is an abbreviated version of the *Monthly Digest of Statistics.*

2 *Monthly Digest of Statistics:* This is published monthly and contains more detailed information of the variables contained in the *Annual Abstract.*

3 *Department of Employment Gazette:* This is an important monthly publication containing statistics on labour and on prices.

4 *Trade and Industry:* This is a weekly publication compiled by the Department of Trade and Industry. It includes the Wholesale Prices Index and other indices measuring industrial production and imports.

5 *National Income and Expenditure Blue Book:* Published by the CSO, this provides details of:

(a) Gross National Product, analysed in various sections such as transport and communication;

(b) Gross National Income, analysed by reference to the self-employed, the employed, company profits, etc;

(c) Gross National Expenditure, analysed into sections relating to capital expenditure, consumer expenditure, imports, etc.

6 *Economic Trends:* Published monthly by the CSO, this provides statistics and graphs on topics such as: production, investment, labour, external trade, prices, and wages.

7 The *FT–Actuaries All-Share Index* is compiled by the *Financial Times*, the Institute of Actuaries in London, and the Faculty of Actuaries in Edinburgh. This is an index

of share prices as quoted on the Stock Exchange. It is made up of 750 different types of ordinary shares.

For examination purposes it is important to be aware of the wealth of published statistics (not all mentioned here), the main sources of published statistics and the content of some of the more important ones.

2.3 Primary data

Suppose precise objectives have been defined, and sources of secondary data have been investigated and none found to be appropriate. At this stage the investigator is faced with the problem of acquiring his/her own (therefore 'primary') data.

2.3.1 Quantitative and qualitative data

Before considering the preparations to be made and the methods to be employed in collecting data, we need to consider the nature of the data likely to be encountered. It may be of several types. It is either qualitative or quantitative.

Qualitative data reflects some quality of what is being observed. Suppose a manufacturer has available packaging in red, yellow and blue and wishes to test the customer-appeal of the different colours of packaging used for a particular product. The investigation provides information about customer preference, which is a qualitative property of the packaging.

Suppose the packages referred to above are assumed to contain 1 kg of the product. If the manufacturer wanted to test this assumption, he could take a number of packages and weigh them. He would be observing the variable (see below) 'weight of package', and the resulting data would be the observed weights of the chosen packages. Unlike qualitative data, quantitative data can be measured (in this case, weighed).

Qualitative data results from observing the attributes of a particular population, in this case 'colour preference'. Quantitative data is often the observed values of a particular variable, in this case 'weight of package'.

2.3.2 Discrete and continuous variables

It is possible to further divide data into that resulting from observing a discrete variable and that resulting from observing a continuous variable.

2.3.2.1 Discrete variables

Discrete variables are variables which can only take a finite or countable number of values in a given range. 'The number of people in a family', 'shoe size' and 'people entering Toy Town Bank on a particular day', are examples of discrete variables. Within the (arbitrarily chosen) range [3, 7] it is possible to observe the data, '4 people in a family' but impossible to observe '4.5 people in a family'. It is possible to observe a '(British) shoe size' of 4½ but not 4⅓. It is possible to observe 6 people entering Toy Town Bank on Monday, but not 6.25 people.

2.3.2.2 Continuous variables

Continuous variables are variables which may take on any value in a given range. Height in metres or the output of a company's production in kilograms, are examples of continuous variables. The output of a company may be measured to the nearest 1000 kg, giving a value of, say, 4000 kg. On the other hand, there is no reason why output should not be measured to the nearest kilogram giving a value of 4475 kg. It depends on the requirements of the measurer.

2.3.3 The population and sampling frame

Once it is decided that primary data is required, the investigator needs to define the relevant population.

2.3.3.1 Population

The population is the group of people or objects of interest to the investigator. For instance, if the investigator is concerned with the level of service British Telecom provides for its customers, then the relevant population will be all British Telecom telephone users. Alternatively, if the investigator is concerned with the views of the 'top users' of British Telecom services, it is necessary to know the population of 'top users'. It is unclear who would be members of the relevant population, unless the term 'top users' is clearly defined. Suppose the 'top users' are defined as the group of users represented by the Telecom Managers Association (TMA); the TMA represents the top 300 users of British Telecom who provide approximately 30 per cent of its profits. The membership of the TMA would be a suitable population. Once the population is known, it is necessary to be able to define its members. The members of the population of customers of British Telecom is defined by the entries in the telephone directories (ex-directory users would need to be identified separately for completeness). Such a list is called a sampling frame.

2.3.3.2 Sampling frame

The sampling frame is a list of the members of the population. Examples are:

1 Stock items which can be identified from the input on the relevant inventory file.

2 Students at a college who can be identified from their enrolment forms.

3 Employees who can be identified from personnel forms.

4 The electoral register.

It is worth noting that many populations do not have a readily available sampling frame. Examples are items coming off a production line or the customers of a restaurant. The method employed to collect data often depends upon the availability of a sampling frame.

2.4 Principles of sampling

2.4.1 Census or exhaustive sampling

In carrying out a survey it seems reasonable to begin by considering the possibility of testing every member of the population. A survey which examines every member of a population of people or households is called a census. A survey which examines every member of a population of objects, such as components produced by a factory process, is an exhaustive sample. If a firm wished to introduce a new bonus scheme, it might judge it prudent to survey all of its workforce; this would be a census. The Government Statistical Service carries out a population census every 10 years, gathering information such as sex, age, relationship to head of household, occupation, number of rooms in the dwelling, etc. This is carried out for the whole United Kingdom. The Government is similarly responsible for other business-related censuses.

In most situations it is not practicable to carry out a census. Also, it is worth noting that it may not be the best thing to attempt.

1 It can be shown mathematically that after a certain sample size has been collected there is very little value in testing further members of the population.

2 A census may require the use of many investigators making it necessary to employ semi-skilled personnel, which can result in loss of accuracy in collected data.

3 If the investigation is restricted to a sample of the population then it is possible to test more aspects of the population.

4 The costs involved in carrying out a census or exhaustive sample may far outweigh any benefits to be gained from the data collected.

2.4.2 Samples

A sample is a subset of the population selected to represent the population. Sampling is one of the most important areas in statistics. It demands careful planning, which does not end once the decision is made to take a sample rather than a census. There are many types of sample, ranging from simple random samples to stratified quota samples. The type used depends on the particular circumstances of the investigation.

The sample should have the same characteristics as the population from which it is chosen.

Sometimes a sample may be selected in such a way as to introduce sampling bias.

2.4.2.1 Sampling bias

A sample is biased if not all members of the population have an equal chance of being chosen. For example, suppose the government were considering replacing the present voting system with proportional representation, and they decided to seek the views of the members of the electorate. If the sample chosen were limited to voters in the south-east of England, it would not be representative of the whole (voting) population and a biased result would be obtained. If a bias-free sample is required, then random sampling techniques must be used.

2.4.2.2 Using random number tables

Before considering the methods one can employ in sampling it is worth being aware of the way random numbers are used in selection procedures.

Random sampling numbers consist of the digits 0 to 9 arranged in groups for reading convenience. They are generated randomly, usually using a computer. This can be taken to mean that any digit is as likely as any other digit to occur in a particular position. A table of random numbers is given in appendix 1. A section is reproduced here:

28 89 65 87 08	13 50 63 04 23	25 47 57 91 13	52 ...
30 29 43 65 42	78 66 28 55 80	47 46 41 90 08	55 ...
95 74 62 60 53	51 57 32 22 27	12 72 72 27 77	44 ...

Suppose we wished to take a sample of 4 from a population consisting of 800 items. As long as it was possible to number the items (0 to 799) we could choose a sample using the numbers above grouped in threes (reading from left to right). The sample would be chosen as follows:

sample (size 4)	ignored (numbers > 799)
288	965
506	870
304	813
232	

(See section 2.4.3.1)

2.4.3 Types of sample

2.4.3.1 Simple random sample

A simple random sample is chosen in such a way that every member of the population has an equal chance of being chosen. To choose a random sample it must be possible to list and number the members of the population, i.e. it must be possible to define a sampling frame. Once the population is numbered, random numbers can be used as described above to select a random sample. However, it will not necessarily produce a perfectly random sample, since some sections of the population may (by chance) have been omitted. Consequently, it is often decided that random sampling is not the best method to employ. There are a number of quasi-random or non-random sampling methods which satisfy the need to have a representative and bias-free sample. They are dealt with in sections 2.4.3.2 to 2.4.3.6.

2.4.3.2 Systematic sample (quasi-random)

Every *n*th item is selected after a random start. Consider the selection of a sample of size 25 from a population of 1000: after a random start, every 40th (1000 ÷ 25) item would be chosen. So, if 28 was chosen as the start, the sample would consist of the 28th, 68th, 108th, ... and 988th items.

The gap of 40 between each sample is the **sampling interval**. The advantage over simple random sampling is that it allows the investigator to move systematically through the population choosing the sample, rather than having to search for randomly selected parts of the population. The investigator must be wary of patterns within the data which could result in a biased sample. Consider the following situation: Five machines, A to E, are producing identical products at the same rate and these are passed onto a single conveyor where they will be sampled. It is decided to sample using a systematic approach with a sampling interval of 25. It could happen that the items on the conveyor form sets of five in the same sequence as they are produced by the machines. With a sampling interval of 25 the sample chosen will contain items all produced by the same machine, resulting in a biased sample. There are ways of avoiding such situations, such as choosing more than one starting point, each randomly selected.

2.4.3.3 Stratified sample (quasi-random)

To obtain a sample that is truly representative it is often necessary to identify sections or strata within the population that will together reflect the population as a whole. For example, in seeking the views of the electorate it is possible to identify certain characteristics of the population which will enable all sections of the population to have their views represented. A simplified stratification is as follows:

sex: male/female	two strata
age: 18–30 and 30+	two strata
geographical area: A, B and C	three strata

Table 2.1 Population breakdown: by sex by age by geographical area

Sex	Size of stratum	Age	Size of stratum	Area	Size of stratum
Male	25 000	18–30	15 000	A	6 000
				B	5 000*
				C	4 000
		30+	10 000	A	4 000
				B	3 000
				C	3 000
Female	35 000	18–30	20 000	A	9 000
				B	6 000
				C	5 000
		30+	15 000	A	6 000
				B	5 000
				C	4 000
Population total					60 000

Information related to the size of each stratum would also be required. A simplified example (population size 60 000) is shown in table 2.1. The data in this table can be used to read the size of each stratum. For example (see the asterisked item), there are 5000 males in the age range 18–30 living in area B. Suppose the sample size we require is 300. The various sample stratum sizes can be calculated to provide information as illustrated in table 2.2.

Table 2.2 Sample strata sizes

Sex	Size of stratum	Age	Size of stratum	Area	Size of stratum
Male	125	18–30	75	A	30
				B	25*
				C	20
		30+	50	A	20
				B	15
				C	15
Female	175	18–30	100	A	45
				B	30
				C	25
		30+	75	A	30
				B	25
				C	20
Sample total (5% of population total 60 000)					300

To calculate the subsample sizes in each stratum it is necessary to refer to the corresponding stratum size in the population. For example, the asterisked item is found as follows:

The population stratum size is 5000 which is 0.083 (3 d.p.) of the total population 60 000 ($5000/60\,000 = 1/12 = 0.083$).

The corresponding sample stratum size is therefore 25:

$$\text{total sample size} \times 0.083 = 300 \times 0.083 = 25$$

The samples are chosen using a random selection or other appropriate process.

This method is particularly useful in situations which require many important categories to be covered by a relatively small overall sample. If a simple random sample were taken there would be a high risk of some categories being omitted, resulting in sample bias. But, as the strata represent all the views within the population, this bias is avoided. However, bias could still arise because of the subjective nature of choosing the strata.

2.4.3.4 Multi-stage sampling (quasi-random)

At times the geographical area to be surveyed is so large that it makes random sampling prohibitively expensive. One solution is to multi-stage sample, which involves dividing the area into sub-areas or regions, randomly selecting a small number of regions, and taking samples from these regions, ensuring that the samples are proportional to the population size.

The sampling method chosen, once the regions have been selected, can be random, systematic or stratified, depending on the nature of the investigation. The extent to which one divides and further subdivides depends on the circumstances. Consider the following: a national survey of mineral water consumption is to be carried out with a proposed sample size of 10 000. The country is divided into 100 areas, from which 10 are chosen at random and the sample is divided amongst them according to their population. From each of these areas 5 towns are chosen and subsamples again divided in proportion to the population. At this stage it may be necessary to stratify the samples: if 80 per cent of the population live in large towns and 20 per cent in villages, then, in choosing the 5 towns, 4 should be large and 1 should be a village to reflect the population as a whole. From each town/village 2 areas are chosen at random and the survey is carried out.

The main advantage of this method is one of cost – saving on travelling expenses, etc.

Disadvantages include:

1 possible bias if only a small number of regions are selected, and

2 the method is not truly random, as once the final sampling areas have been selected the rest of the population cannot be in the sample.

Ideally the population should be heterogeneous and the areas chosen should reflect this. If all of the areas chosen are significantly different from one another, then choosing some and excluding others (even if it is randomly done) will result in a bias sample.

The sampling methods considered so far have necessitated the existence of a sampling frame. It is often impossible to identify a satisfactory sampling frame and when this is the case other sampling methods have to be employed.

2.4.3.5 Cluster sampling (non-random)

This method involves selecting a small number of areas and sampling *all* the members of the target population. Suppose information is sought concerning the use of solar-powered cells in domestic heating. It is unlikely that the population of users could be readily identified. Likely areas could be selected and the residents surveyed to establish who falls into the target population: those using solar power for their domestic heating. Once identified, they can be interviewed.

Cluster sampling has the advantages that it is:

1 a good alternative to multi-stage sampling if a satisfactory sampling frame does not exist;

2 inexpensive to operate as little organization or structure is involved.

The main disadvantage is the potential for considerable bias because of the non-random method of choosing clusters.

2.4.3.6 Quota sampling (non-random)

Quota sampling is much used in market research; it involves interviewers being given a fixed number of subjects (their quota) to interview. The population is usually stratified and this is reflected in the interviewer's quota. This method allows the interviewer to choose the subjects (there is no preselection of the sample), which are specified according to a number of criteria, such as sex, age, social class, etc. The first two controls (sex and age) are easy to implement, but the third may prove to be more problematic. The problems inherent in this method extend beyond difficulties of identification of suitable subjects, which can be partly solved by careful and detailed definition of the controls. Bias can be introduced because of such matters as the location in which the interviewer chooses to base herself/himself for the purposes of the survey and the time of day the interviewer chooses to commence the survey (see disadvantages).

Advantages are:

1 low cost;
2 no non-responses, as each interviewer continues until his/her quota is complete;
3 stratified sampling is usually employed.

Disadvantages include:

1 sampling is non-random and considerable selection bias could result;
2 much depends on the skill of the interviewers and the use of semi-skilled personnel could result in serious interviewer bias.

2.4.4 Sample size

Whatever sampling method is employed, the size of sample selected is a very important aspect of the procedure. There is no universal rule to determine how large a sample should be, though certain special cases will be considered later in the text. General principles about the choice of sample size are:

1 the larger the sample size the more precise will be the results, the ultimate precision being gained from a census; and
2 after a certain sample size, there is little to be gained by further increasing the sample size.

In practice, the decision on how large a sample to aim for is determined by considering certain factors:

1 Money and time available.
2 Degree of precision required. If the aim of the survey is to obtain an approximate idea of customer reaction to a new product, then a small sample (which would therefore be imprecise) would do. On the other hand, if the intention was to ascertain the safety level of a new brake component for the car industry, it would be necessary to gain information that is as precise as possible.
3 Number of subsamples required. If the survey involves a complex stratified sampling procedure, then it will be necessary to have a large overall sample size to ensure each stratum is adequately represented. The problems related to sample size will be returned to later in the text.

2.5 Methods of collecting data

Having gone through the preliminary stages involved in an investigation, we need to decide how to collect the data needed to satisfy the specified objectives. Secondary data and its relevance have already been mentioned and involve referring to the appropriate documents. Whether secondary data is used or primary data is sought depends on the likely costs and the desired accuracy. We will now consider the collection of first-hand or primary data.

2.5.1 Observation

The observation method involves monitoring production line output or traffic movement, etc. It is accurate but can be labour-intensive. Also, its application is limited; it is possible to observe the quantity of an item purchased but not the customer's views on the quality of that item.

The following methods of data collection will tend to involve the use of questionnaires, the design of which will be considered in section 2.6.

2.5.2 Individual interview

An individual interview means a personal and structured interview, rather than the informal 'street interview'. It is probably the most expensive method, and is often resource-intensive, as it requires:

1 interviewers to be trained;
2 interviews to be arranged;
3 for an unbiased sample, a random sampling procedure to be employed in the selection procedure.

It tends to be questionnaire-based and is useful as the basis for a wider survey. It allows the 'professional' interviewer, where necessary, to guide the respondent. When used as a 'pre-survey' survey it is an example of a pilot survey and is useful in establishing the success or otherwise of one's survey planning (see 'Sample questions' at the end of this chapter).

2.5.3 Telephone and street interviews

Telephone and street interviews are widely used in all types of investigative work and are different from the personal interview described earlier in that they often have an in-built bias: it may or may not be important that one's sample only include telephone owners. Street interviews usually involve many interviewers, not all of whom are 'professional', and consequently one can lose the conciseness of the personal interview.

2.5.4 Postal questionnaires

Postal questionnaires are much cheaper than the personal interview method, primarily because the labour factor is much reduced (an expensive part of survey work). However, it is necessary to take much more care with the design of the questionnaire as, once the questionnaires have been distributed, it will be impossible, without some form of pilot study, to know if questions have been understood. There are advantages and disadvantages connected with this type of approach.

Advantages are:

1 the convenience of being able to reach a dispersed population;
2 not needing to make interview appointments.

Disadvantages are:

1 there is likely to be a poor response rate;
2 there is a risk that the questionnaire will be answered by an inappropriate respondent.

2.6 Questionnaires

Most of the methods discussed involve the use of questionnaires and the success or failure of an investigation can rest on the design of the questionnaire. If all the preliminary stages have been successfully completed, it makes sense to ensure that the survey is served by a well-designed questionnaire.

The questionnaire as a basis for data collection can be a powerful and useful tool. Unfortunately it can also prove to be a hindrance if the design, the compilation, the composition and the number of its questions, are not given adequate attention. At first it would seem a simple matter to list a number of questions referring to the area of investigation in which one is interested; but it requires careful planning and attention to some basic rules. These include:

1 Keep questions short to allow for quick and easy understanding. It could prove prohibitive for a respondent to have to 'unravel' a lengthy question before being able to offer an answer.
2 Keep the language simple. The use of unnecessarily complicated or 'technical' language could deter people from responding.
3 Avoid leading questions, particularly if it is the honest intention of the investigator to obtain unbiased data. The following example illustrates the different ways information can be sought:

   Question A: 'Which television channel do you feel offers the best choice of Saturday-night viewing?'
   (i) BBC1 (iii) ITV
   (ii) BBC2 (iv) Channel 4

Question B: 'Do you agree that Channel 4 offers unparalleled Saturday-night viewing?'
(i) Yes
(ii) No

Question B puts pressure on the respondent to agree, unlike Question A which allows the respondent to decide for him/herself.

4 Check that the information is available. It could prove to be time-wasting for the investigator, and distressing or aggravating for the respondent, if a question proves to be unanswerable because of lack of information. For example:

Question C: 'Do you think that government policy on monetary control will have a positive or negative influence on the quantity of M3 in the economy?
(i) Yes
(ii) No

Trained economists might have problems answering such a question; certainly the 'man in the street' is likely to experience difficulty.

5 Avoid sensitive issues. Sometimes the nature of the enquiry means that issues concerning a respondent's health, sex life or financial circumstances are a necessary part of the investigation. Even if this is the case, the questions should be carefully compiled and the respondent treated with sensitivity. In other circumstances questions of a sensitive nature should be avoided altogether or left until the end of the questionnaire, perhaps with an optional status. By the end of an interview a skilled interviewer will have been able to gain the confidence of the respondent and consequently such issues should prove to be less problematic.

6 Go for 'the whole truth.' 'Sensitive' issues are not just those mentioned above. For instance, individuals may not wish to reveal that they watch certain television programmes, believing such an admission would undermine their integrity. It is necessary to incorporate a checking procedure in the questionnaire where there is a risk of getting 'misinformation'. This is done by seeking information about the 'sensitive issue' in more than one part of the interview and incorporating a cross-checking procedure.

7 Get the order right! The order in which questions are presented can be vital. It is a useful technique to start with easy 'confidence-building' questions, enabling the questioner and respondent to establish a working relationship. It is also important to ask questions in a logical order:

Question D: 'Are you at present employed?'
(i) Yes
(ii) No

Question E 'Does your employment require you to work on Sunday?'
(i) One or more Sundays each month.
(ii) Less than one Sunday a month, e.g. one Sunday every three months.
(iii) Never.

It would be illogical to ask Question E before ascertaining whether the respondent is employed (Question D).

8 Get the facts! The questionnaire should contain an adequate number of questions in order to obtain the desired information.

9 Keep it brief! After all the above points have been adhered to, one still has to be careful not to make the questionnaire too lengthy. Respondents often want to be sure they will not have to spend hours answering questions before they will consider starting!

10 Carry out a pilot survey. There's no better way of checking the validity of something than by testing it. This is exactly what a pilot survey does: a small sample is chosen to test the various aspects of the questionnaire and any necessary changes are made as a result of the findings.

11 Be aware of the various types of questions one can ask. These can range from those requiring simple 'yes/no/don't know' responses to 'open-ended' questions.

(a) If it is required that respondents answer a question with a simple 'yes', 'no' or 'don't know', then it will be necessary to keep the question simple and free of ambiguity. Some questions appear straightforward but can cause difficulties. For example:

Question F: 'Do you think it is permissible for a company to extend its range of products?'

Yes No Don't know

There are many problems here:

(i) What company is being referred to?
(ii) What type of company is it?
(iii) What products are being referred to?
(iv) Are these products potentially dangerous?

The question needs to be changed quite a lot before it can be used in any serious sense. An improvement follows:

Question G: Hazchem Co Ltd has existed in the area for 20 years. Its record of safety is excellent; only one minor accident in the period. The company wishes to extend the range of its horticultural products; this will involve no new processes, but will require extra storage space. The local council is at present objecting to these proposals. Do you think the company should be allowed to go ahead with its plans?

Yes No Don't know

(b) Multiple choice questions give a range of possible answers to a given question. They provide a quick way to collect information, as the respondent only has to tick a box or make a choice. For example:

Question H: 'Why did you vote in the last general election?'

(i) I wished to exercise my democratic right.
(ii) I believed my vote was needed by the candidate I supported.
(iii) My husband/wife was voting, so I decided to do so too.
(iv) Other reason – please state.

Underline your answer.

This type of question design introduces some ambiguity: it is uncertain if one can choose more than one answer. Also, research has shown that the order of questions influences the respondent's choice, introducing more uncertainty about the validity of the resulting data.

(c) Open-ended questions leave a lot of responsibility to the respondent. They tend to prompt useful, but qualitative, data as the respondent is generally allowed to answer as he/she wants to. For example:

Question I: 'What is your opinion of the college canteen?'

2.7 Data: accuracy, error and bias

Whether it is a secondary source of data that is being considered or data collected as the result of one's own investigation, the way it is presented and the level of precision necessary has to be considered carefully. One must also check that one's interpretation of the information, so-presented, is the correct interpretation.

It is worth noting (particularly relating to later sections on measures of centre, etc) that absolute statistics are often meaningless unless compared to other statistics or to some predetermined measure. However, it is sometimes difficult to compare two sets of statistics because of a difference in units: for example, sales data might exist in 'number of units sold', while profits might be measured in monetary value. It is common in such cases to use percentages as a means of analysis. When used properly this is a very powerful method, but one needs to state clearly where one's data comes from and what the percentages are percentages of. The following example illustrates the potential for information to be misleading.

Consider table 2.3. The message is that Department B has convincingly outmatched Department A in generating profits for the company.

Table 2.3 Profits generated by two departments of Credit-On Co Ltd (1988 profits are given as percentages of 1987 performances)

Department A:	3%
Department B:	25%

Table 2.4 Profits generated by two departments of Credit-On Co Ltd (profits, in £000s, are given for 1987 and 1988)

	1987	1988	increase (%)
Department A:	250	257.5	3
Department B:	12	15	25

Table 2.5 Sales and profits generated by two departments of Credit-On Co Ltd (sales and profits, in £000s, are given for 1987 and 1988)

	1987		1988		increase (%)	
	Sales	Profits	Sales	Profits	Sales	Profits
Department A:	10 000	250	20 000	275.5	100	3
Department B:	150	12	200	15	33	25

Table 2.4 suggests we should look again. Even though Department A did 'only' increase profits by 3 per cent, it started from a much higher base. However, even that is not the full story, as table 2.5 reveals. Department A produced £257 500 profit from £20m worth of sales (1.29 per cent). By comparison, Department B produced £15 000 profit from £200 000 worth of sales (7.5 per cent). Clearly, before drawing conclusions about published statistics we should be sure we understand exactly what is being said, and whether the 'implied' information is acceptable. This is particularly true of data given as percentages.

Another source of error in 'reported' data is illustrated by the following example. The result of an auction held in the USA prompted the following report in a British newspaper: 'Impressionist painting sells for £156 250'. The painting had not been sold for any amount of pounds sterling, but had been reported by an American news agency as having been sold for 'about $250 000'. The British newspaper reporter had used an approximation of the current rate of exchange (£1 = $1.6) to convert the sum to pounds sterling for his British readership. The painting had actually sold for $250 600 and the exchange rate at the time was £1 = $1.6350. A more accurate sterling equivalent would have been £153 272.17, a difference of £2977.83, almost £3000.

Rounding of data, measurement errors and reference to bias and unbiased error will be considered in chapter 9. If you feel you need to revise basic mathematical background material, you should refer to chapter 9 before continuing.

The next chapter deals with the stages immediately following data acquisition, classification, frequency distributions, tables, graphs, etc.

Before moving on you should attempt the following past examination questions.

Sample questions

1 An accountant is sampling from a computer file. The first sample number is selected randomly and is item 41; the rest of the sample is selected automatically and comprises items 91, 141, 191, 241, 341, ...

This type of sample is termed:

A simple random **B** stratified **C** multi-stage **D** quota **E** systematic

CIMA; Foundation stage, Section B; Mathematics and statistics; May 1986.

Answer: E

2 (a) Sampling methods are frequently used for the collection of data. Explain the terms simple random sampling, stratified random sampling and sampling frame.

(b) Suggest a suitable sampling frame for each of the following in which statistical data will be collected:

(i) an investigation into the reactions of workers in a large factory to new proposals for shift working;

(ii) a survey of students at a college about the relevance and quality of the teaching for their professional examinations;

(iii) an enquiry into the use of home computers by schoolchildren in a large city.

(c) Explain briefly, with reasons, the type of sampling method you would recommend in each of the three situations given above.

CIMA; Foundation stage, Section B; Mathematics and statistics; May 1986.

Solution/hints

(a) See main text.

(b) (i) Company employment register or similar.
(ii) College registers or class lists of students on professional courses.
(iii) The city's school registers.

(c) Refer to text. Consider the costs involved, the convenience, the time factors, the reasons for referring to a sample rather than taking a census, etc.

3 The Health Ministry of your government wishes to know the percentage of people in the country who have a broken leg on, say, 1st June this year. A large random sample is to be chosen from one of the following five populations. Which population is likely to provide the least biased sample?

A Those on a (GP) doctor's list.
B Those shopping in a large department store.
C Those watching a football match.
D Those returning from a skiing holiday.
E The in-patients of a large general hospital.

CIMA; Foundation stage, Section B; Mathematics and statistics; May 1985.

Answer: A

4 A large organization has a sports and leisure complex (the facilities include tennis and squash courts, football, rugby and hockey pitches, gymnasium, sports hall and swimming pool) which is available for the use of members of staff, administrative, clerical and production, and their families, at a very low subscription.

The organization has most of its administrative and clerical staff located in the city centre; the production staff are located in three factories on the outskirts of the city. The sports and leisure complex is close to one of the factories but some distance from the other two factories. The management of the organization are concerned by the high costs of this complex and also that the facilities are very much under-used. The management are considering two alternatives: (i) selling the complex; (ii) retaining the complex but trying to encourage use by existing staff and/or making the facilities available to people other than staff and their families. Such people would have to pay an economic subscription.

The management has asked you to conduct two surveys: (i) to find whether employees currently use the complex and/or are likely to use the complex in the future; and (ii) to find out whether people other than staff and their families living in the area would wish to use the complex.

(a) Briefly explain how you would conduct BOTH surveys.

(b) Draw up a suitable questionnaire or recording schedule for use with the survey of EMPLOYEES.

CIS; Part 1; Quantitative studies; June 1987.

Solution/hints

Part (a)

(i) Staff survey: as there are distinct subgroups within the overall number of staff, each with distinct characteristics (most of the administrative staff are based in the city; the other staff are distributed around three factory sites, only one of which is close to the centre) it would seem appropriate to stratify the sampling frame, which in this case is likely to be the company's list of staff employed. The stratification would need to take into account location and type of employee, as these characteristics could influence the company's final decision. (It could be that distance from the centre is prohibitive and/or that only a particular type of staff member uses the centre. The latter could prompt the company to investigate the possibility of providing different facilities.) See 'Stratified samples' (section 2.4.3.3).

(ii) Survey of non-employees in the area of the centre: A ready-made sampling frame would probably not exist. The electoral register would not be suitable, as the centre would probably be suitable for use by young people not old enough to vote (and consequently not registered). A quota sample of a target group in the area would probably suffice. See 'Quota sampling' (section 2.4.3.6) and 'Stratified samples' (section 2.4.3.3).

In both cases suitable questionnaires would need to be designed. They could be posted (in the case of local residents and possibly to employees also) and/or distributed around the company's factories and offices (in seeking employee opinion).

Part (b)
See main text for examples and guidelines.

5 Briefly explain FOUR of the following:

(a) questionnaire design;
(b) multi-staged stratified sampling;
(c) non-response;
(d) quota sampling;
(e) sampling frame;
(f) interviewer bias.

CIS; Part 1; Quantitative studies; June 1986.

Solution/hints

Refer to main text. *Note*: You are asked to explain only four of the six items. Answering more would earn you no more marks.

6 Explain what is meant by:

(a) stratified sampling; (b) quota sampling; (c) cluster sampling;

supporting your answer by brief examples. Comment in each case on the advantage(s) and disadvantage(s) of each method of sampling.

The Polytechnic of North London; Accounting Foundation Course.

Solution/hints

Refer to main text.

7 Your company is anxious to monitor the use of the company carpark, which has parking space for 75 cars. It is aware that the number of employees is greater than this figure and is consequently considering the implementation of a special pass scheme. If this is done, it is believed that not all pass holders will want to (or be able to) use the car parking facilities every day: this may be because of illness, business commitments elsewhere, etc.

Required:
In order to advise on the number of passes that need to be issued, explain how you would:

(a) carry out the necessary survey; and
(b) analyse the resulting data.

The Polytechnic of North London; Accounting Foundation Course; 1984.

Solution/hints

Part (a) The objectives need to be clearly understood. A number of factors should be considered:

(i) the number of employees using the carpark each day. If, for example, there are days when it is not fully utilized, then these could be designated as 'No pass needed' days;

(ii) the identities of the users: if some employees only use the available facilities occasionally, then the company might consider this to be a good reason for not issuing a pass at all. Alternatively, some employees might be considered worthy of special treatment because of their status or the importance of their job.

Referring to the above (and any other factors you feel are relevant) you have to decide on how to carry out the survey.

Factors hinted at are:

- when to survey;
- who to survey; and
- what information to seek.

Methods which you should consider are:

- sampling *all* users for a predetermined period (one week, say);
- randomly selecting days to carry out the survey and randomly selecting a sample from the users.

Part (b) Having decided on the method of sampling, you should consider how the results could be used. In particular, what would be the effect on the precision of the results and/or the ease with which the survey could be conducted if you adopted your chosen method.

8 Comment on each of the following:

(a) Postal questionnaires are a cheap and reliable way of collecting data.

(b) One of the main problems with surveys using interviewers is the risk of interviewer bias.

(c) Writing a questionnaire is just a matter of listing the questions you want to ask.

(d) Carrying out a pilot survey is a waste of time and a waste of money.

The Polytechnic of North London; BA, Business studies; 1985.

Solution/hints

(a) Refer to main text (section 2.5.4).

(b) Bias as a result of the interviewer is certainly a risk in such surveys. However, appropriate training should be given to interviewers to minimize this. The type of bias that can occur include:

(i) misinterpretation on the part of the respondent because of some particular emphasis the interviewer places on asking certain questions;

(ii) sample bias as a result of an interviewer choosing (intentionally or otherwise) only a particular type of respondent, e.g. all male or all female.

(c) Refer to the main text.

(d) The view that pilot surveys are a waste of time and money is a very shortsighted one. Some surveys involve a big investment of time and money and to ensure that the planning has been well done it is worth investing a little more time (and money) in a pilot survey; it does not need to be particularly precise, as it is designed to be a testing mechanism. As long as a reasonable sample is chosen (to reflect the planned population) then it will usually save time and money in the long run.

9 Your fellow students have shown concern about the lack of available space to do private study. You have been asked to represent them in approaching the college management committee in order to press for some improvement in appropriate library space. Before you do this you want to be sure that you are representing a majority view, not just the feelings of a few 'complaining' individuals.

Required:

Describe how you would survey the students to gain the required information.

The Polytechnic of North London; BTEC HND, Business Studies; 1986.

Solution/hints

The intention here is to obtain a representative sample of student opinion relating to the problem of study areas. Before sampling, it may be necessary to divide the student population into different courses (see stratified sampling) in order to ensure a representative sample. If it is possible to obtain a list of all students (see sampling frame), this will prove a useful starting point in selecting a randomized sample (see random sampling). Alternatively, it may be decided that such a level of precision is not necessary and that it is sufficient to take random samples from a few groups. Having decided on that, a very simple questionnaire is all that is required. Nevertheless attention should be paid to how questions are asked. For example, the question 'Do you feel there is inadequate study provision in this college?' is a 'leading question', therefore inappropriate.

Recommended reading

Chapman, M., in collaboration with Mahon, B., *Plain Figures* (HMSO, 1986).

Key Data 87, Central Statistical Office (HMSO, 1987).

Owen, F. and Jones, R., *Statistics* (Pitman, 1988), chapters 1, 14, 15, 16 and 23.

Whitehead, P. and Whitehead, G., *Statistics for Business* (Pitman, 1984), chapters 1, 2, 3 and 12.

Curwin, J. and Slater, R., *Quantitative Methods for Business Decisions* (Van Nostrand Reinhold, 1986), chapter 1.

3 Data II: Presentation tables, charts, graphs

3.1 Introduction

In chapter 2 we considered the stages and methods involved in collecting data. This chapter concentrates on ways of presenting that data, once it has been collected.

Before considering the various pictorial forms (tables, charts, graphs, etc) of data presentation, we will look at the ways in which data can be classified. You will be expected to be familiar with many different types of data presentation and consequently you should, as well as practising many different types of examination questions, refer to data presentation in newspapers, magazines, etc.

3.2 Classification of data

Data, once collected, has to be arranged in such a way as to make it possible to communicate the information efficiently to the desired audience. As explained in the last chapter, information or data can be qualitative or quantitative. The type of data will influence how it is presented. Consider the results of an analysis of staff in a large department in a polytechnic. The 'status' of staff is qualitative information. In such cases it is possible to present the information in a table. However, before data is tabulated it must be collected together or classified in some way. The polytechnic information was probably collected in the following form:

Ann Smith, Clerical assistant
John Brown, Principal Lecturer
Fred Jones, Head of Department
etc

Before tabulating the data or preparing it for presentation in a chart or diagram (see later), the number of classes or categories must be decided upon. In this case it could have been decided to have just two classes:

Teaching staff
Non-teaching staff

This would have resulted in losing a lot of the detail from the available data. Alternatively, if a wide range of classes were used, e.g.

Clerical Grade 1	Research assistant
Clerical Grade 2	Research Fellow
etc	Lecturer Grade II
Technician 1	etc
Technician 2	
etc	

it would be difficult for the reader to assimilate all the available information. Neither form is satisfactory. The final version of such an analysis carried out by the author is shown in table 3.1.

A quantitative aspect of the staff of this department is the salaries they earn. Again, before one reaches the stage of presenting data in tabular form, one must decide on a relevant classification or grouping of the data. As in the previous example, two classes could have been chosen:

Annual salaries: under £10 000
£10 000 and above

The information thus presented would be brief and tell us little about the distribution of salaries in the department. A general rule is to aim for 8 to 10 categories or classes, enabling one to strike a balance between having too little and too much information, both extremes making it difficult or impossible to interpret the data. The table produced for the polytechnic survey is shown in table 3.2.

Table 3.1 Analysis of staffing levels in the Department of Business Studies, the Polytechnic of North London

Employment description	Frequency Male	Frequency Female	Total frequency
Administrative/clerical	1	10	11
Technician	1	0	1
Research assistant	0	1	1
Research fellow	1	0	1
Lecturer II	20	10	30
Senior lecturer	10	6	16
Principal lecturer	12	1	13
Head of Department	1	0	1
Totals	46	28	74

Table 3.2 Analysis of salary levels of staff employed within the Department of Business Studies, the Polytechnic of North London

Salary (£ per annum)	Frequency Male	Female	Total
under 10 000	0	4	4
10 001–11 000	10	10	20
11 001–12 000	5	3	8
12 001–13 000	5	2	7
13 001–14 000	3	0	3
14 001–15 000	1	6	7
15 001–16 000	2	2	4
16 001–17 000	6	0	6
17 001–18 000	12	0	12
18 001–19 000	1	1	2
19 001 and above	1	0	1
Totals	46	28	74

Table 3.3 Analysis of salary levels of staff

Salary	Frequency
Under 10 000	4
10 000–11 000	20
11 000–12 000*	8
12 000–13 000†	7
13 000–14 000	3
⋮	⋮
Total	74

3.2.1 Limits of classification

When presenting data, in any form, great care should be taken to avoid ambiguity. One common error is failing to explain to the reader the intended limits of particular class intervals. Consider table 3.2 again and table 3.3. The presentation of data in table 3.3 has a number of faults; in particular, it is unclear which salaries should go in which interval. For example, if an individual earns exactly £12 000 per annum, would that information be recorded in the interval marked *, or the interval marked †. Table 3.2 shows more clearly the method that should be employed in such a case.

3.3 Presentation of data: tables

There are a variety of methods one can employ in presenting data, including a number of pictorial methods that will be considered in some detail later. Before moving on to these, we will give further consideration to a common form of data presentation: tables (including frequency tables).

There are a number of guidelines one should follow when presenting data in table form:

1 The table should have clear labels, titles, etc.

2 Where appropriate sub-totals and totals should be clearly shown.

3 There should not be so much data as to make the whole table unclear.

4 Aim to have 8 to 10 class intervals. This is a guide, not a rigid rule – table 3.2 had 11 intervals.

5 If possible, make the interval widths equal.

3.3.1 Frequency distributions

In cases where particular values occur more than once, it is common practice to list the values making a note of the number of occurrences, the frequency, of each value. The resulting information, set out in tabular form, is called a frequency distribution (see tables 3.1, 3.2 and 3.3).

3.3.1.1 Ungrouped frequency distribution

In the following example the variable is 'hours of television watched'. A random sample of 20 individuals responded to the question 'How many hours of television do you watch each week?', as follows:

29	35	28	31	28	35	28	29	30	32
29	30	28	32	33	35	32	28	35	35

Some values occur more than once, making it desirable to count the frequency of each value. The result is a simple frequency distribution. The lowest value is 28 and the highest is 35: a range of 8. The frequency distribution can be established by listing the full range of values and 'tallying' them as shown in table 3.4.

Table 3.4 Frequency table showing the amount of time spent watching television during a randomly selected one-week period

Hours of TV	Tally	Frequency
28	𝍸	5
29					3
30				2
31			1
32					3
33			1
34		0
35	𝍸	5
		Total frequency 20

(Source of data: sample of 20 students of North London Polytechnic, 1986.)

3.3.1.2 Grouped frequencies

If the quantity of data is large, it is often more convenient to group frequencies into class intervals. This results in some loss of precision. However, it is often worth it, if it makes it easier to read the data. Suppose the data from the above 'television' survey was as follows:

27	26	35	34	29	28	40	40	24	33
22	25	21	20	31	30	37	38	22	23
15	12	29	16	10	36	29	17	39	28
25	41	29	40	18	37	25	18	27	27
25	40	33	31	34	26	40	20	12	15

Producing an ungrouped frequency distribution table of this data as above would result in a table with 31 rows (values 10 to 41). This would be little more helpful than having the data written out in its 'raw' format. In such a case it is useful to group the data into classes or intervals, as follows. First find the range of data values:

Maximum data value = 41
Minimum data value = 10
Range of values = 41 − 10 = 31

The problem now is to decide how to group the data. If one is aiming for 8 to 10 class intervals, then classes of 4 hours 'width' will suffice, as in table 3.5.

Table 3.5 Grouped frequency table showing the amount of time spent watching TV during a period of one week

Hours of TV	Actual data values	Frequency
10–13	12, 10, 12	3
14–17	15, 16, 17, 15	4
18–21	21, 20, 18, 18, 20	5
22–25	24, 22, 25, 22, 23, 25, 25, 25	8
26–29	27, 26, 29, 28, 29, 29, 28, 29, 27, 27, 26	11
30–33	33, 31, 30, 33, 31	5
34–37	35, 34, 37, 36, 37, 34	6
38–41	40, 40, 38, 39, 41, 40, 40, 40	8
	Total frequency	50

One would not usually write out each value as shown in the middle column of table 3.5. That has been done to illustrate how the specific values have been located in their class intervals.

3.3.1.3 Cumulative frequency table

The number of observations (or frequency) above or below a specified value is often required. For example, we might wish to know the number of people who watch less than 30 hours of television per week or, perhaps, more than 21 hours per week. From the cumulative frequencies in table 3.6, it can be seen that the answers are 31 (*) and 38 (†) people, respectively.

Table 3.6 Grouped frequency table showing the amount of time spent watching TV during a period of one week

Hours of TV	Frequency	Cumulative frequencies (less than)	Cumulative frequencies (greater than)
≥ 10 and < 14	3	3	50
≥ 14 and < 18	4	7	47
≥ 18 and < 22	5	12	43
≥ 22 and < 26	8	20	38†
≥ 26 and < 30	11	31*	30
≥ 30 and < 34	5	36	19
≥ 34 and < 38	6	42	14
≥ 38 and ≤ 41	8	50	8
Total frequency	50		

Notes to table 3.6

1 '≥ 10 and < 14' is equivalent to '10–13', and is read as 'values equal to or greater than 10 *and* less than 14'.

2 Interpretation of cumulative frequency data

(a) 'Cumulative (less than)' shows that:

- 3 people watched less than 14 hours of television that week;
- 7 people watched less than 18 hours of television that week;
- 12 people watched less than 22 hours of television that week, and so on.

(b) 'Cumulative (greater than)' shows that:

- 50 people watched 10 hours or more of television that week;
- 47 people watched 14 hours or more of television that week, and so on.

One important reason data is presented in table form is that it provides a summary of the collected information, allowing the reader to get an immediate 'picture' of that information. However, in a frequency table the data is still in numerical form and it is undeniable that some people will avoid wading through a mass of numbers to acquire information. Therefore it is essential to have other, more easily accessible, means to communicate the information contained in collected data.

3.4 Pictorial representation of data

We will now consider the many ways of presenting data in a visual form, ranging from simple bar charts to the various graphs used in special cases. In all cases it is important to bear in mind that, though charts and graphs play an important role in data presentation, they can give misleading information. Compare the two bar charts in fig. 3.1. They show the same information but, because of different scales and an arbitrary starting point in chart (b) (which is not made explicit), the impression is given that the category Lec. II is even larger than the other categories than is in fact the case.

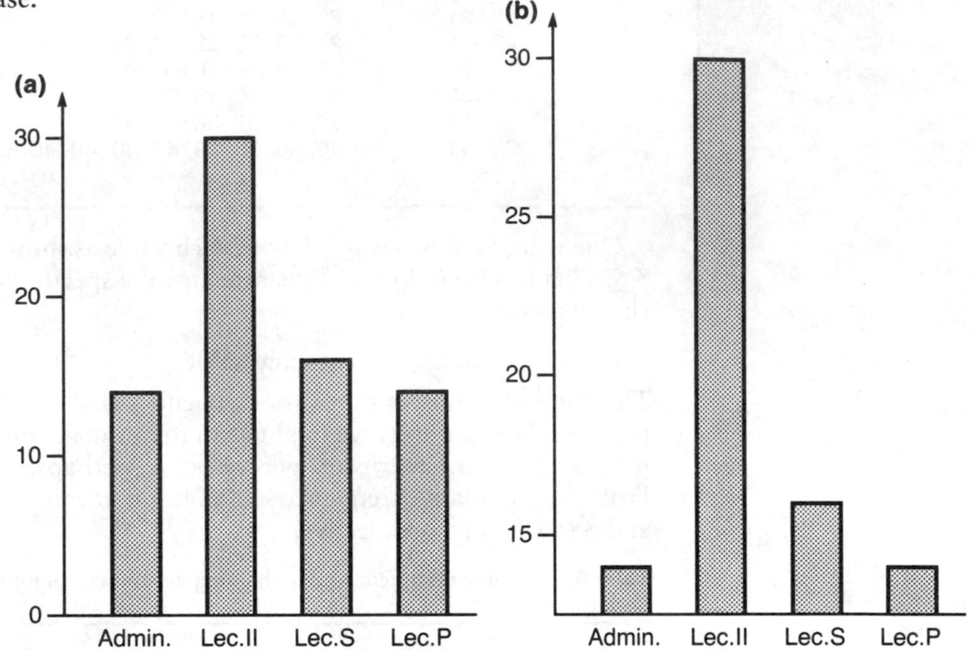

Fig. 3.1 Analysis of the staffing levels in the Department of Business Studies, the Polytechnic of North London.
Admin.: this category contains administrative, clerical, technical and research staff.
Lec. II: this category contains lecturers, grade II.
Lec. S: this category contains lecturers, senior grade.
Lec. P: this category contains lecturers, principal grade, and heads of department.

Pictorial presentation of data is an important means of communication, but one must be conscious of the need for the information given to be accurate and honest as well as easily understood. The following are examples of some of the more commonly used charts and graphs.

3.4.1 Bar charts

Bar charts represent the data in bars of equal width. The height of each bar shows the value of the data. There are a number of ways of using bar charts:

3.4.1.1. Simple bar charts

Figure 3.1(a) is a good example of a simple bar chart. It is in fact a bar chart representing the data given in table 3.1.

3.4.1.2 Percentage bar charts

Again, using the data shown in table 3.1, fig. 3.2 illustrates the use of the percentage bar chart. It provides an easy way of comparing the proportions of, in this instance, male and female post-holders. It does, however, result in some loss of detail: the actual number of employees could not be obtained from the chart if one was not given the information in the form of a footnote.

3.4.1.3 Component bar charts

Figure 3.3(a) shows the sales record of a hypothetical company over a period of four years. Figure 3.3(b) illustrates the effect of adding extra information concerning the different components of sales (products A, B and C). This second chart, a component bar chart, allows the reader to gain information about the principal point of interest, the company's overall sales performance, while allowing access to additional information; for instance, it can be seen that product C is the major contributor to sales. While it is clear that product C is the major component of sales, it is not easy to see the exact size of C's contribution, or to monitor its change over time. New emphasis is required and the following is an ideal method to employ. (The method of constructing a component bar chart is explained in the solution to sample question **1** at the end of this chapter.)

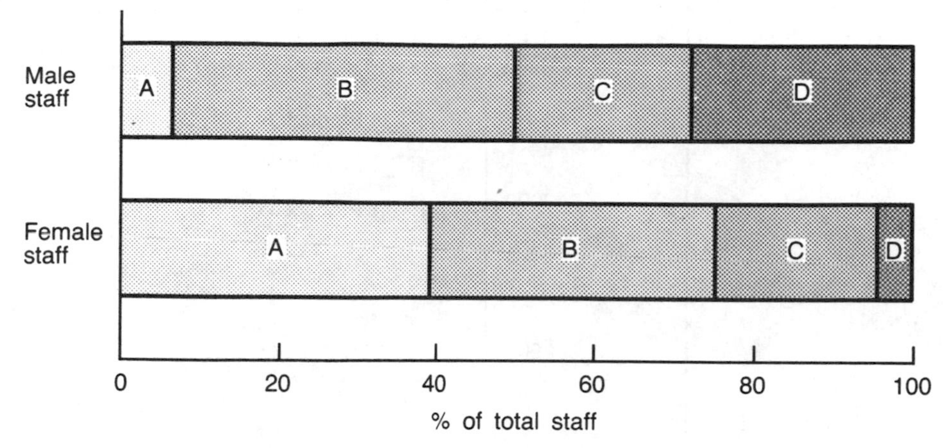

Fig. 3.2 Analysis of staffing levels in the Department of Business Studies, the Polytechnic of North London, given as percentage of the total male (female) staff level.
A: this category contains administrative, clerical, technical and research staff.
B: this category contains lecturers, grade II.
C: this category contains Lecturers, senior grade.
D: this category contains lecturers, principal grade and heads of department.
Note: Total male staff, 46; total female staff, 28.

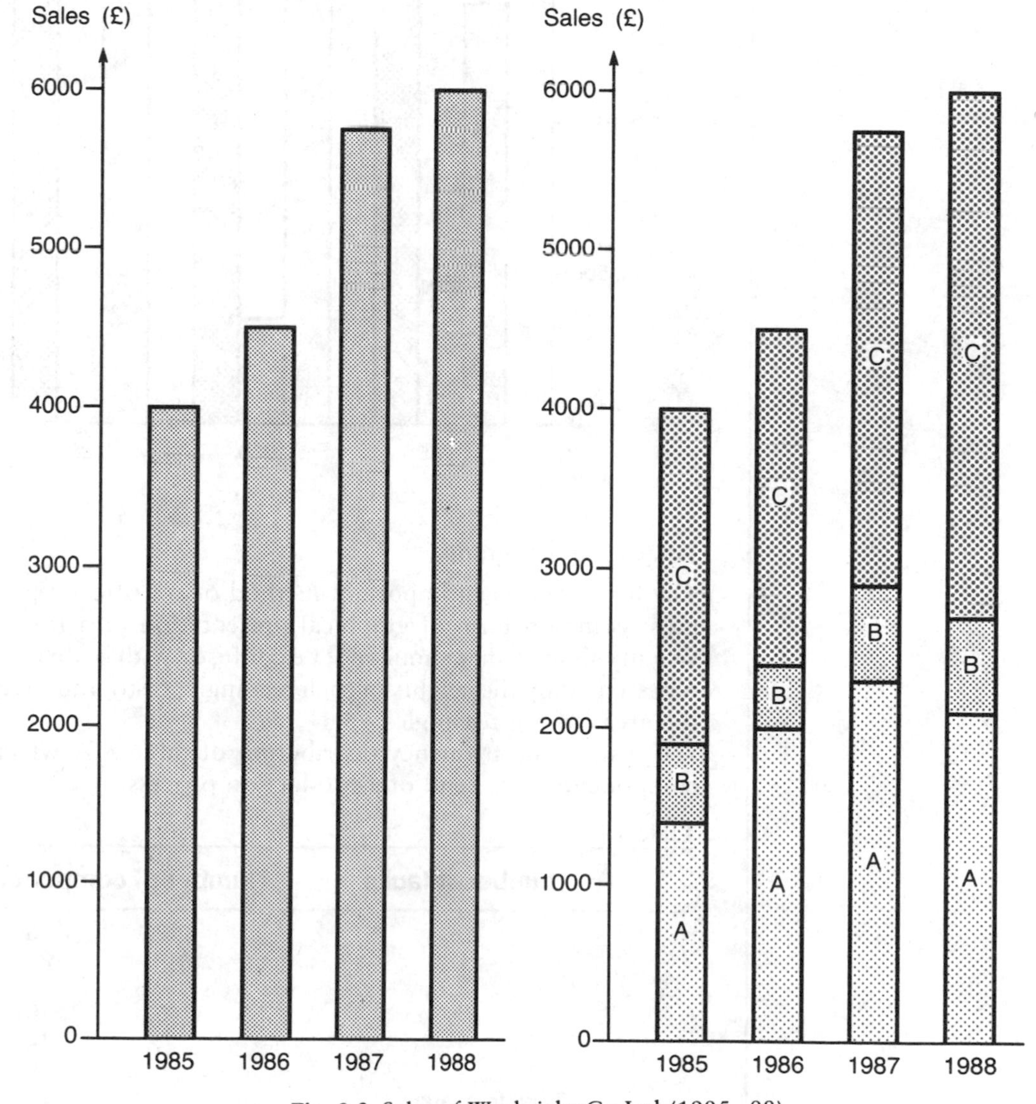

Fig. 3.3 Sales of Workright Co Ltd (1985–88)

3.4.1.4 Multiple bar charts

As with component bar charts, multiple bar charts allow one to draw attention to a number of features of a set of data on one chart. An example of such a chart (using the same data as fig. 3.3) is shown in fig. 3.4.

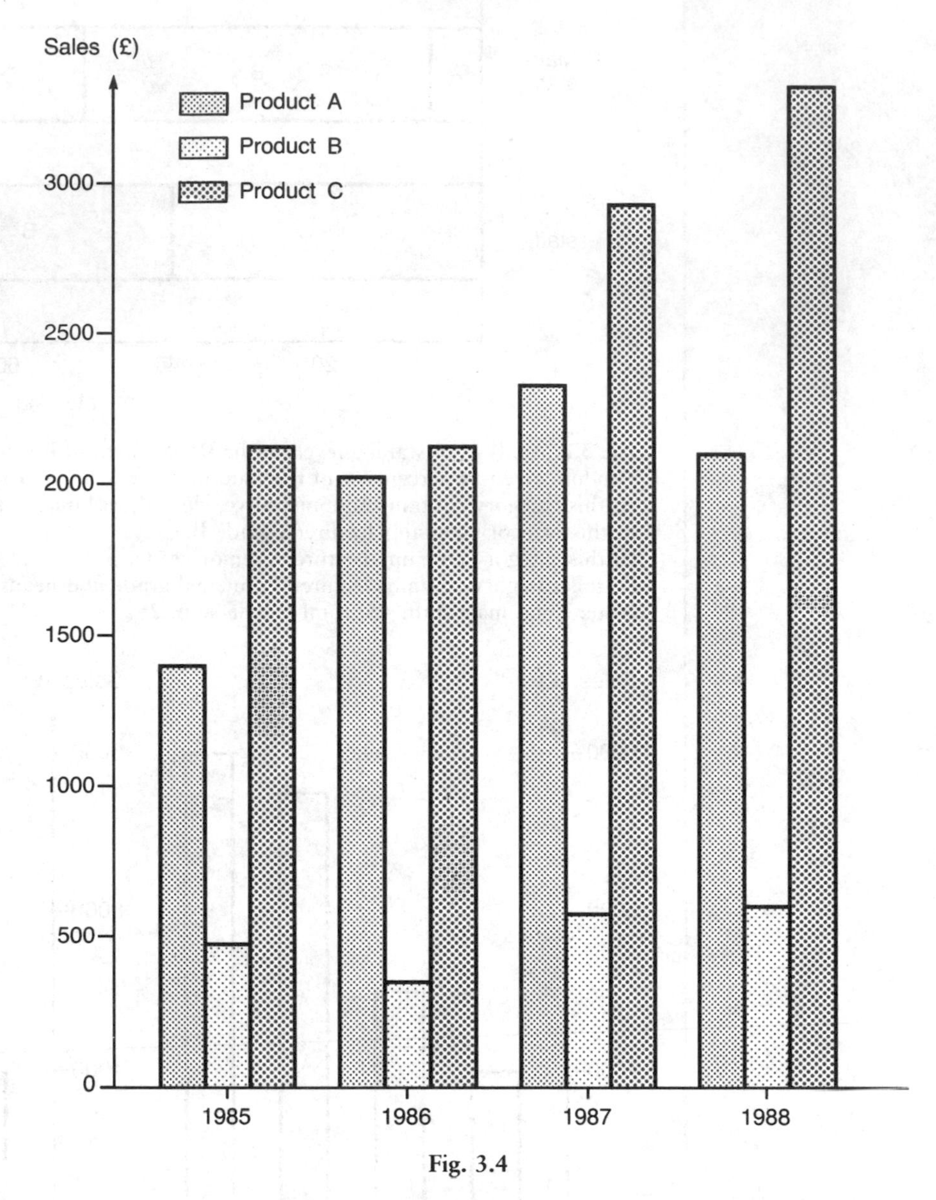

Fig. 3.4

3.4.1.5 Histograms

The histogram is an important method of statistical representation of data. It is the most common form of graphical presentation of a frequency distribution. Though visually similar, they should not be confused with bar charts. Bar chart frequencies are represented by the heights of each rectangle; histogram frequencies are represented by the areas of the rectangles.

Consider the frequency distribution of table 3.7, which results from monitoring components at the end of a production process:

Table 3.7

Number of faults	Number of components (frequency)
0	5
1	25
2	21
3	16
4	12
5 or more	0
Total frequency	79

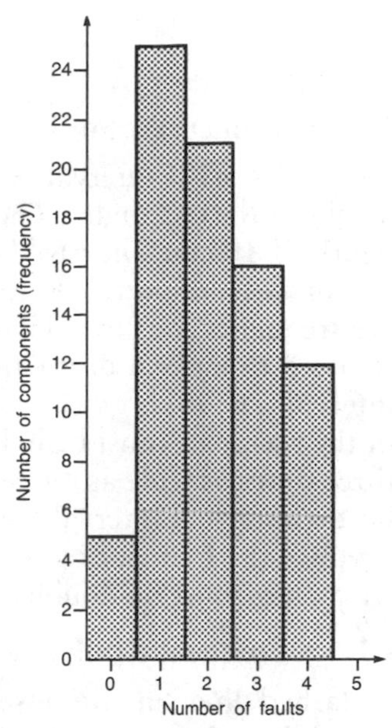

Fig. 3.5 Histogram showing the frequency of component faults

Figure 3.5 shows the resulting histogram. Note that the frequency rectangles are positioned centrally on the observed number of faults. This is to avoid any ambiguity and is typical of the method of construction of a histogram for the observations of a discrete variable.

Consider another example, the frequency distribution which results from observations of component diameters (table 3.8). This (continuous) data is shown as a histogram in fig. 3.6.

Table 3.8 Diameters of 79 components

Diameters (cm)		Frequency
10 but less than	20	3
20	30	7
30	40	10
40	50	18
50	60	23
60	70	8
70	90	7
90	120	3
Total frequency		79

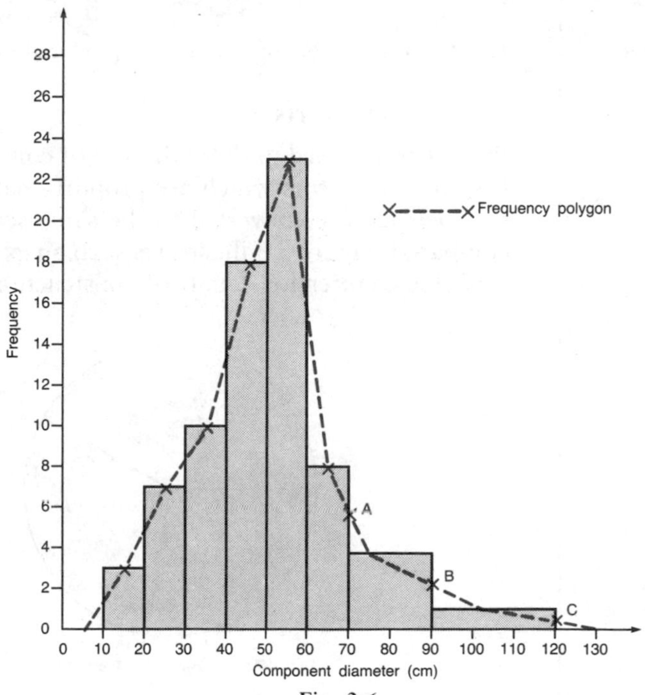

Fig. 3.6

Points to note:

1 The class intervals can be described in a number of ways.

2 The limits of the intervals are not equal, calling for great care when deciding on the height of the rectangles. The intervals [10, 20] to [60, 70] are all the same, having a width of 10 cm. The problems arise with the last two intervals, which are 20 and 30 cm wide, respectively. Recall that it is the area of the rectangle which represents the frequency, not the height of the rectangle. If all the intervals are equal, there is no need to make a distinction. The adjustments are easily made. For instance, the interval [70, 90] is twice as wide as the preceding intervals. Consequently the height of the rectangle must be half the interval frequency. The next interval, [90, 120], is three times as wide as the other intervals. Hence the height of the rectangle is found by dividing the interval frequency by three. In general, doubling the class width requires the frequency to be halved and, conversely, if the class width is halved the frequency must be doubled to obtain the rectangle height.

3.4.1.6 Frequency polygons

The dashed line superimposed on the histogram in fig. 3.6 is called a frequency polygon. It is drawn so that the area under it is equal to that under the histogram.

If the class intervals are equal, the frequency polygon is drawn by joining the mid-points at the top of each rectangle. When the intervals are of various widths it is more difficult to handle. The (vertical) mid-points are joined – points A, B and C on the figure. Note that the polygon extends beyond the range of the histogram to meet the base axis. This is to ensure that total areas (of the bars and the polygon) are the same.

3.4.2 Line charts

Line charts are employed in much the same way as simple bar charts; see fig. 3.7.

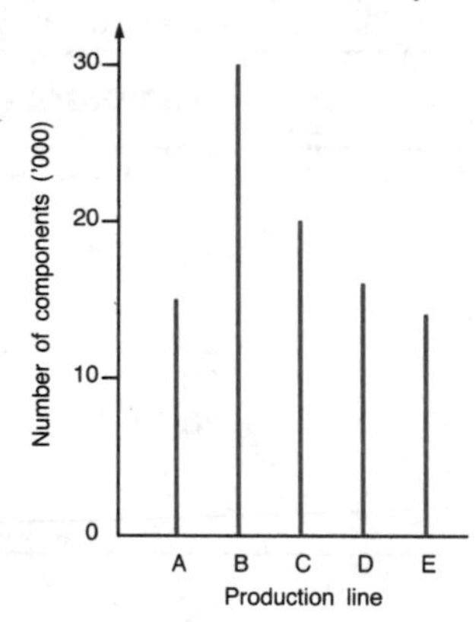

Fig. 3.7 The number of components produced by Sedan Chairs Ltd; analysed by production line

3.4.3 Pie charts

Pie charts are used to show the size of constituent parts relative to the whole. A circle is split into sectors which are proportional to the part each represents. If data from two periods are converted to the same scale, pie charts allow the data to be easily compared. Figure 3.8 illustrates such an example. See the solution to sample question 1 of this chapter for details of construction.

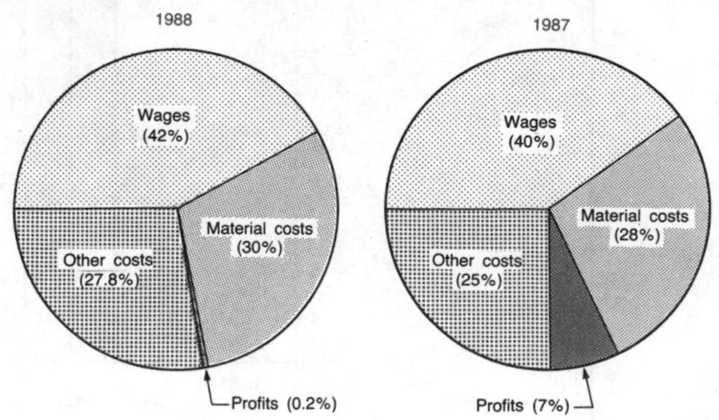

Fig. 3.8 Workright & Co Ltd: Distribution of revenue, 1987 and 1988

3.4.4 Pictograms

Pictograms are probably the most imprecise form of data presentation. They are used to catch the reader's attention and are used in advertising, company reports, etc. In fig. 3.9, the main intention is to present a picture which combines the focus of interest, oil rigs, and the feature of that interest, the number of oil rigs in operation. The pictures of oil rigs are scaled to reflect the change in variable, the number of oil rigs in operation. Problems could arise because it is not clear whether it is the height of the oil rig which is significant or the 'area' covered by the oil rig diagram.

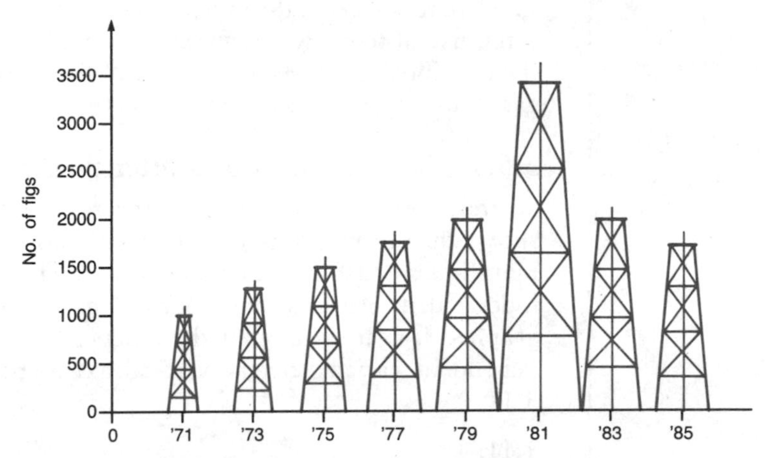

Fig. 3.9 Oil rigs operating in the United States of America (average number per annum)

3.4.5 Statistical maps

Data classified by geographical area is the obvious candidate for statistical maps, sometimes called cartograms. In order to make the communication of information relatively easy, a colour code or system of shading is used to differentiate the areas. Figure 3.10 shows the rates of unemployment in different regions of the United Kingdom, in broad bands.

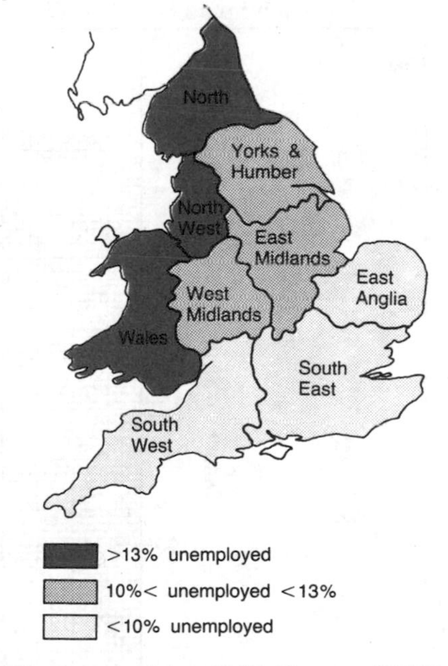

Fig. 3.10 Regional unemployment rate: May 1987 (*Source: KEYDATA 87, HMSO, 1987*)

Examination questions often ask for charts and/or graphs to be drawn. You should be aware of the methods involved and the appropriateness of each type of pictorial representation. It is important to remember to:

1 provide clear labelling for table and charts;

2 give full information concerning the source of data;

3 avoid building ambiguity or misrepresentation into the presentation of data (e.g. fig. 3.1(b) and table 3.3).

3.5 Graphical representation of data

A number of pictorial representations of data have been considered. In each case the aim was to present data in a form which allowed the reader to visualize the statistical information. Graphs also do this, but are different in that they show the relationship between two variables, for instance, sales and time or profits and advertising revenue. In such cases it is often useful to consider one variable as independent and the other, dependent. For example, if we were to monitor the profits of a company as related to the sales of that company, we could reasonably claim that profits were dependent on sales; that is, sales is the independent variable and profits, the dependent variable. It is often useful to show a number of related variables on the same graph. For instance, the relationship between sales and level of production could be illustrated on the same graph as wages and level of production.

3.5.1 Scattergraphs or scatter diagrams

Scattergraphs are used to look for relationships between two variables. Table 3.9 shows the output (in thousands of units) over a period of 20 weeks and the expended man-hours (×10) over the same period. The variable 'man-hours' is interpreted as the independent variable and 'output' as the dependent variable. When the data is graphed it is usual to plot the independent variable on the horizontal axis and the dependent variable on the vertical axis. The resulting scattergraph is shown in fig. 3.11.

Table 3.9

week	**1**	**2**	**3**	**4**	**5**	**6**	**7**	**8**	**9**	**10**
output ('000s)	12	12	10	10	16	15	12	20	25	10
man-hours (×10)	40	42	36	35	45	44	41	58	76	34
week	**11**	**12**	**13**	**14**	**15**	**16**	**17**	**18**	**19**	**20**
output ('000s)	21	20	10	15	16	14	10	18	16	18
man-hours (×10)	65	66	33	47	47	43	35	55	49	54

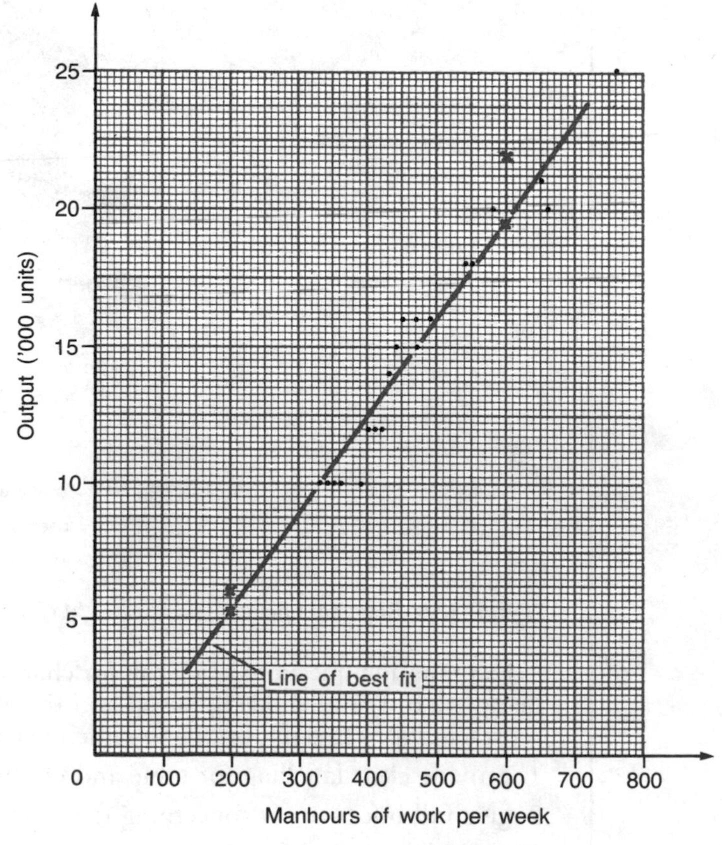

Fig. 3.11 Scattergraph of production level (thousands of units output) against man-hours of work over a period of 20 weeks

The aim is often to identify a linear relationship between the variables. If such a relationship is identifiable, then a 'line of best fit' can be added to the scattergraph as has been done in this case. The line of best fit is that line which best represents the (linear) relationship existing between the variables. Establishing the line of best fit involves correlation and regression which will be considered in chapter 6.

3.5.2 Ogives

Ogives, or cumulative frequency curves, are used to show the number of observations less than or equal to a given value. Alternatively, the number of observations more than a given value can be shown. For grouped data, the less than (or greater than) cumulative frequencies are plotted against the upper (or lower) limit of each interval. Consider the data presented in table 3.6. Figure 3.12 shows the less than and greater than ogives, respectively, of that data.

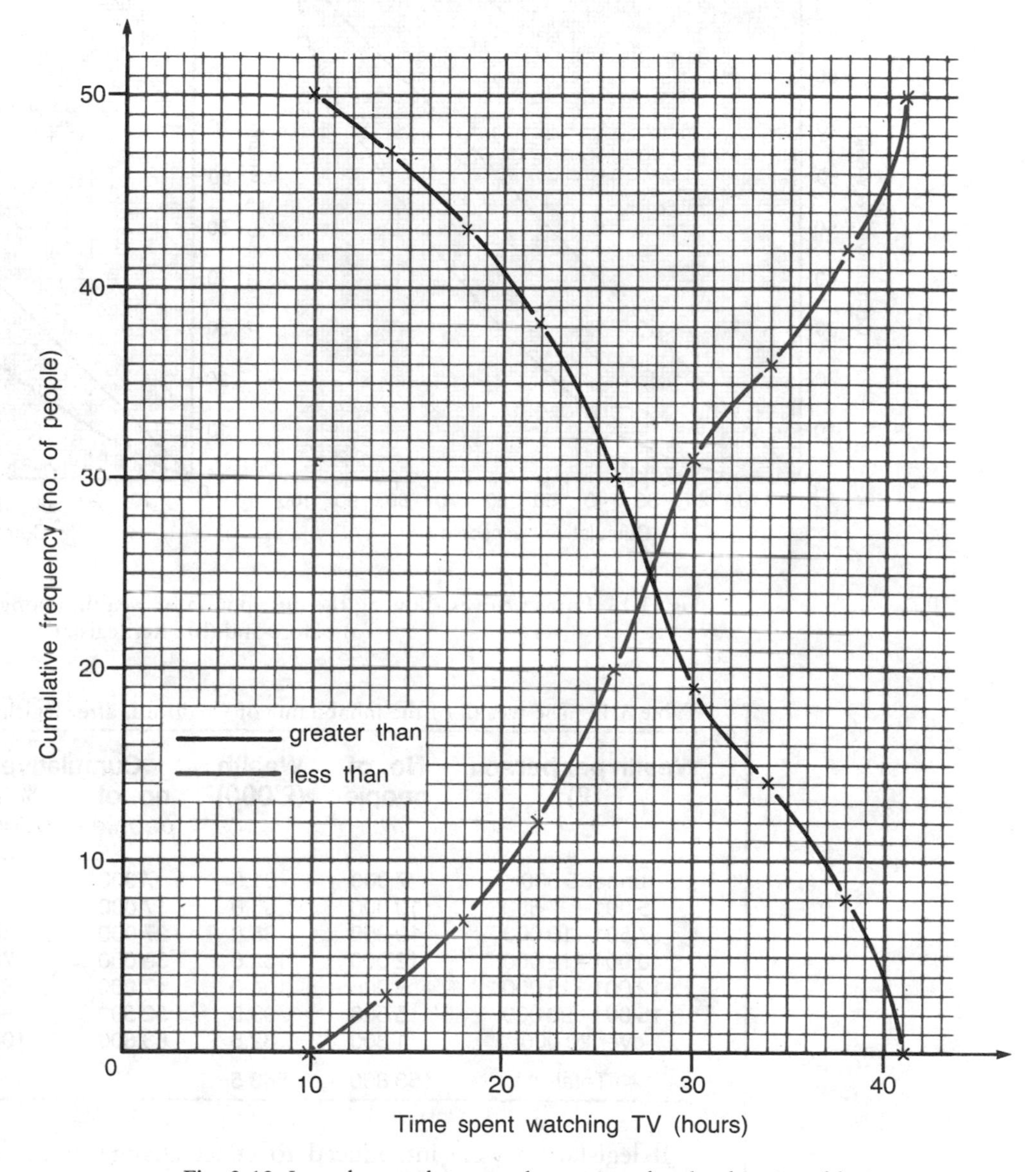

Fig. 3.12 Less than and greater than ogives for the data in table 3.6

3.5.3 Lorenz curves

A Lorenz curve is a particular form of cumulative frequency curve which compares two sets of cumulative data. The *x*- and *y*-axes are often both percentages scales, though this is not essential. Consider the distribution of wealth of the inhabitants of Crediford in the year 1987 and the relevant percentage cumulative frequencies, shown in table 3.10 (overleaf). Plotting the cumulative percentage wealth against the cumulative percentage population results in the Lorenz curve in fig. 3.13(a).

Table 3.10 The wealth of the inhabitants of Crediford, 1987

Wealth per person (£)	No. of people	Wealth (£'000)	Cumulative no. of people	Cumulative % of people	Cumulative wealth (£'000)	Cumulative % wealth
under 5 000	19 000	35.0	19 000	35	35.0	9
5 001– 7 500	10 000	60.0	29 000	54	95.0	25
7 501–10 000	12 000	96.0	41 000	76	191.0	50
10 001–12 000	7 000	77.0	48 000	89	268.0	70
12 001–15 000	4 000	56.0	52 000	97	324.0	85
15 001–20 000	1 000	19.0	53 500	99	343.0	90
over 20 000	800	40.0	53 800	100	383.0	100
Total	53 800	383.0				

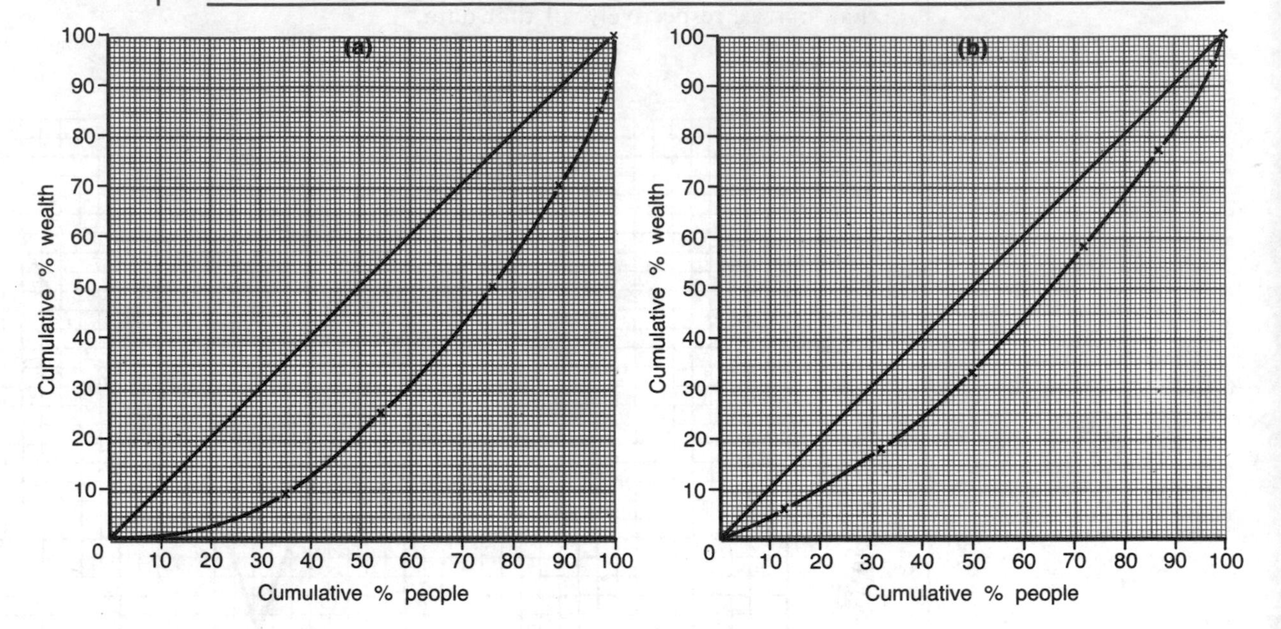

Fig. 3.13 Lorenz curves showing the distribution of wealth among the inhabitants of Crediford, (a) before and (b) after legislation

Table 3.11 The wealth of the inhabitants of Crediford, after legislation

Wealth per person (£)	No. of people	Wealth (£'000)	Cumulative no. of people	Cumulative % of people	Cumulative wealth (£'000)	Cumulative % wealth
under 5 000	7 000	31.5	7 000	13	31.5	6
5 001– 7 500	10 000	65.0	17 000	32	96.5	18
7 501–10 000	10 000	85.0	27 000	50	181.5	33
10 001–12 000	12 000	132.0	39 000	72	313.5	58
12 001–15 000	8 000	104.0	47 000	87	417.5	77
15 001–20 000	5 500	93.5	52 500	98	511.0	94
over 20 000	1 300	32.5	53 800	100	543.5	100
Total	53 800	543.5				

If legislation were introduced to effect change in the distribution of wealth, as shown in table 3.11, a different picture would result: fig. 3.13(b). For instance, the number of people with wealth below £5000 in 1988 is 7000, compared to 19 000 in 1987. In this same category, the total wealth is £31 500 in 1988, compared with £35 000 in 1987, an increase in average wealth (in the under-£5000 category) from £1842 per person (£35 000/19 000) to £4500 per person (£31 500/7000).

The Lorenz curve in fig. 13.13(b) is closer to the diagonal line, reflecting a change in the distribution of wealth of the inhabitants of Crediford. Lorenz curves show the degree of concentration of a variable (in this example, wealth). The further from the diagonal line the curve is, the greater the concentration of the variable. The closer to the diagonal, the more even the spread of the variable. In 1987, the wealth of the people of Crediford was concentrated in a small section of the community (50 per cent of the population owned only 21 per cent of the wealth). The legislation seems to have resulted in a move towards a more equitable distribution of wealth.

3.6 Graphs showing data changes over time

In business, it is frequently necessary to monitor data over time. Examples range from observing sales affected by seasonal influences, to monitoring the changes in a company's sales brought about by increasing the advertising revenue.

3.6.1 Strata graphs

An example of the use of strata graphs is shown in fig. 3.14. This shows the population of children and of elderly people at different points in time, with the information given in age groups.

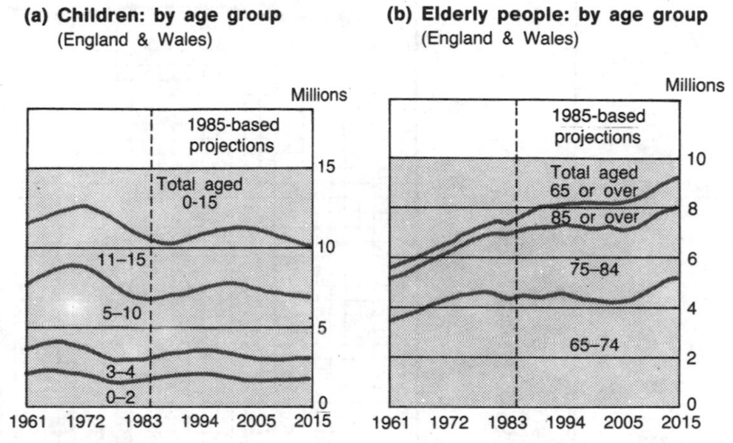

Fig. 3.14 (a) Children and (b) elderly people in England and Wales, by age group; 1961–1985 and 1985-based projections to 2015

(Office of Population Censuses and Surveys; Government Actuary's Department; Social Trends, 1987, Charts 1.4 and 1.5)

3.6.2 Time series

A series of statistical values showing the performance of data over a period of time is called a time series. If one were budgeting for raw materials or labour for a planned new project, it would be advisable to refer to existing data on past costs, in particular to investigate such data for any signs of possible increases. Figure 3.15 shows the level of unemployment over the period 1980 to 1987. Analysis of time series will be considered in more detail in chapter 5.

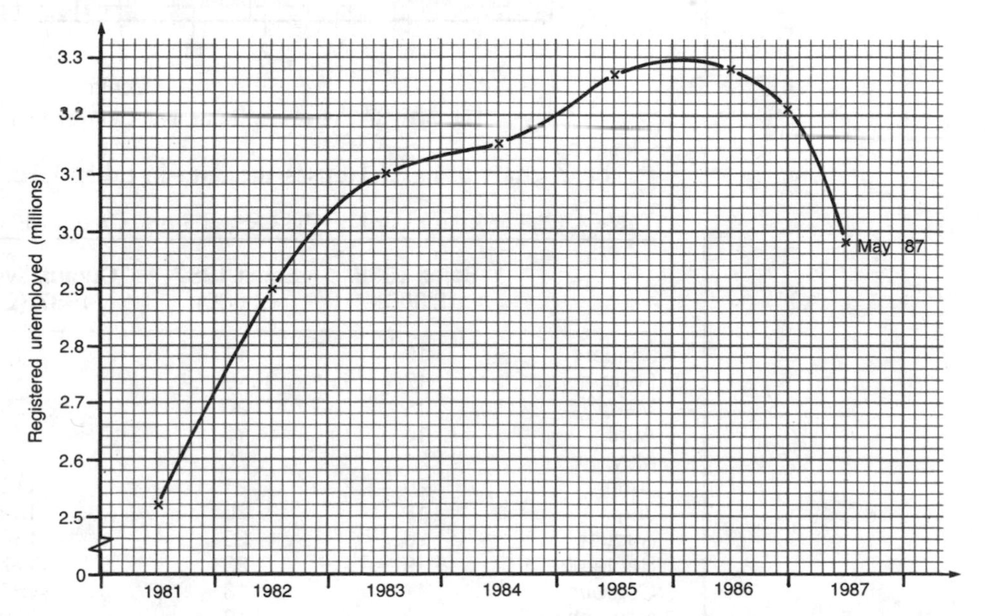

Fig. 3.15 Registered unemployed (millions), seasonally adjusted, excluding school-leavers

(Key Data '87; Central Statistics Office.)

3.6.3 Z charts

The Z chart allows one to interpret a time series in three different ways using the one graph:

1 The base line of the 'Z' represents a 'standard' time series showing the variable under consideration, for example, sales or profits.

2 The diagonal line of the 'Z' represents the cumulative total of this variable, for example cumulative sales, cumulative profits, etc.

3 The top line of the 'Z' represents the total of sales or profits for the year 'to that point'; in the example shown in fig. 3.16 the January figure would be the total sales from the beginning of February 1986 to the end of January 1987.

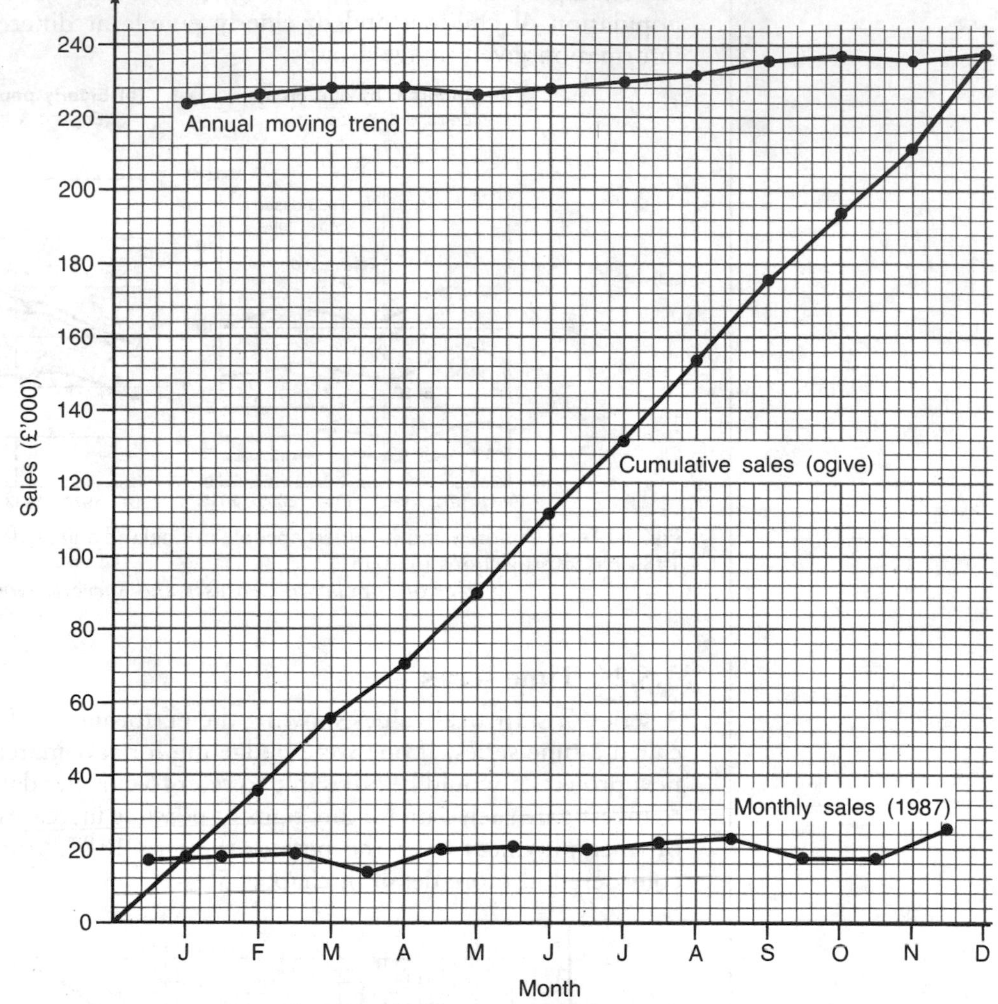

Fig. 3.16 Z chart of sales of The Central Store, Crediford, 1987

The Z chart in fig. 3.16 uses the data in table 3.12.

Table 3.12 The sales record of The Central Store, Crediford: 1986 and 1987

	Sales 1986 (£'000)	Sales 1987 (£'000)	Cumulative sales 1987 (£'000)	Annual moving total (£'000)
January	16	17	17	224
February	17	19	36	226
March	18	20	56	228
April	15	15	71	228
May	21	20	91	227
June	20	21	112	228
July	18	20	132	230
August	20	22	154	232
September	19	23	177	236
October	17	18	195	237
November	18	18	213	237
December	24	26	239	239

Sample questions

1 You have been appointed personal assistant to the finance officer of Workright Co Ltd. It is planned to present information to the company employees in an attempt to encourage greater interest in the company's business operations. You have been asked to contribute to this by providing suitable data presentations. The relevant data is as follows:

Year:	1988 £'000	1987 £'000	1986 £'000	1985 £'000
Sales	6.0	5.8	4.5	4.0
Wages	2.5	2.2	1.9	1.3
Production costs	0.8	0.6	0.5	0.4
Material costs	1.8	1.6	1.3	1.0
Taxation	0.0	0.3	0.1	0.2
Other costs	0.8	0.7	0.5	0.5
Profit	0.1	0.4	0.2	0.6
Price index	170	150	115	100

There are three products produced by the company and their share or sales is given as follows:

Product Year:	1988 %	1987 %	1986 %	1985 %
A	35	40	45	35
B	10	10	8	12
C	55	50	47	53
	100	100	100	100

Required:

Prepare visual displays to show each of the following:

1 sales by product;

2 a comparison of the last two years, analysing total sales into costs, taxation and profits, highlighting the influence of wages;

3 total sales over the full period, taking inflation into account.

The Polytechnic of North London; Accounting Foundation Course.

Solution/hints

Parts 1 and 2

Component bar charts and pie charts are appropriate methods of displaying this data. The methods involved in constructing these charts are as follows:

Component bar chart

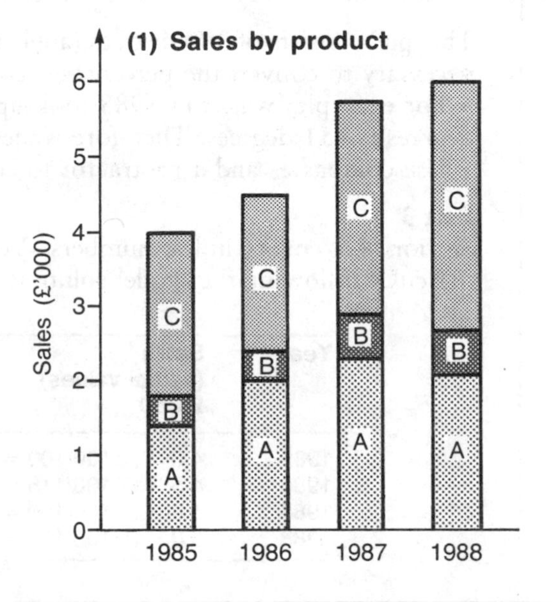

This is quite straightforward in its construction.

First calculate the absolute values of the sales, using the percentages given and the total sales figure. For example, the 1988 bar:

Information given in the question: total sales = £6000
A = 35% of total sales
B = 10% of total sales
C = 55% of total sales

Therefore: sales of A = 35% × £6000 = £2100
sales of B = 10% × £6000 = £600
sales of C = 55% × £6000 = £3300

The 'portions' of sales are then 'stacked' on top of one another to form the complete bar of height (6000).

Pie chart

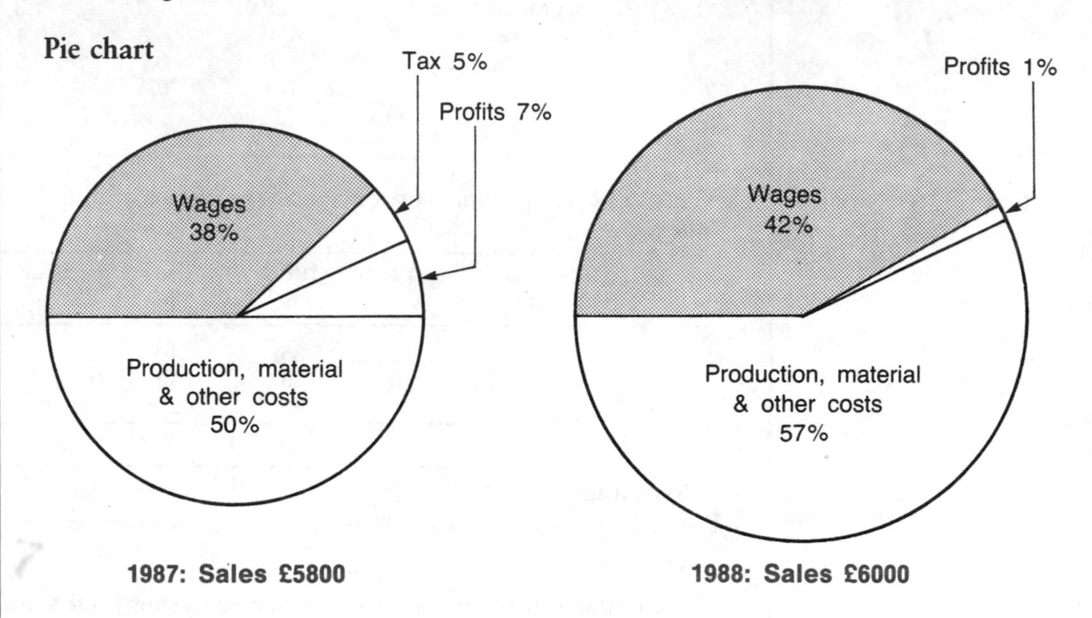

1987: Sales £5800 **1988: Sales £6000**

To compare the two years' figures, costs, taxation and profits must first be expressed as percentages of the years' total sales revenue. You are asked to highlight the influence of wages, so calculate that figure separately from the other costs. For example, for 1987:

total sales = £5800
taxation = £300 = 5% of total revenue
profits = £400 = 7% of total revenue
wages = £2200 = 38% of total revenue
production, materials and other costs = £600 + £1600 + £700 = £2900 = 50% of total revenue

The 'pie' consists of 360 degrees (angle measurement). To construct the pie chart it is necessary to convert the percentages to degree equivalents.

For example: wages in 1988 took up 42 per cent of revenue; 42 per cent of 360 degrees is 151 degrees. Therefore wages takes a 151 degree 'portion' of the 'pie'.

Use compasses and a protractor to construct pie charts.

Part 3

Section 4.5 covers index numbers. You will find that section useful if you have difficulty following the model solution to this part of the question.

Year	Sales (actual values) £'000	Sales (scaled to 1985 price level using price index)
1985	4.0 (× 100/100 = 1)	1 × 4.0 = 4.0
1986	4.5 (× 100/115 = 0.87)	0.87 × 4.5 = 3.92
1987	5.8 (× 100/150 = 0.67)	0.67 × 5.8 = 3.88
1988	6.0 (× 100/170 = 0.59)	0.59 × 6.0 = 3.54

OR

Year	Sales (actual values) £'000		Sales (scaled to 1988 prices)
1985	4.0	(×170/100 = 1.7)	1.7 × 4.0 = 6.80
1986	4.5	(×170/115 = 1.48)	1.48 × 4.5 = 6.66
1987	5.8	(×170/150 = 1.13)	1.13 × 5.8 = 6.55
1988	6.0	(×170/170 = 1.0)	1.0 × 6.0 = 6.00

2 Polyproducts Ltd have produced annual reports which include the following information:

(i) Divisional profits:

	%
Computer division	40
Robotics	20
Publications	8
Consultancy	7
Research	10
Components	15

(ii) Cash flow, with comparative data from previous year:

	1988 £'000	1987 £'000
Salaries	30	25
Taxation on profits	10	8
Interest on loans	5	5
Renewal of equipment	9	8
Investment costs	12	10
Repairs	4	4
Totals	70	60

Required:
You have been requested by the editor of the company magazine to present the above data in a form suitable for inclusion in the company's annual report. She wants the data to be attractive and easy to understand. In order to provide the maximum flexibility for inclusion in the report, she wants TWO versions for the data shown in (i).

For the data shown in (ii), show the change in circumstances from one year to the next:

(a) in absolute terms, and

(b) in relative terms; as a proportion of the total cash flow each year.

The Polytechnic of North London; Accounting Foundation Course.

Solution/hints

Part (i)

Divisional profits of Polyproducts Ltd

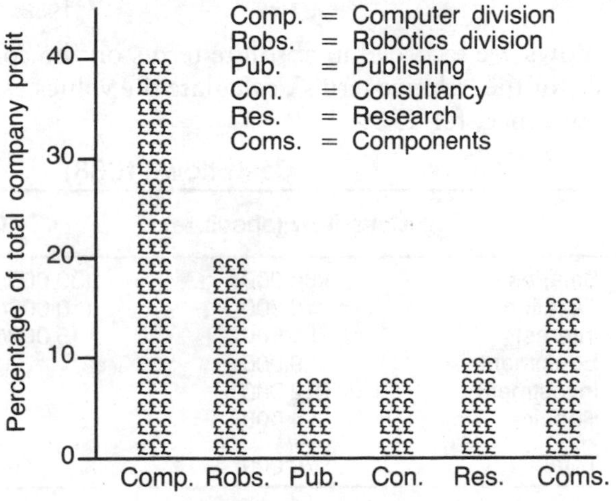

Polyproducts Ltd: Profits by division

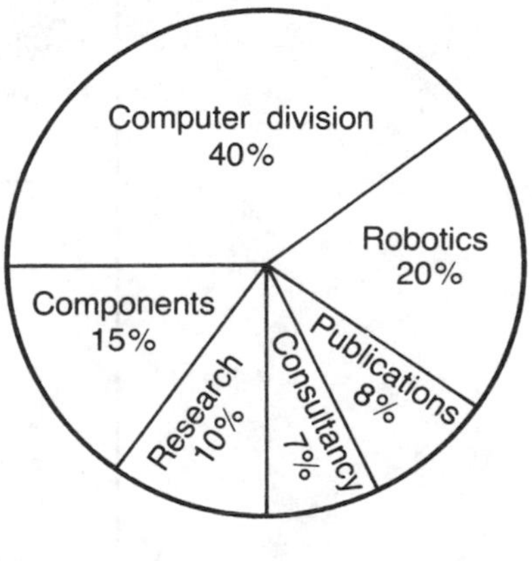

A simple pictogram, which in this case is not significantly different from a bar chart (the 'bars' are simply filled with £ signs), has been chosen to represent the divisional profits. Each column represents the divisional profits (as a percentage).

The second pictorial representation is a pie chart. The calculations are shown below:

Division	Proportion of profit (%)	Pie chart angle (degrees)
Computer	40	40/100 × 360 = 144
Robotics	20	20/100 × 360 = 72
Publications	8	8/100 × 360 = 29
Components	15	15/100 × 360 = 54
Research	10	10/100 × 360 = 36
Consultancy	7	7/100 × 360 = 26

Part (ii)

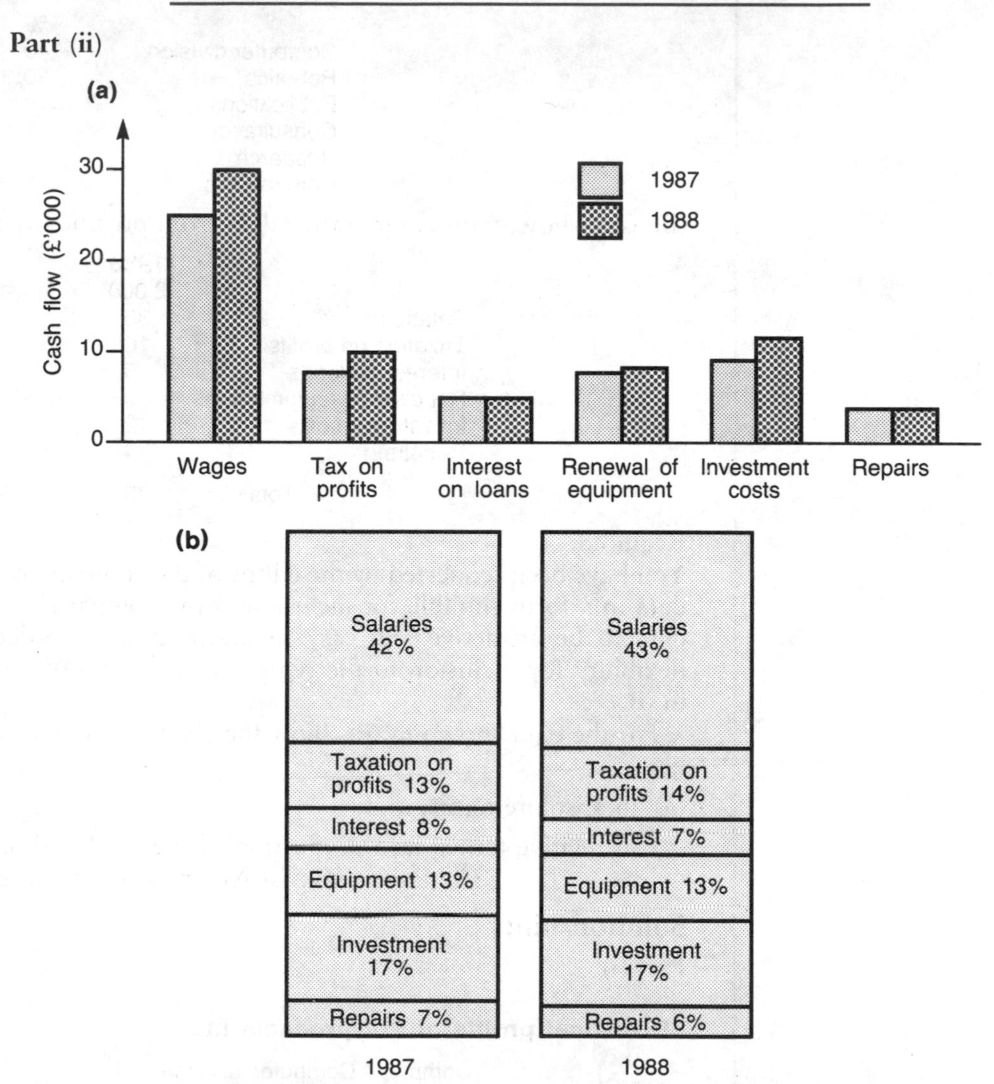

The cash flows are recorded in absolute terms on the component chart. For part (**b**), which calls for the 'relative terms', calculate the values as percentages of the total cash flow. For instance, for 1988:

Cash flow (1988)

	Cash flow (absolute)	Cash flow (%)
Salaries	£30 000	(30 000/70 000 × 100 =) 42.86
Taxation	£10 000	(10 000/70 000 × 100 =) 14.29
Interest	£5 000	(5 000/70 000 × 100 =) 7.14
Equipment	£9 000	12.86
Investment	£12 000	17.14
Repairs	£4 000	5.71
Total	£70 000	100.00

3 Produce a histogram from the following data:

Sales (£'000)	Frequency
under 3	2
3 and under 6	10
6 and under 9	12
9 and under 12	15
12 and under 18	10
18 and under 21	9
21 and over	5
Total	63

(Source of data: Record of the sales achieved by staff of Sellmore Co Ltd for 1988)

The Polytechnic of North London; Accounting Foundation Course.

Solution/hints

As stressed in section 3.4.1.5, the construction of a histogram demands more planning than a simple bar chart. The values of the data (the frequency) are proportional to the *areas* of the bars and not, as is the case with bar charts, the heights. The bars will not necessarily be of equal width: this happens only when the intervals are equal, which is not the case here.

The first step is to determine the interval widths, in particular the width that occurs most often. In this case it is £3000 ('3 and under 6', '6 and under 9', and so on). Choose this width as the 'standard' interval. The interval '12 and under 18' is *twice* the size of the standard; consequently its height is *halved* (from 10 to 5).

In this example there are also open intervals – ones which do not have a lower (or higher) limit. To construct the chart one is forced to choose limits for these intervals. It is reasonable to assume that sales could have had a *zero* value, so the first interval is taken to begin at 0. The upper interval (21 and over) is not so straightforward. In such a case it is usual to refer to the existing interval widths, which in this case results in choosing an upper limit of 24.

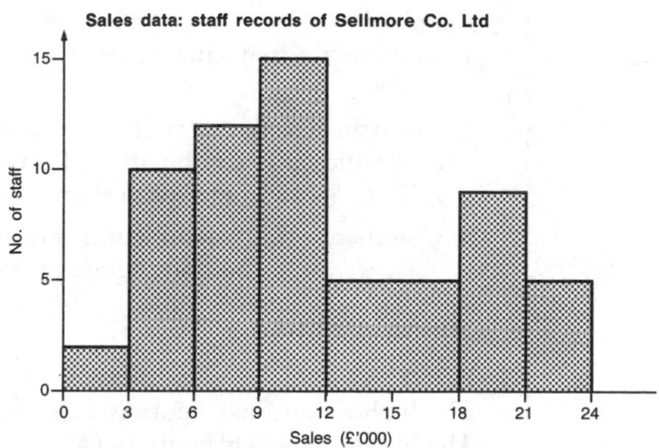

4 (a) Briefly explain, using suitable sketches, four different methods of presenting data diagrammatically.

(b) Distribution of personal incomes before tax 1979–80:

Percentage of people		Percentage of total income before tax
Top 10		24.8
10 and under 20		15.6
20	30	12.8
30	40	11.0
40	50	9.2
50	60	7.8
60	70	6.6
70	80	5.3
80	90	4.0
Bottom 10		2.9

(Source: Social Trends)

(i) Draw a Lorenz curve to represent the above data.

(ii) Comment briefly on your Lorenz cuve.

CIS, Part 1; Quantitative studies; June 1987.

Solution/hints

See section 3.5.3.

5 Social scientists have devised a means to classify people according to certain social and economic variables. This classification, the socio-economic grouping, is divided into the following groups: A,B,C1,C2,D,E, in descending order.

The following data represents the weekly earnings of a sample of 100 individuals in relation to their socio-economic class:

Socio-economic/earnings data. Sample size 100.

135 (B)	60 (E)	48 (D)	183 (C2)	54 (E)	96 (C2)	66 (C2)	192 (C1)	186 (C1)
99 (D)	147 (D)	84 (C1)	180 (C2)	222 (B)	150 (C2)	147 (C2)	192 (B)	99 (D)
87 (C2)	63 (E)	72 (C1)	144 (C2)	69 (D)	81 (C2)	78 (E)	78 (D)	111 (C2)
111 (C2)	255 (A)	54 (C1)	126 (C2)	129 (C1)	201 (B)	57 (E)	66 (D)	51 (C2)
51 (C2)	57 (D)	69 (D)	150 (C2)	198 (C1)	222 (C1)	51 (E)	111 (C2)	165 (C1)
156 (C2)	82 (D)	111 (C2)	126 (C1)	120 (C2)	237 (B)	69 (E)	72 (D)	93 (D)
132 (C1)	45 (E)	54 (E)	120 (C2)	195 (C1)	51 (E)	135 (B)	60 (E)	183 (C2)
54 (E)	66 (C2)	192 (C1)	99 (D)	147 (D)	188 (C2)	222 (B)	147 (C2)	192 (B)
87 (C2)	63 (E)	144 (C2)	69 (D)	78 (E)	78 (D)	111 (C2)	255 (A)	126 (C2)
129 (C1)	57 (E)	51 (C2)	57 (D)	69 (D)	198 (C1)	222 (C1)	111 (C2)	156 (C2)
82 (D)	126 (C1)	120 (C2)	237 (B)	69 (E)	72 (D)	132 (C1)	45 (E)	54 (E)
195 (C1)								

Required:

1 Construct a frequency distribution of EARNINGS, choosing intervals of a suitable width.

2 Construct a two-way table showing the frequencies in each earnings/socio-economic class combination. Consider earnings in the THREE groups: less than £75; £75–135; and more than £135.

3 Comment on the main features of the data, as viewed in your answer to 2 above.

The Polytechnic of North London; Accounting Foundation Course.

Solution/hints

Part 1

The highest earnings figure is £255 per week.
The lowest earnings figure is £45.
This gives a range of £255–£45 = £210.

Aiming for 8–10 intervals, appropriate interval widths would be 21 to 26. A convenient interval width of £25 gives the following frequency distribution.

Weekly earnings (£)	No. of individuals (frequency)
45– 70	30
71– 95	14
96–120	13
121–145	12
146–170	9
171–195	11
196–220	3
221–245	6
246–270	2
Total	100

Part 2

Wages and Social Class

Weekly earnings (£)	Socio-economic class A	B	C1	C2	D	E	Total
less than 75			2	5	10	16	33
75–135		2	7	15	8	2	34
more than 135	2	7	10	12	2		33
Total	2	9	19	32	20	18	100

Part 3

There is a tendency for high earnings to be in the 'high' socio-economic groups; even though the 'low', 'medium' and 'high' earnings are distributed in roughly equal proportions throughout the sample.

Recommended reading

Key Data 87, Central Statistical Office (HMSO, 1987).

Chapman, M., in collaboration with Mahon, B., *Plain Figures* (HMSO, 1986).

Owen, F. and Jones, R., *Statistics* (Pitman, 1988), chapters 2 and 3.

Whitehead, P. and Whitehead, G., *Statistics for Business* (Pitman, 1984), chapters 4–7.

Reichmann, W. J., *Use and Abuse of Statistics* (Pelican, 1978).

Curwin, J. and Slater, R., *Quantitative Methods for Business Decisions* (Van Nostrand Reinhold, 1986), chapter 2.

4 Descriptive Statistics

4.1 Introduction

If you have just started a course, you might be interested in obtaining information on the starting salaries of past graduates of your course. One possible source of relevant information would be the careers office. Suppose they presented a sample of past graduates' starting salaries in the form of a table of data (see table 4.1).

Table 4.1 Starting salaries for graduates of the Business School (annual figures; £)

Main subject area Accounting	Salary	Main subject area Personnel	Salary
1	9 850	1	8 100
2	8 750	2	8 000
3	9 650	3	8 800
4	11 250	4	8 250
5	10 500	5	10 800
6	12 750	6	12 400
7	14 000	7	9 200
8	13 000	8	10 000
9	10 000	9	10 900
10	7 900	10	9 750
11	8 750	**Main subject area Business Studies**	**Salary**
12	8 500	1	9 500
13	9 300	2	10 500
14	9 550	3	12 500
15	10 500	4	10 800
16	10 600	5	11 200
17	11 550	6	25 250*
Main subject area Marketing	**Salary**	7	9 550
1	8 200	8	9 100
2	9 500	9	9 550
3	11 750	10	11 400
4	11 250	11	12 500
5	9 000	12	13 600
6	12 300	13	10 900
7	10 500	14	9 750

Table 4.2 Frequency distribution of graduate salaries (all subject specialisms)

Annual salary (£)	Number of graduates (frequency)
⩽ 8 000	2
> 8 000 but ⩽ 9 000	8
> 9 000 but ⩽ 10 000	14
> 10 000 but ⩽ 11 000	9
> 11 000 but ⩽ 12 000	6
> 12 000 but ⩽ 13 000	6
> 13 000 but ⩽ 14 000	2
> 14 000	1
Total frequency	48

It is difficult to absorb the information in such a form, but we have seen in the preceding chapters how data can be summarized as, for example, a frequency distribution (table 4.2), or presented in some diagrammatic form, for example, a histogram (fig. 4.1). While these are undeniably useful, it is often desirable to summarize data in the form of a numerical value: a descriptive statistic such as the arithmetic mean or the standard deviation, which provide **measures of centre and dispersion**, respectively. This chapter considers the various forms of descriptive statistics in use and the relevant applications.

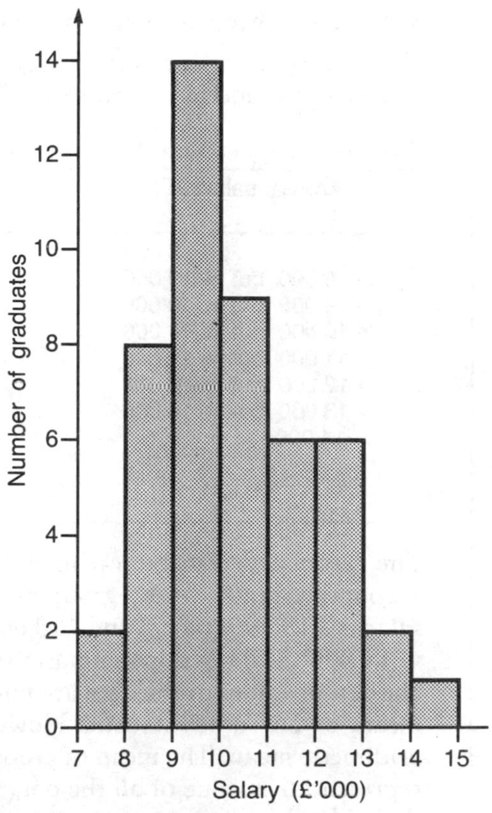

Fig. 4.1 Graduate salaries (all subject areas)
(*Note:* End classes are 'open': ≤ 8 and > 14. In such cases the missing limits must be estimated.)

4.2 Measures of centre

Measures of centre (or **measures of location**) are statistics which give an indication of the centre of a set of data. The arithmetic mean is such a measure. If calculated using all of the population data, it is called the **population mean**. If calculated using a sample taken from the population, then the resulting statistic is called a **sample mean**. For instance, referring to the data in table 4.1, the average salary of personnel graduates, £9620, is a sample mean, as the data used is not the whole population of graduate salaries, but a sample. This statistic would enable one, for example, to compare salary offers made by prospective employers and to judge the relative merit of such offers. The mean is just one example of a number of measures of central location that we will consider.

4.2.1 Arithmetic mean

The arithmetic mean is the measure of centre most people are referring to when they use the term 'mean' or 'average'. It is probably the most important measure of central location.

Consider again the data in table 4.1. Suppose we are interested in the starting salary of graduates whose main subject area is marketing. The relevant subset of data can be seen in table 4.1. The mean is calculated by summing the data and dividing by the total frequency. So, by considering the individual data as $x_1, x_2, \ldots, x_n$, (i.e. $x_1 =$ £8200, $x_2 =$ £9500, ..., $x_7 =$ £10 500), the mean ($\bar{x}$) is calculated as follows:

$$\bar{x} = \frac{x_1 + x_2 + \ldots + x_n}{n}$$

$$= \frac{(£8200 + £9500 + \ldots + £10\,500)}{7}$$

$$= £10\,357.14$$

This is usually summarized symbolically in the form:

$$\bar{x} = \frac{\Sigma x_i}{n}$$

(Σ is read as 'the sum of'.)

4.2.2 Arithmetic mean: grouped data

When data is presented in classes or intervals the situation is slightly different. We will now consider a typical example.

Table 4.3 Frequency distribution of graduate salaries (main subject area: accounting)

Annual salary (£)	Midpoint (£) m	Frequency f	$m \times f$
⩽ 8 000	7 500	1	7 500
> 8 000 but ⩽ 9 000	8 500	3	25 500
> 9 000 but ⩽ 10 000	9 500	5	47 500
> 10 000 but ⩽ 11 000	10 500	3	31 500
> 11 000 but ⩽ 12 000	11 500	2	23 000
> 12 000 but ⩽ 13 000	12 500	2	25 000
> 13 000 but ⩽ 14 000	13 500	1	13 500
> 14 000	14 500	0	0
	Totals	17	£173 500

The frequency distribution in Table 4.3 has simplified the data presentation for accounting graduates by grouping the salaries into class intervals. For example, the salaries £10 500 (twice) and £10 600 are contained within the interval '> 10 000 but ⩽ 11 000'. Data in this summarized form is certainly easier to read, but detail is lost—the actual values of the data are not known. This makes it impossible to calculate the mean, or any other statistic. However, it is still possible to **estimate** a value of the arithmetic mean: The mean of grouped data uses the mid-value of each interval as the representative value of all the data in the interval. The mean is estimated by treating the mid-values as if they were the actual observations. For instance (referring again to table 4.3), £8500 is the mid-value of the interval [8000, 9000] and is interpreted as occurring 3 times (in fact, the observations within these limits are £8750, £8750 and £8500). Once the mid-values have been established, an estimate for the total salary (£173 500) can be found and the arithmetic mean is estimated as before (by dividing by the total frequency):

$$£173\,500 \div 17 = £10\,205.88$$

(The true sample mean for accounting graduates is £10 376.47. The estimate has an error of £170.59 or 1.644 per cent of the true value.)

As an exercise, find estimates of the means of the data for the graduates in the other subject areas (refer to table 4.1 for data). You will need to group the data as shown in table 4.3. By comparing your results with the actual means you should find that estimates calculated in this way will compare favourably with the true values even though the sample sizes are small. As the sample size increases one would expect thc precision of the estimate to improve. This method of estimation relies on the assumption that data will be uniformly distributed within each class interval, that is, that data is not 'bunched' within the interval.

Other factors which will influence the accuracy of any estimate are:

1 The presence of extreme values, such as the salary of £25 250 of the graduate specializing in business studies (marked with an asterisk (*) in table 4.1).
2 The choice of limits for open-ended class intervals. In the example considered above it was assumed that the end intervals had the same width as the other intervals. Hence, '⩽ £8000' was taken as [£7000, £8000], giving the mid-point £7500, and '> £14 000' was taken as [£14 000, £15 000], giving the midpoint £14 500.

4.2.3 The mode

The mode is that data value which occurs the most frequently. The mode of the salaries of the graduates specializing in business studies is £9550, which occurs twice. The other values all occur once only. The mode of the subset of salaries for those graduates specializing in accounting is not unique. The values £10 500 and £8750 each occur twice, making that subset of data **bimodal**. In such a case it would not be sensible to use the mode as a measure of centre. In general, if the sample size is small it is inadvisable to do so. The median would be a better choice.

4.2.4 The median

The median is that value below (or above) which 50 per cent of the data falls. The median can be thought of as the value of the item with rank $(n+1)/2$, when n items of data are ranked in order of magnitude. If there is an odd number of items, then the median is the middle value. If there is an even number of data values then the median is the arithmetic mean of the middle two data values.

Referring again to table 4.1, the median salary for the marketing graduates is the underlined value in the following listing:

£8200 £9000 £9500 <u>£10 500</u> £11 250 £11 750 £12 300

The median salary for the personnel graduates is the mean of the two middle terms:

8000 £8100 £8250 £8800 <u>£9200 £9750</u> £10 000 £10,800 £10 900 £12 400

i.e. Md = median = (£9200 + £9750)/2 = £9475

The arithmetic mean is the most commonly used measure of centre but there are situations where the median is more suitable. For instance, one salary figure (£25 250) is noticeably larger than the rest of the data in the same subset. It would be considered an extreme value and its inclusion in the calculation of the mean would give a distorted view of the 'average' salary available for graduates in this area. One solution to this problem is to exclude any extreme values. In this instance (business studies specialists), we have a mean value of £11 864.29 and an amended mean value (excluding the value £25 250) of £10 834.62. Referring to fig. 4.2, it can be seen that the amended value is more representative of the salaries earned by graduates in that area. An alternative would be to consider using the mode or the median. In this case the median, £10 850.00, is more appropriate (there are two modes).

In general, the median is the 'average' one would choose as a statistic to illustrate typical income.

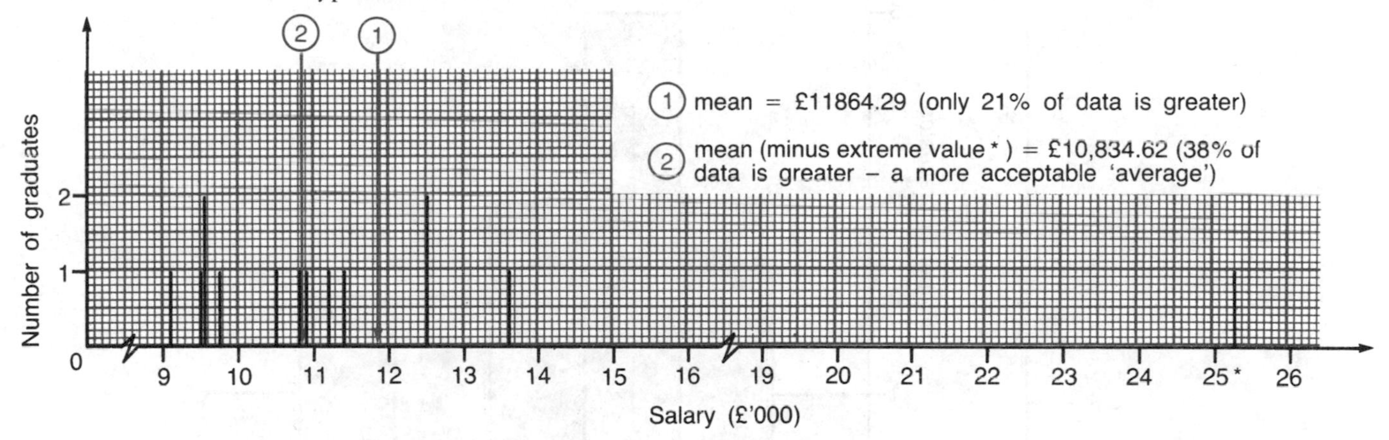

Fig. 4.2 A line chart showing the distribution of salaries of business studies graduates

4.2.5 The median: grouped data

Table 4.4 Frequency distribution and cumulative frequency of graduate salaries (main subject area: business studies)

Annual salaries	Frequency	Cumulative frequency
⩽ 8 000	0	0
> 8 000 but ⩽ 9 000	0	0
> 9 000 but ⩽ 10 000	5	5
> 10 000 but ⩽ 11 000	3	8*
> 11 000 but ⩽ 12 000	2	10
> 12 000 but ⩽ 13 000	2	12
> 13 000 but ⩽ 14 000	1	13
> 14 000	1	14

*median (7th item) occurs in the interval '> 10 000 but ⩽ 11 000'

You are likely to encounter questions which require you to find the median of grouped data, such as that illustrated in table 4.4. In such a case it is impossible to list the data in ascending order, because the actual values of the data are unknown.

Consequently the median must be estimated. To estimate the median we need to consider the cumulative frequency of the data (cumulative frequencies and ogives were considered in chapters 2 and 3). When estimating the median we are looking for the $n/2$ term. In table 4.4, the interval that contains the $n/2$ term is marked with an asterisk (*). As in the estimation of the arithmetic mean from grouped data, we assume the items are uniformly distributed throughout the interval. We can then estimate the median as follows:

$$\mathbf{Md = Median} = \boldsymbol{L + (n_1/n_2) \times l}$$

where: L is the lower limit of the median interal;
$n_1 = \frac{1}{2}N - (\Sigma f)_L$, where N is the total frequency and $(\Sigma f)_L$ is the cumulative frequency up to the median interval (in the example (*), this gives $n_1 = 14/2 - 5 = 7 - 5 = 2$);
n_2 is the frequency of the median interval; and
l is the width of the median class.

For example, the estimated median, for the business studies graduates, is calculated as follows:

$$\text{median salary} = £10\,000 + (2/3) \times £1000 = £10\,666.67$$

By referring to the original, ungrouped data (table 4.1), you will find that the true median is £10 850. The estimate can be seen to be a good one (an absolute error of £183.33 or an error of 1.69 per cent relative to the true value).

4.2.6 Mode and median: graphical methods

The mode or median can also be obtained from graphs.

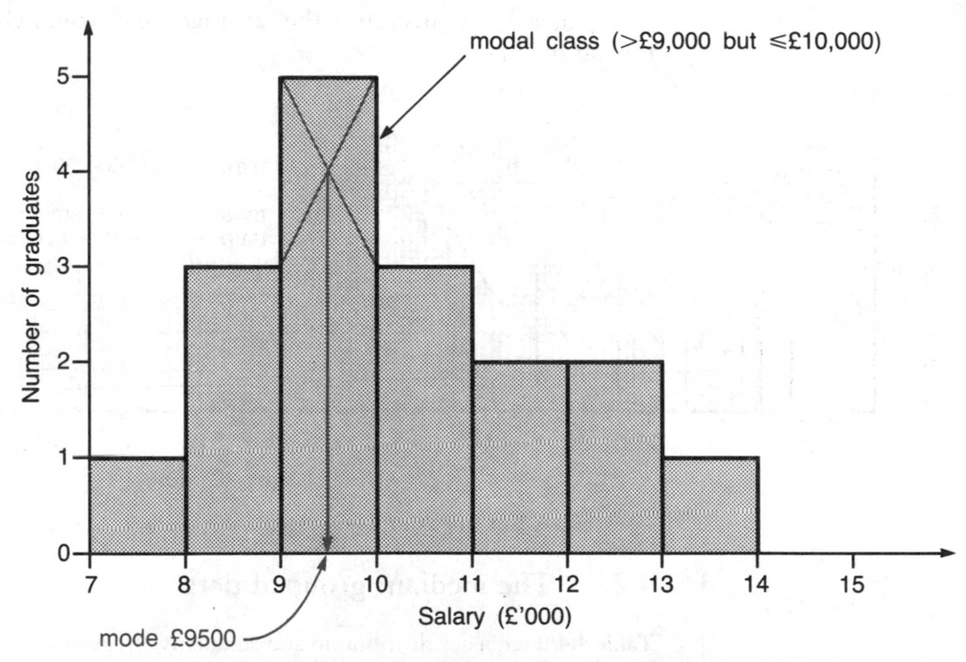

Fig. 4.3 Graduate salaries (main subject area: accounting)

For an example of obtaining the mode using graphical techniques, consider the subset of graduates' salary data referring to accounting specialists (grouped in table 4.3). Figure 4.3 shows the histogram of this data. First, the modal class (or interval) is found. In this case there is only one modal class [£9000, £10 000]. The mode is then found by joining the corners of the modal class to the intersecting points of the adjacent classes, as shown on the histogram. The estimate of the mode is £9500.

The assumptions are that the true mode is likely to be in the interval or class containing most values (the modal class), and that within that class the mode is likely to be proportionally closer to the adjacent class containing more observations. If both adjacent classes have the same frequency (as in this example), the mode is estimated as the mid-value of the modal class.

An ogive is used to estimate the median (see fig. 4.4, drawn from data in table 4.2). It is worth noting that the estimation of the median is not influenced by the lack of knowledge concerning the values of the extreme class intervals: both upper and lower limits are estimated. In such cases the median is a good statistic to use if we wish to establish a measure of location. The mode is similarly unaffected by the values of the end intervals. It would have to be a peculiarly skewed distribution (see section 4.4) for this not to be so.

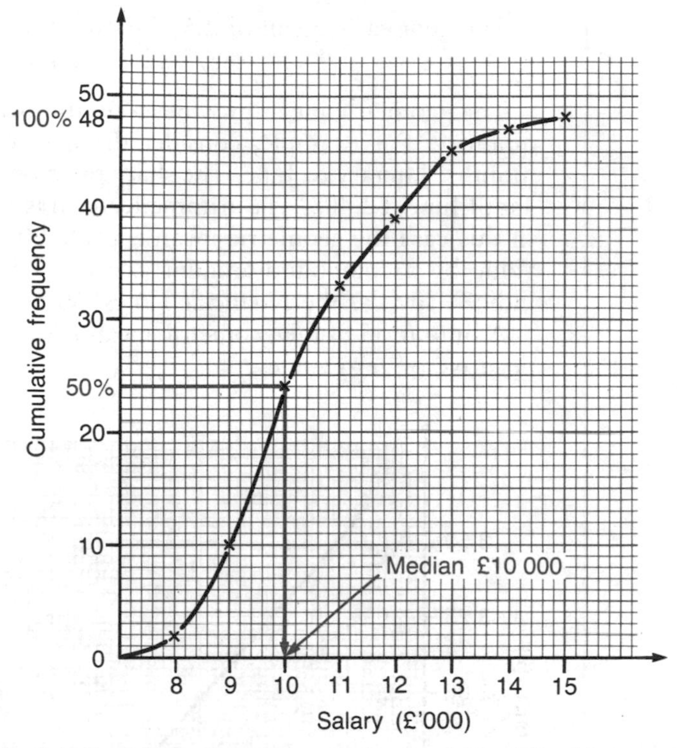

Fig. 4.4 Cumulative frequency – graduate salaries (all subject specialisms)

4.2.7 The best measure of centre to use

The arithmetic mean, the mode and the median are the measures of location you are most likely to encounter in an examination.

The **mean** is the most commonly used, and most easily understood by the general public – the term 'average' tends to be used exclusively for the mean rather than for the mode or median. The **advantages** of the mean are:

1 most people understand what it is describing; and
2 unlike the mode and the median, the mean uses all of the available information. That is, all of the data is used in its computation.

The **disadvantages** are that:

1 the mean can be distorted by extreme values; and
2 the mean will rarely correspond to an actual value contained in the data (the mean salary in our example (all subject areas) is £10 650.00, which no single graduate earned).

The **mode** is used in situations which demand that we know the most common value of an item. For instance, a house builder would probably find it more useful to know that the most commonly sought after house was three-bedroomed, rather than that the mean number of bedrooms wanted by purchasers was 3.235.

The **median** is a useful way of analysing some types of data – e.g. life expectancy and typical income. In monitoring salaries over time, the median will be unaffected by extreme values, unlike the mean.

Neither the mode nor the median take every value into account. This can be an advantage in that extreme values do not distort the statistic obtained, but it can also be a disadvantage, in that dispersion of data is ignored. Two less commonly used averages are the geometric mean and the harmonic mean.

4.2.8 The geometric mean

The geometric mean of n items of data is calculated by multiplying the data together, then taking the nth root of the product.

i.e. The geometric mean of $x_1 x_2 \ldots, x_n$ is equal to

$$\sqrt[n]{x_1 \times x_2 \times \ldots \times x_n}$$

e.g. The geometric mean of 2.5 and 5.6 is equal to

$$\sqrt{2.5 \times 5.6} = 3.74 \text{ (2 d.p.)}$$

The geometric mean of 2.5, 5.6 and 4.8 is equal to

$$\sqrt[3]{2.5 \times 5.6 \times 4.8} = 4.07 \text{ (2 d.p.)}$$

The geometric mean provides a useful measure of the rate of change in a value or quantity. For example, consider the market price of a share over a period of six months, January to June. The share price on the 1st of January was £3.57 and on the 1st of June, £2.405. The arithmetic mean is £2.9875. The geometric mean is £2.9302. In this case the geometric mean is probably more accurate because shares change in value by a compound amount (see fig. 4.5). It is used in the preparation of some indices, such as the *Financial Times'* 30-Share Index.

A limitation of the geometric mean is that it cannot be applied if the data contains negative or zero values.

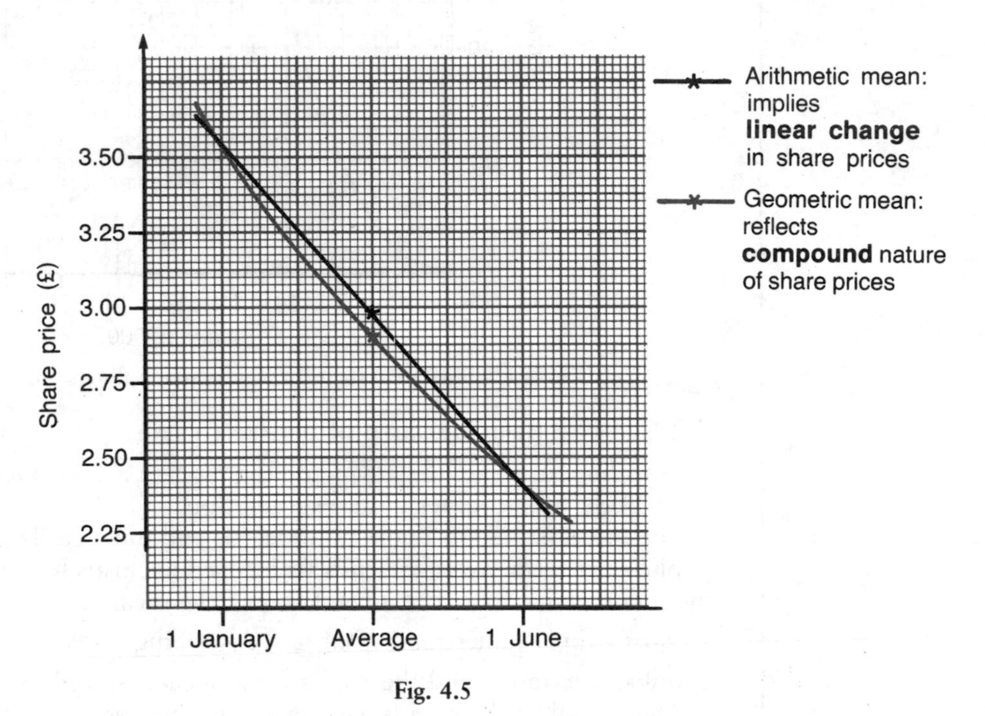

Fig. 4.5

4.2.9 The harmonic mean

The harmonic mean of n items of data ($x_1, x_2, \ldots, x_n$) is equal to

$$\frac{n}{1/x_1 + 1/x_2 + \ldots + 1/x_n}$$

For example, a car travels from A to B at 50 miles per hour (m.p.h.) and from B to A at 30 m.p.h. The distance from A to B is 18 miles. At first it might seem reasonable to calculate the arithmetic mean in order to find the average speed for the round trip – this would be 40 m.p.h. Further investigation shows this to be incorrect. Note the following:

Distance travelled	Time taken
18 miles (at 50 m.p.h.)	21.6 minutes
18 miles (at 30 m.p.h.)	36.0 minutes

Total distance travelled = 36 miles
Total time taken = 57.6 minutes

This gives an average speed of $60/57.6 \times 36 = 37.5$ m.p.h.

Using the formula given above the harmonic mean is equal to:

$$\begin{aligned}\text{harmonic mean} &= 2/(1/50 + 1/30) = 2/(0.02 + 0.03333)\\ &= 2/0.05333\\ &= 37.5 \text{ m.p.h.}\end{aligned}$$

4.3 Measures of dispersion

Averages provide information about the middle or centre of the range of data under investigation. However, it is often not enough to know the average of a set of data. Consider the sets of data illustrated in table 4.5 and shown in the form of histograms in fig. 4.6.

Table 4.5 Frequency distributions of the salaried staff of departments A, B and C of Polyproducts Co Ltd

Annual salary (£'000)	Number of employees		
	Department A	**Department B**	**Department C**
> 11 but ⩽ 12	7	22	0
> 12 but ⩽ 13	10	8	13
> 13 but ⩽ 14	15	0	23
> 14 but ⩽ 15	9	0	9
> 15 but ⩽ 16	6	2	5
> 16 but ⩽ 17	3	18	0
Total frequency	50	50	50
Mean salary	£13 620	£13 620	£13 620

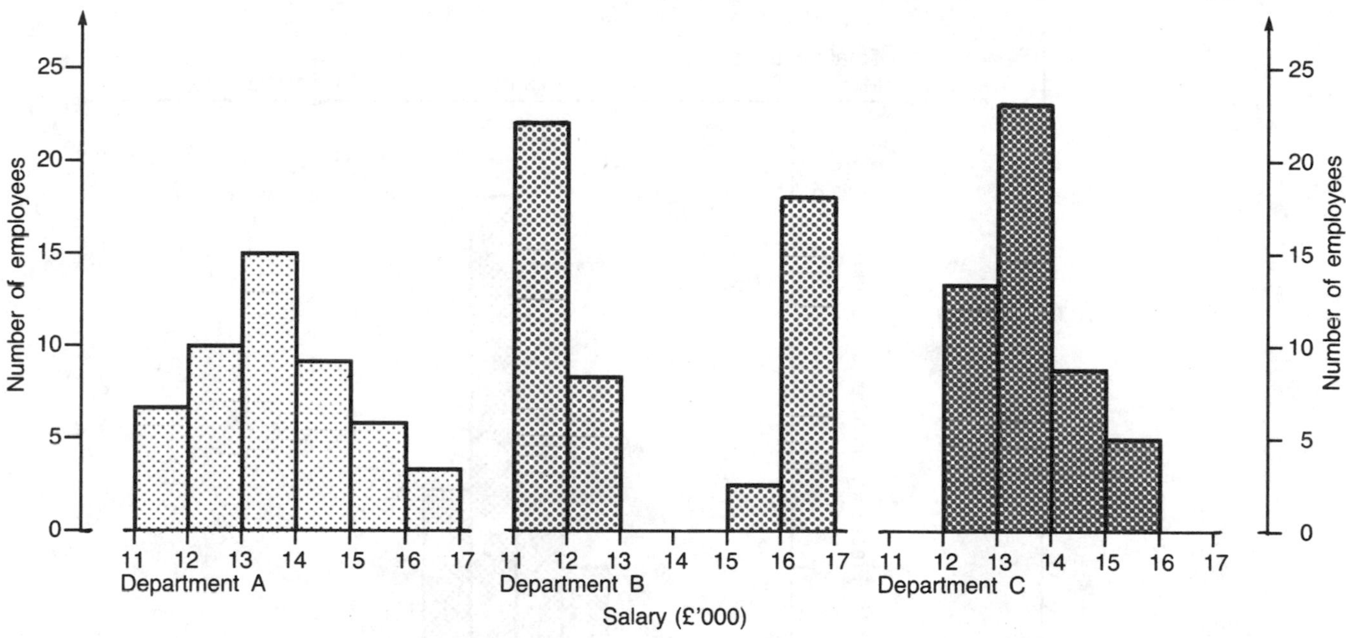

Fig. 4.6 Polyproducts Co Ltd – staff salaries, Departments A, B and C

Suppose you were considering taking up an appointment with Polyproducts Co Ltd and you were in a position to choose which department to join. One area of concern would be the expected level of salary and it would be reasonable to investigate the levels of salaries enjoyed by the present members of the departments. One method of doing that would be to look at the average salary in each department. In this case the mean value is the same in each case, £13 620. You could, on this basis, accept that it made no difference which department you worked in, or you could consider an alternative method of comparing salary levels between the departments.

The histograms in fig. 4.6 show that salaries are distributed, or spread, differently within each department. The salaries of members of Department A range from £11 000 to £17 000, as do those in Deparment B. There is, however, a dramatic difference. If you joined Department B it would be impossible for you to start with the average salary of £13 620. Also, while more staff in that department are earning relatively high salaries, there are also more earning relatively low salaries. Do you gamble on being a high earner or choose Department A, the relatively safe path to a middle salary level? Perhaps the best choice is Department C: in this department no one is paid less than £12 000, though no one earns more than £16 000.

It must be obvious by now that the mean fails to give the fullest information about a set of data. In the example just considered, further information was obtained by observing how the data was spread about its mean. There are a number of ways of measuring variability (or spread) and the different measures we will consider vary from simple-to-compute but limited-in-use (the range), to the relatively more complicated-to-compute but much more useful (the standard deviation).

Table 4.6 Data showing the number of days taken to respond to requests for component orders. Delivery service: Earlybird Delivery Service and Ontime Transporters

Delivery time (days)	Frequency	
	Earlybird	Ontime
5		3
6		7
7		0
8	10	15
9	20	30
10	51	35
11	19	20
12		10
13		12
14		10
15		8
Total frequency	100	150
Mean time	9.79 days	10.35 days

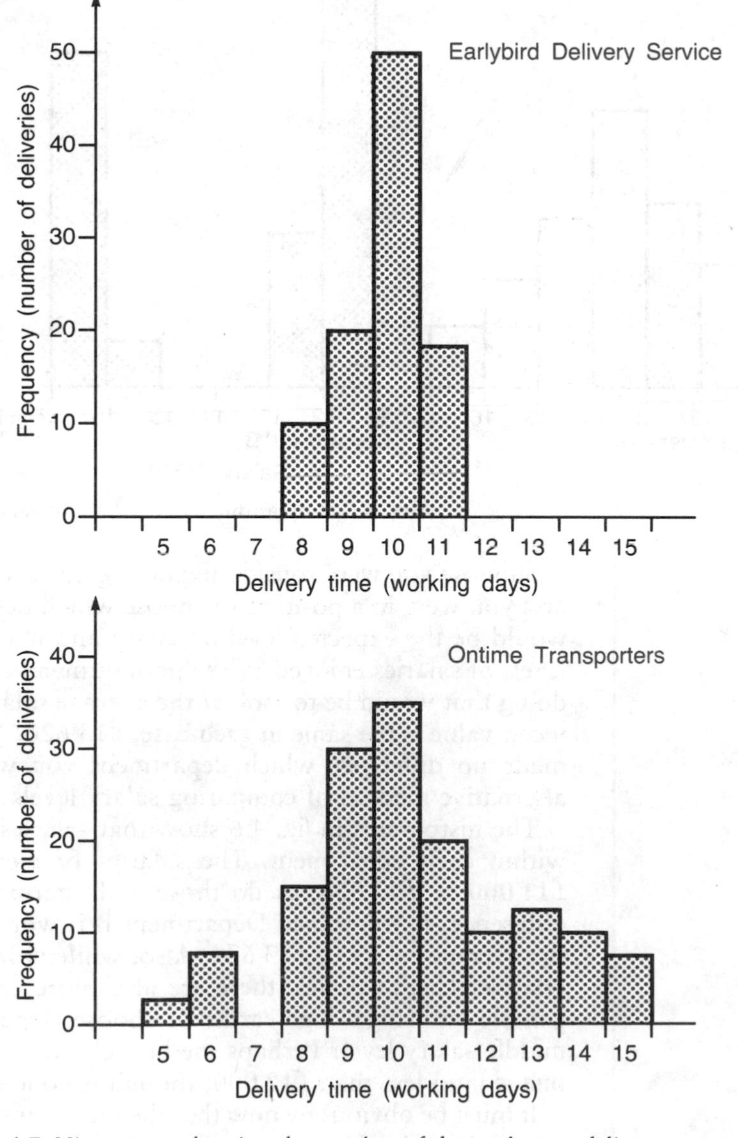

Fig. 4.7 Histograms showing the number of days taken to deliver component orders

The following examples refer to the data illustrated in table 4.6 and fig. 4.7, showing the service records of two delivery companies, Earlybird Delivery Service and Ontime Transporters. They both claim to deliver goods in 10 working days. The average delivery time (arithmetic mean) shows that both companies did (approximately) adhere to their claims of 10-day delivery. However, reliability is an important factor of business life and it does appear that Ontime have been the less reliable. At times they took as little as 5 days to deliver, but often took more than 12 days. The following methods can be used to measure this variability in performance.

4.3.1 The range

The range is probably the simplest way of measuring the variability or spread of data. It is simply the difference between the largest and the smallest data values:

$$\textbf{Range} = \textbf{maximum value} - \textbf{minimum value}$$

In our example:

$$\text{Range (Earlybird)} = 11 - 8 = 3 \text{ days}$$
$$\text{Range (Ontime)} = 15 - 5 = 10 \text{ days}$$

The range, though simple to compute, is seldom used as it uses only two pieces of data. This makes it sensitive to extreme values. For example, if, on one occasion, Earlybird were to deliver in one day, then the range for this company would become $(11 - 1)$ 10 days, suggesting that the two companies had similar variability in delivery time, an unjust comparison. One way around this problem of extreme values is to ignore them, as is done in the following measure of dispersion.

4.3.2 The interquartile range

The lower quartile (Q_1) is that value below which 25 per cent of the data falls. The middle quartile (Q_2), more usually called the median, is that value below (or above) which 50 per cent of the data falls. The upper quartile (Q_3) is that value above which 25 per cent of the data falls. The interquartile range is simply the difference between the upper quartile and the lower quartile:

$$\textbf{Interquartile range} = Q_3 - Q_1$$

To find the quartile and interquartile range, calculate the cumulative frequencies as percentages (see table 4.7).

Table 4.7 Delivery service: Earlybird Delivery Service and Ontime Transporters – cumulative frequency data

	Earlybird			Ontime		
Delivery time (days)	**Frequency**	**Cumulative frequency**	**%**	**Frequency**	**Cumulative frequency**	**%**
5				3	3	2
6				7	10	6.67
7				0	10	6.67
8	10	10	10	15	25	16.67
9	20	30	30 (Q_1)	30	55	36.67 (Q_1)
10	51	81	81 (Q_3)	35	90	60
11	19	100	100	20	110	73.33
12				10	120	80 (Q_3)
13				12	132	88
14				10	142	94.67
15				8	150	100
Mean time	9.79 days			10.35 days		

For Earlybird, the lower quartile is 9 days. The upper quartile is 10 days. These values give an interquartile range of 1 day $(10 - 9)$. For Ontime, the lower quartile is also 9 days, while the upper quartile is 12 days, giving an interquartile range of 3 days $(12 - 9)$. This measure again reflects the greater variability in Ontime's delivery record when compared to Earlybird's.

For grouped data you would use a similar method to that employed in calculating the median. It is possible that an examination question would require you to show the median, lower and upper quartiles on an ogive. Figure 4.4 showed an ogive of graduate salaries, with the median indicated. This same ogive is shown in fig. 4.8 with upper and lower quartiles indicated. The interquartile range in this case is £2300 (£11 500 – £9200).

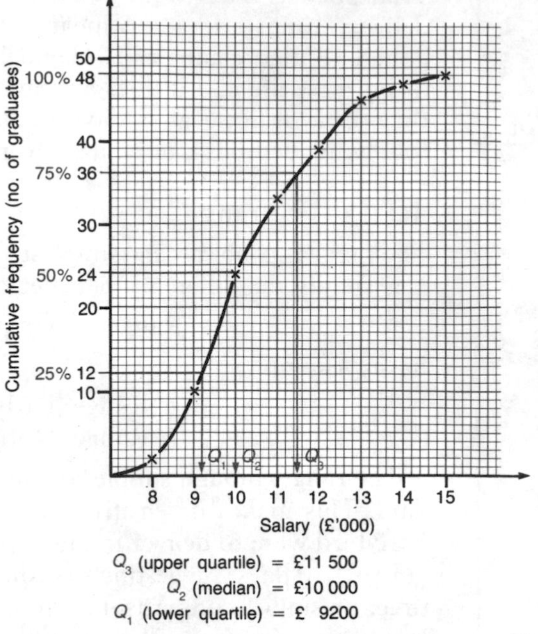

Fig. 4.8 Graduate salaries – lower, middle and upper quartiles

4.3.3 The mean deviation

The mean deviation measures the average difference between the mean and each item of data. Symbolically this is expressed as:

$$\textbf{Mean deviation} = \Sigma|x_i - \bar{x}|/n$$

where: $|x_i - \bar{x}|$ is the absolute difference between the mean, $\bar{x}$, and the *i*th piece of data x_i (e.g. $|12-8| = 4$ and $|8-12| = |-4| = 4$);

$\Sigma|x_i - \bar{x}|$ is the sum of the differences. If we are dealing with a grouped data, this is expressed as $\Sigma f_i|x_i - \bar{x}|$; and

n (or Σf_i) is the number of items of data (or total frequency).

(*Note:* For grouped data the mid-values of each class would replace 'x_i' in the calculations.)

The symbols and terminology should be clear after the following example.

Table 4.8 Delivery time and the mean deviation calculation for Ontime Transporters

Delivery time (days) x	Deviation from the mean $d = \|x - \bar{x}\|$	Frequency f	$f \times d$	Examples of calculations
5	5.35*	3	16.05	*d = $\|x - \bar{x}\|$ = $\|5 - 10.35\|$ = 5.35
6	4.35	7	30.45	
7	3.35	0	0.00	
8	2.35	15	32.25	†d = $\|x - \bar{x}\|$ = $\|13 - 10.35\|$ = 2.65
9	1.35	30	40.50	
10	0.35	35	12.25§	
11	0.65	20	13.00	§f × d = 35 × 0.35 = 12.25
12	1.65	10	16.50	
13	2.65†	12	31.80	
14	3.65	10	36.50	
15	4.65	8	37.20	
Totals		150	269.50	

Mean time = 10.35 days
Mean deviation time = 269.5/150 = 1.797 days

The mean deviation of delivery time (Ontime Transporters) is calculated as shown in table 4.8. The value is 1.797 days. This compares with 0.674 (67.4/100) days, the mean deviation of delivery time for Earlybird Delivery Service. (The calculation for the Earlybird data is left for you to do as an exercise.)

The mean deviations of the two companies (Earlybird 0.674 days and Ontime 1.797 days) allow us to compare the variability of their delivery times and judge that Earlybird offers a more reliable delivery service.

The mean deviation is seen as an improvement on the range and the interquartile range, as it uses all the available data and yet is not affected to any significant amount by extreme values. This is because the deviations of each data value from the mean are averaged.

The calculation of the mean deviation can be cumbersome (especially if the number of observations, n, is large). Though the mean deviation has intuitive appeal, the fact that the signs are ignored (absolute values are used) limits its applicability in further statistical work.

4.3.4 The variance and the standard deviation

The variance, like the mean deviation, looks at the difference between the data values and the arithmetic mean. Instead of ignoring the signs though, calculating the variance involves squaring these differences (or deviations from the mean). Symbolically, the **variance** is expressed in the following way:

$$\sigma^2 = \Sigma f_i(x_i - \bar{x})^2 / \Sigma f_i \qquad (1)$$

Because the variance results in 'squared' units, its square root, the standard deviation, is more useful. For instance, if analysing data measured in kilograms, the variance would have units of square kilograms, which have no meaning. By calculating the square root of the variance we retain the initial base unit. Symbolically, the **standard deviation** is:

$$\sigma = \sqrt{\Sigma f_i(x_i - \bar{x})/\Sigma f_i} \qquad (2)$$

(This is the formula for grouped population data.)

Consider the example illustrated in table 4.9. The columns of table 4.9 represent the stages in calculating the variance and the standard deviation using formulae (1) and (2).

Table 4.9 Frequency distribution of graduate salaries (main subject area: accounting) – calculation of variance and standard deviation

Annual salary (£'000)	Mid-point x	Frequency f	$x-\bar{x}$	$(x-\bar{x})^2$	$f(x-\bar{x})^2$
$\leqslant$ 8	£7 500	1	−2 705.88	7 321 786.6	7 321 786.6
> 8 but $\leqslant$ 9	£8 500	3	−1 705.88	2 910 026.6	8 730 079.7
> 9 but $\leqslant$ 9	£9 500	5	−705.88	498 266.6	2 491 332.9
> 10 but $\leqslant$ 11	£10 500	3	294.12	86 506.6	259 519.7
> 11 but $\leqslant$ 12	£11 500	2	1 294.12	1 674 746.6	3 349 493.1
> 12 but $\leqslant$ 13	£12 500	2	2 294.12	5 262 986.6	10 525 973.1
> 13 but $\leqslant$ 14	£13 500	1	3 294.12	10 851 226.6	10 851 226.6
> 14	£14 500	0	4 294.12	18 439 466.6	0.0
Totals		$\Sigma f = 17$			$\Sigma f(x-\bar{x})^2 = 43\,529\,411.7$

Mean = $\bar{x}$ = £10 205.88 (see section 4.2.2 and table 4.3)
Variance = σ^2 = 43 529 412.0/17 = 2 560 553.6
Standard deviation = $\sigma = \sqrt{2\,560\,553.6}$ = £1 600.17

4.3.4.1 Calculation short-cuts

The calculations involved in the example in table 4.9 involved large numbers. Usually this will not be a problem – most calculators can cope with the numbers you are likely to encounter. If not, there are short-cuts you can adopt to make the calculations simpler:

The formulae (1) and (2) can be rewritten as follows:

$$\sigma^2 = \frac{\Sigma f(x-\bar{x})^2}{n} = \frac{\Sigma fx^2}{n} - \left(\frac{\Sigma fx}{n}\right)^2$$

$$= \frac{\Sigma fx^2}{n} - (\bar{x})^2 \qquad (3)$$

$$\sigma = \sqrt{\frac{\Sigma fx^2}{n} - (\bar{x})^2} \qquad (4)$$

We can also take account of the large numbers involved by reducing all the data to a more manageable scale and adjusting the standard deviation accordingly. The amended formula for the standard deviation would then read:

$$\sigma = c\sqrt{\frac{\Sigma fd^2}{n} - \left(\frac{\Sigma(fd)}{n}\right)^2} \qquad (5)$$

where c is a scaling factor, and $d = x/c$.

The calculations can be made even simpler by assuming (estimating) a mean (particularly useful if the mean is unknown). The formula for the standard deviation then looks the same as (5) but $d = (x - m)/c$, where m is the assumed mean.

This probably seems very complicated, so consider another example. The short-cuts will be applied to the accounting graduate salary data (tables 4.3 and 4.9). The calculations are shown in table 4.10. As can be seen, the results are not significantly different (£1600.20 compared with £1600.17), but the numbers are much easier to deal with.

Table 4.10 Frequency distribution of graduate salaries (main subject area: accounting) – mean and standard deviation calculated using short cut techniques
Assumed mean, m = £10 500; scaling factor, c = 1 000.

Mid-point x	Frequency f	$(x-m)/c$ d	fd	fd^2	Notes:
£7 500	1	−3*	−3	9	$^*d = \frac{(7500 - 10\,500)}{1000} = -3$
£8 500	3	−2	−6	12	
£9 500	5	−1	−5	5†	
£10 500	3	0	0	0	
£11 500	2	1	2	2	$\dagger fd^2 = 5 \times (-1)^2 = 5$
£12 500	2	2	4	8	NB: This is not to be confused with $(5 \times -1)^2 = 25$
£13 500	1	3	3	9	
£14 500	0	4	0	0	
	$\Sigma f = 17$		$\Sigma fd = -5$	$\Sigma fd^2 = 45$	

Standard deviation, $\sigma = 1\,000\sqrt{45/17 - (-5/17)^2}$
$= 1\,000\sqrt{2.6471 - 0.0865}$
$= 1\,000\sqrt{2.5606}$
$= 1\,000 \times 1.6002$
$=$ £1600.20 (allowing for rounding errors)

The mean is $m + (c \times d)$ = £10 500 + (1000 × (−5/17))
= £10 500 − 294.12
= £10 205.88

4.3.4.2 Sample statistics

In all statistical analysis it is important to distinguish between population statistics and statistics of samples taken from a population.

The formulae given above for the variance and the standard deviation are valid when a population is being investigated. The accounting graduates' starting salaries (initially encountered in table 4.1) has thus far been considered as the population of graduate salaries. If the data is considered to be a sample taken from a population, the variance ($\hat{\sigma}^2$) and standard deviation ($\hat{\sigma}$) statistics are calculated in essentially the same way as σ^2 and σ, with one important difference: Sample statistics are **estimates** of the relevant population statistics and, though the reason is not important for this text, better estimates of the population variance and the standard deviation are found if a correction (called **Bessel's correction factor**) is applied.

Bessel's correction factor is $\frac{n}{n-1}$.

(The derivation of this factor is beyond the scope of this text.)

Hence the formula for the sample variance becomes:

$$\hat{\sigma}^2 = \left[\frac{\Sigma fx^2}{n} - (\bar{x})^2\right] \times \frac{n}{n-1}$$

and the formula for the standard deviation is:

$$\hat{\sigma} = \sqrt{\left[\frac{\Sigma fx^2}{n} - (\bar{x})^2\right] \times \frac{n}{n-1}}$$

The standard deviation may seem complicated to compute, but with practice (and a reliable calculator) you should not encounter too many difficulties. Many examination papers provide the formulae, making it unnecessary to remember the different versions. However, you should practice using them until you feel confident about your skills.

The standard deviation is an important measure of dispersion because it is based on all of the available data and it can be used for further statistical analysis, as we will see later in the text. Like the mean deviation, the standard deviation can be used to compare the dispersion of two or more distributions.

4.4 Measures of skewness

We have considered some of the more commonly used measures of centre and measures of dispersion. For instance: the median is a measure of centre concerned with the point around which data is concentrated and standard deviation is a measure of dispersion concerned with the extent that data varies. It is unlikely that you would be asked to find any other measure, but skewness is an important statistical property which is worth noting.

Measures of skewness are concerned with the extent to which data is symmetrically distributed. A skewed distribution is one which is not symmetrical.

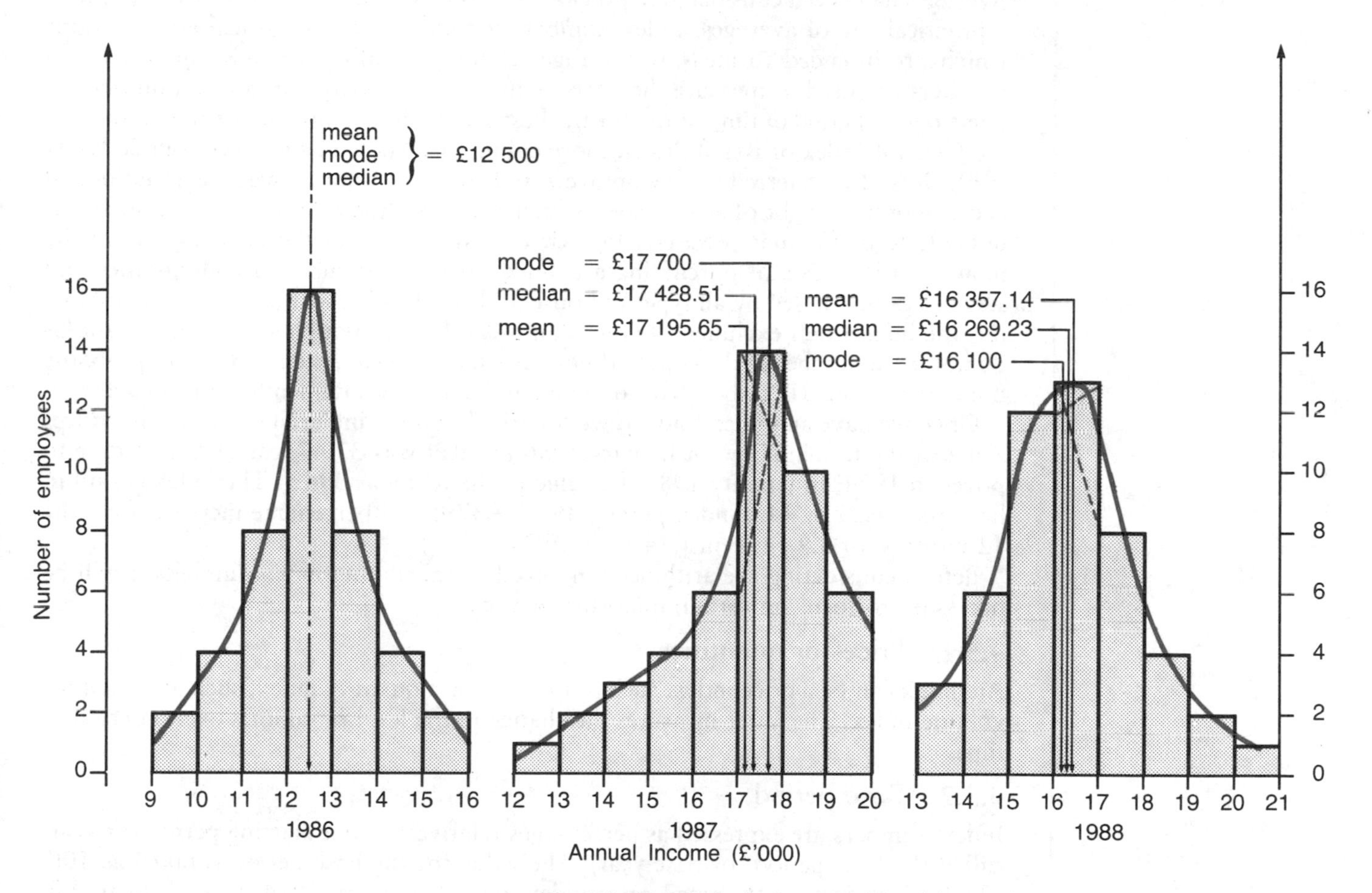

Fig. 4.9 ABC Co Ltd – income received by employees 1986–88

Figure 4.9 shows the distribution of employee incomes of ABC Co Ltd. The income data for 1986 is symmetrically distributed, that for 1987 is negatively skewed (the 'tail' is on the left) and that for 1988 is positively skewed (the 'tail' is on the right; see section 4.4.1), though not excessively so.

Because the mean is affected by extreme values, it tends to be pulled in the direction of those values. In contrast, the median will be positioned in the middle of the distribution and the mode will be positioned where the greatest concentration of data is. The relative positions of the mean, the mode and the median are indicated for each of the distributions in fig. 4.9. The charts show the typical influence of skewness on a uni-modal distribution.

4.4.1 Pearson's coefficient of skewness

Pearson's coefficient of skewness measures the degree of asymmetry in a data set by considering the difference between the mean and the median, relative to the standard deviation. It is defined as follows:

$$S = \frac{3(\bar{x} - \text{Md})}{\sigma}$$

where $\bar{x}$ is the mean, Md is the median and σ is the standard deviation.

S will be positive if the mean exceeds the median and the mode and negative if the mean is less than the median and the mode. Hence a distribution is said to be positively skewed (e.g. fig. 4.9, 1988 data) or negatively skewed (e.g. fig. 4.9, 1987 data).

As an example, the coefficient for the 1988 data in fig. 4.9 is calculated as follows:

$$S(1988) = \frac{3(£16\,357.14 - £16\,269.23)}{£1549} = +0.17$$

[$S(1986) = 0$; $S(1987) = -0.41$]

4.5 Index numbers

Averages have been considered in previous sections; index numbers are an example of a practical use of averages. Index numbers (or indices) are statistical constructions which are intended to measure (average) change, usually economic change. Index numbers are used to measure the changes in the prices or quantities of a number of items over a period of time. Probably the best known index (and most often quoted) is the General Index of Retail Prices, more commonly known as the Retail Price Index (RPI). It is often referred to by employees and employers during wage negotiations. It is commonly thought of as the index which measures 'the cost of living', though it is difficult to define that precisely. In fact, the RPI measures the changes, month by month, in the costs of purchasing a defined (average) number of commodities and services as purchased by all types of households in the UK, with certain high and low income households excluded. If you as an individual do not have a mortgage and/or do not use a number of the commodities chosen as 'typical', then you would probably not consider the RPI to be a 'cost of living' index for you. Remember it is an average.

Once we have an index, how do we use it? The following brief example introduces some of the terminology. In January 1986 the RPI was 379.7, calculated relative to prices in 1974. In January 1987 the value of the RPI was 394.5. The index is said to have increased by 14.8 index points (394.5 − 379.7), a percentage increase, over the 12 months, of 3.9 per cent [(14.8 × 100)/379.7].

Before considering the arithmetic involved in calculating index numbers it will be necessary to focus on certain important aspects.

4.5.1 Prices or quantities

An index can be a price index, such as the RPI, or a quantity index, such as an import volume index. The latter measures the change in the level of imports over a period of time.

4.5.2 Base period

Index numbers are expressed as percentages relative to some starting period (or year) called the base period (or base year). The value for the base period is taken as 100. The base period for the retail price index was 15 January 1974. It was updated in January 1987. There is no special procedure involved in choosing a base period except that it should be representative. The base period should not be one where prices (or quantities) were extreme. Some examples of simple index numbers follow.

4.5.3 Single item index numbers

Table 4.11 shows the values of the price and quantity (weight) of a large loaf of bread over a period of three years. The relative price in 1988 for a loaf of bread, with 1987 as the base year, is calculated as follows:

$$\textbf{Price index} = p_n/p_0 \times 100$$

where p_0 = the price in the base year;
and p_n = the price in the current year (in this case, 1988).

Table 4.11 Changes in the price and weight of a large loaf during the period 1986 to 1988. (Source of data: a Crediton bread shop)

	1986	1987	1988
Price (£)	0.45	0.55	0.60
Quantity (weight; kg)	0.75	0.80	0.80

Therefore, in our example the 1988 price index is $60/55 \times 100 = 109.1$, indicating that in 1988 bread has increased in price by 9.1 per cent (relative to 1987 prices). If 1986 is taken as the base year, the resulting index is $60/45 \times 100 = 133.3$. That is, a large loaf of bread has increased in price by just over 33 per cent between 1986 and 1988.

A quantity index can be calculated similarly:

$$\textbf{Quantity index} = q_n/q_0 \times 100$$

where q_0 = the quantity in the base year;
and q_n = the quantity in the year of interest.

The 1988 quantity index in our bread example is $0.80/0.80 \times 100 = 100$, indicating that there has been no change in the weight of a large loaf in 1988 (relative to 1987). In practice, it is unlikely that anyone would want to calculate an index number based on one item only. It would be enough to simply compare prices (or quantities) to see whether they had changed, though the percentage change would not be obvious.

Before considering more complex, and useful, index numbers it is worth further explaining **index points**. An example should help to illustrate the idea:

The tax and price index (TPI) measures changes in both direct and indirect taxes, as well as changes in retail prices. This is in contrast to the RPI which measures changes in purchasing power after taxation. The TPI for the period July 1986 until January 1987 (base period: January 1978) was as follows:

July	192.1
August	192.9
September	194.0
October	194.3
November	196.3
December	197.1
January	198.0

The index has risen by 5.9 'index points' over this period ($198.0 - 192.1$). As a percentage, this is a 3.1 per cent increase ($5.9/192.1 \times 100$). An index point is equal to 1 per cent of the base year index. Index points are used to compare index values because it is arithmetically easier to make this comparison than to calculate the percentage change each time.

4.5.4 Multi-item index numbers

Index numbers tend to be based on a collection of items or 'basket of goods'. An index based on one item would have limited applications. The RPI is calculated from a 'typical' 'basket of goods', said to be representative of the purchasing habits of the 'average' citizen. Suppose the RPI were based on only four commodities: bread, coffee, chocolate and champagne. Admittedly this is not an average collection, but it will suffice for illustrative purposes. We will further assume that the prices for 1983 and 1987 were as follows:

	1983	1987
Bread	35 p per loaf	55 p per loaf
Coffee	220 p per packet	180 p per packet
Chocolate	40 p for 125 gm	75 p for 125 gm
Champagne	950 p per bottle	1200 p per bottle

A simple index could be calculated by adding the prices of single items in 1987 ($35 + 220 + 40 + 950 = £12.45$) and dividing by the corresponding sum relating to 1983 ($55 + 180 + 75 + 1200 = £15.10$). The index for this example would be 121.29 ($15.10/12.45 \times 100$).

This type of index is called a **simple aggregate index.** There are a number of reasons why such a simple approach may not be appropriate. Before attempting to calculate a 'better' index number, the problems involved in carrying out such a calculation should be considered. Referring to the above information, it can be seen that not all of the prices have increased – the price of coffee has actually gone down. Those prices that have increased have done so at different rates – this would not affect the validity of the simple aggregate index. Also, the prices of the commodities are given in different units and the relative importance of each commodity is not obvious – this would affect the validity of the simple aggregate index. These points suggest that however the index is calculated, it will result in a compromise. The **difficulties,** which are common to most index calculations, can be summarized as follows:

1 Changes in prices of goods are usually at different rates and the changes are not necessarily all increases or all decreases.

2 Different commodities are frequently referred to in different units – e.g. milk in litres, coffee in kilograms.

3 The relative importance of each commodity is not always made obvious.

The first difficulty is a consequence of the desire to get some overall view of price changes and is unavoidable. Common experience might lead one to argue that the prices of most commodities seldom decrease, making it unnecessary to be too concerned about this difficulty. The other difficulties can be overcome by weighting the items to be used in constructing the index.

4.5.5 Weighting used in index construction

Weighting reflects the importance of each item. One way of establishing the relative importance of the items would be to carry out a household survey to find out the proportion of income spent on each of the items each month. Assume that in 1983 such a survey revealed the information summarized in table 4.12. The proportion of income spent on each item is given as a percentage – e.g. 41 per cent of income was spent on bread, 43 per cent on coffee, etc. These proportions represent the relative importance of each items or level of weighting which should be attributed to each item. Assume further that a similar survey carried out in 1987 revealed the information summarized in table 4.13.

Table 4.12 Survey of purchasing habits: 1983

Commodity	Price per unit	Quantity (no. of units)	Average monthly spending (£)	Percentage of total spending
Bread	35 p	24	8.40	41
Coffee	220 p	4	8.80	43
Chocolate	40 p	2	0.80	4
Champagne	950 p	0.25	2.38	12
Totals			20.38	100

Table 4.13 Survey of purchasing habits: 1987

Commodity	Price per unit	Quantity (no. of units)	Average monthly spending (£)	Percentage of total spending
Bread	55 p	28	15.00	48
Coffee	180 p	4	7.20	23
Chocolate	75 p	4	3.00	10
Champagne	1200 p	0.5	6.00	19
Totals			31.20	100

It is now possible to calculate an index that is both easy to calculate and easy to understand.

4.5.6 An expenditure index

Comparing the average monthly expenditures for 1983 and 1987, an index can be found using the total expenditure figures from tables 4.12 and 4.13:

$$\text{Expenditure index} = 31.20/20.38 \times 100$$
$$= 153.09$$

This shows an increase in expenditure over the period 1983 to 1987 of 53.09 per cent.

With base prices and quantities, p_0 and q_0, and current year price quantities, p_n and q_n, as defined above, the equation describing this index is:

$$\frac{\Sigma p_n q_n}{\Sigma p_0 q_0} \times 100$$

It seems straightforward to conclude that the reason for increased expenditure is (on average) higher prices and/or greater quantities of commodities being purchased. It is usual, however, to attempt to isolate either the contribution made by changing prices or the contribution made by changing quantities, resulting in a price index or a volume index, respectively.

4.5.7 Laspeyres' price index

Referring to table 4.12, in 1983 the total average monthly expenditure was £20.38. If the quantities purchased in 1987 were the same as the 1983 quantities, the total expenditure in 1987 would have been £24.90 [$(55\times24)+(180\times4)+(75\times2)+(1200\times0.25)$].

An index can be found using this information:

$$24.90/20.38\times100 = 122.18$$

Because the quantities have been kept constant, the index must show change caused by price fluctuation only. In this example, prices have increased (on average) by 22.18 per cent.

The formula for this index, called Laspeyres' Price Index, is:

$$\textbf{Laspeyres' price index} = \frac{\Sigma p_n q_0}{\Sigma p_0 q_0} \times 100$$

The quantities purchased in the *n*th year are assumed to be the same as the quantities purchased in the base year. This is an improvement on the index calculated in 4.5.4, the simple aggregate index, because the relative importance of commodities purchased have been taken into account. Also, because the quantities purchased have been fixed, the index measures change relative to prices only, unlike the expenditure index from which, because of its construction, it was impossible to tell if changes were caused by price fluctuations or changes in purchasing habits. Laspeyres' index is often calculated using price relatives. The following section explains how this is done.

4.5.8 Price relatives

The price relative of each item is found by calculating the value of the price in the 'new' year (p_n), relative to the price in the base year (p_0) and multiplying by 100:

$$\textbf{price relative} = p_n/p_0 \times 100$$

The price relatives of our current examples are shown in the following table:

Commodity	Price 1983	Price 1987	Price relative (PR)
Bread	35	55	55/35 × 100 = 157.1
Coffee	220	180	81.8
Chocolate	40	75	187.5
Champagne	950	1200	126.3

To calculate Laspeyres' price index, it is necessary to weight the price relatives using the base year weights or expenditure. The formula for this index is:

$$\text{Price index} = \frac{\textbf{price relative} \times \textbf{weight}}{\textbf{total weight}}$$

For our example the calculations are as follows:

Commodity	1983 Expenditure (weight)	PR	Weight × PR
Bread	8.40	157.1	1319.6
Coffee	8.80	81.8	719.8
Chocolate	.80	187.5	150.0
Champagne	2.38	126.3	300.6
Totals	20.38		2490.0

The index for 1987 is 2490.0/20.38 = 122.18.

It may seem easier to calculate Laspeyres' price index using ordinary prices (p_n and p_0) and base year quantities (q_0), rather than first calculating price relatives then using levels of expenditure. In practice, however, expenditure data is more easily obtained, for example, by reference to the *Survey of Household Expenditure.* Referring to quantities of a commodity may make sense when discussing bottles of champagne or jars of coffee, but problems of interpretation could arise when considering quantities of public transport.

Summary

Index numbers can be difficult to understand and complicated to calculate, so it is worth summarizing what we have covered so far.

1 An index number is a statistical construction which allows the user to measure changes in some economic variable, such as price, over a period of time.

2 Index numbers are calculated relative to some chosen period called the base period.

3 A simple index of one commodity with $p_0 = 50$ p (base year price) and $p_n = 75$ p (current year price) can be found as follows:

$$p_n/p_0 \times 100$$
$$= 75/50 \times 100$$
$$= 150$$

4 It is usual for index numbers to use data from a number of items.

5 It is usually necessary to have information relating to the relative importance of a number of items (weights). The weights attributed to items can be in the form of the amount spent on the items (expenditure).

6 An expenditure index is found by comparing base year expenditure and current year expenditure. Because both prices and quantities are changing, an expenditure index cannot be used to monitor price changes or quantity changes exclusively.

7 In order to measure the changes in prices only it is necessary to keep quantities constant. Laspeyres' price index, does this by using only base year quantities.

So, in Laspeyres' price index, we have an index which measures changes in prices over a period of time. It takes the relative importance of each commodity into account (using base period quantities or expenditure as weights) and keeps quantity constant so that we can be sure it is only price change that is being measured. Unfortunately there are still difficulties. In practice, purchasing habits change. Not only is it reasonable to suppose that the quantity or volume of a commodity purchased will change, but the actual commodities purchased will probably change. This is particularly true today when society is experiencing such rapid change, technological change as well as other change.

4.5.9 The Paasche price index

To respond to changes in purchasing habits, some statisticians favour using **current weights** (which may still be expenditure) in calculating an index. The Paasche price index uses current weights. In every other respect it is calculated in the same way as Laspeyres' index:

$$\textbf{Paasche price index} = \frac{\Sigma p_n q_n}{\Sigma p_0 q_n} \times 100$$

Refer back to table 4.13. The expenditure in 1987 ($\Sigma p_n q_n$) was £31.20; if the same quantities had been purchased in 1983 as were purchased in 1987, then expenditure in 1983 would have been £24.95 $[(35 \times 28) + (220 \times 4) \times (40 \times 4) + (950 \times 0.5)]$.

The Paasche price index can now be found:

$$31.20/24.95 \times 100 = 125.05$$

Unlike the calculation of Laspeyres' price index, calculating the Paasche price index for a particular period involves updating the changes in quantities as well as prices. It is therefore not a pure price index. This necessitates calculating the weights every time the price index is required, which could be an expensive and time-consuming job. The RPI is published monthly but the weights are only adjusted once a year, a compromise between the Paasche and Laspeyres' indexes.

4.5.10 Volume or quantity index numbers

The examples considered so far have involved changes in prices. It is sometimes necessary to monitor the changes in the volume or quantity of items over a period of time. The problems are similar and the calculations of the same type.

$$\textbf{Laspeyres' volume index} = \frac{\Sigma q_n p_0}{\Sigma q_0 p_0} \times 100$$

$$\textbf{Paasche volume index} = \frac{\Sigma q_n p_n}{\Sigma q_0 p_n} \times 100$$

Consider the following example: Solid Building Company uses three main components in their work. The prices and quantities over a period of time are shown in table 4.14.

Table 4.14 The average daily material requirements (in appropriate units) and prices (£ per unit) of Solid Building Company (1987 and 1988)

	1987		1988					
	Price p_0	**Quantity** q_0	**Price** p_n	**Quantity** q_n	p_0q_0	p_nq_0	p_0q_n	p_nq_n
Sand per bag	0.80	125	0.90	175	£100	£112.50	£140	£157.50
Cement per 50 kg	4.80	25	5.00	40	£120	£125.00	£192	£200.00
Bricks per tonne	100.00	4	105.00	6	£400	£420.00	£600	£630.00
Totals					£620	£657.50	£932	£987.50

Referring to table, the volume index numbers are:

Laspeyres' volume index = £932/£620 × 100
= 150.32

Paasche volume index = £987.50/£657.50 × 100
= 150.19

4.5.11 Fisher's index

Because Laspeyres' index uses base period weights, it tends to overstate change in prices or quantities. This is because when prices increase there is usually a reduction in the volume of the affected commodities consumed. The result is that the index numerator will be too large. Similarly, when prices decrease there is usually an associated increase in consumption. In this case not enough weight will be given to the prices which have decreased, again resulting in an overstatement of change. It can be shown that the Paasche index will tend to understate change. Because these differences arise, statisticians have tried to obtain a better index. Fisher's index is found by taking **the geometric mean of Lapeyres' index and the Paasche index,** giving what is claimed to be an ideal index.

$$\textbf{Fisher's volume index} = \sqrt{\textbf{Laspeyres' index} \times \textbf{Paasche index}}$$

For the Solid Building Co example used in the previous section, Fisher's volume index is equal to

$$\sqrt{150.32 \times 150.19} = 150.26$$

Fisher's index is ideal in that it correctly predicts expenditure. Consider the various price and volume index numbers for the Solid Building Company: (The appropriate calculations are left as an exercise.)

Laspeyres' price index = 106.05
Paasche price index = 105.95
Laspeyres' volume index = 150.32
Paasche volume index = 150.19
Fisher's price index = 106.00
Fisher's volume index = 150.26

The expenditure index can be found as follows (remember: expenditure = price × quantity):

1 Using Laspeyres' index numbers:

$$(106.05 \times 150.32)/100 = 159.41$$

This overstates the expenditure index.

2 Using Paasche index numbers:

$$(105.95 \times 150.19)/100 = 159.13$$

This understates the expenditure index.

3 Using Fisher's index numbers:

$$(106.00 \times 150.26)/100 = 159.28$$

This gives the correct figure for the expenditure index. This can be checked very simply by calculating the expenditure index ($\Sigma p_n q_n / p_0 q_0 \times 100$; left as an exercise).

4.5.12 Index numbers: chain base method

The choice of weights is an important aspect of index number construction. It is usually based on the expenditure involved (in a price index) or the quantities used (in a volume index). If the weights are subject to frequent change, reflecting, for instance, the introduction of new items into the index, then relating the data to a fixed base over a long period of time will probably not make sense – the composition of the index will have changed so much as to make any meaningful comparison difficult. In such a case it is probably better to compare a particular index number value with the value for the previous year. In doing this we are using the chain base method. An example follows, using the RPI (housing) for January 1986 to January 1987.

Base period 15 January 1974: index = 100			**Base period re-defined January 1986**
1986	January	463.7	100.00
	February	465.7	100.43 (= 465.7/463.7 × 100)
	March	467.5	100.82 (= 467.5/463.7 × 100)
	April	483.5	104.27 (= 483.5/463.7 × 100)
	May	482.7	104.10
	June	471.6	101.70
	July	472.8	101.96
	August	475.2	102.48
	September	477.3	102.93
	October	478.4	103.17
	November	497.4	107.27
	December	501.1	108.07
1987	January	502.4	108.35

If the increase is calculated relative to the original indices (base period January 1974) we get the same result over the 12 month period:

$$(502.4 - 463.7)/463.7 \times 100 = 8.35\%$$

That is, the index has increased by 8.35 per cent over the January to January period.

The chain based index over the same period is as follows:

1986	January	100.00	
	February	100.43	
	March	100.39	(= 100.82/100.43 × 100)
	April	103.42	(= 104.27/100.82 × 100)
	May	99.84	(= 104.10/104.27 × 100)
	June	97.69	(= 101.70/104.10 × 100)
	July	100.26	
	August	100.87	
	September	100.44	
	October	100.18	
	November	103.97	
	December	100.74	
1987	January	100.26	

The chain index shows the rate of change in housing prices from one month to the next, detecting drops in prices (April to May and May to June) as well as rises.

It may seem that the information from an index number will depend on which base period and/or method of calculation is used. The information is, of course, essentially the same. For instance: relative to the base period January 1974, the change from December 1986 to January 1987 is 1.3 index points (502.4 – 501.1); relative to the base period January 1986, the change over the same period is 0.28 index points. In both cases, the percentage change is 26 per cent (1.3/501.1 × 100; 0.28/108.7 × 100), as shown using the chain base method. Whilst the information is essentially the same, this should indicate the need for care in interpretation.

4.6 Practical index numbers and their uses

Clearly, index numbers are complex constructions and careful thought must be given to the factors involved in their construction: choice of base period, weights, etc. Before looking at a selection of sample examination questions, we will consider an example of index numbers used in practice.

Index numbers, such as the Retail Prices Index, are frequently referred to by politicians, employers and trade union representatives, particularly when trying to monitor inflation. Inflation has been a main political topic for a number of years, with great credit being taken by governments for keeping it at a low level. The Retail Price Index is commonly used as a measure of inflation and though the current rate of inflation, as measured by the RPI, is low and has actually dropped in recent years, this cannot be interpreted to mean that prices and/or quantities purchased are falling. What is happening is that prices and/or quantities purchased are increasing, but not as rapidly as they were earlier. For instance, during the period January 1985 to January 1986 the RPI (for all items) increased by 19.8 index points (379.7 – 359.9), a percentage increase of 5.5 per cent. During the following 12-month period, January 1986 to January 1987, the RPI increased by 14.8 index points (394.5 – 379.7), a percentage rise of 3.9 per cent. So, even though the increase was smaller (3.9 per cent cf 5.5 per cent), the RPI has still increased. You will recall that the data collected to construct the RPI is based on the expenditure of the 'average citizen'. When applied to a large number of people, the RPI is a useful measure; for the individual it may not be quite as useful, as the following example shows.

4.6.1 Deflating a series using a price index

Consider the case of Fred Smith who is employed by Slave Labour Co Ltd. Over a five-year period he has received an annual increase in salary of £200. He feels unhappy about this but has been told by his employer not to complain as £200 a year is a reasonable increase in wages. Even though Fred agrees that £200 sounds a large amount, he still finds it difficult to maintain the standard of living he knew when first employed by the company. He consulted a friend who happened to be a statistician. Table 4.15 shows the information as set out by the statistician.

Table 4.15 Record of the wages of Fred Smith, an employee of Slave Labour Co Ltd and of the RPI for a five-year period

Year	RPI		Wages (£)	Real wages Year 1 value (£)	Real wages Year 5 value (£)
	(a)	(b) base year = year 1	(c)	(d)	(e)
1	250	100	10 000	10 000.00	12 600.00
2	260	104	10 200	9 807.69	12 357.69
3	275	110	10 400	9 454.54	11 912.73
4	295	118	10 600	8 983.05	11 318.64
5	315	126	10 800	8 571.43	10 800.00

The first thing that Fred noticed was that the pay increase he received in year 2 was £200, 2 per cent of his year 1 wage level (£10 000), while the RPI has increased by 4 per cent – twice as much as his increase in wages. Consider the information in table 4.15 column by column:

Column (a) shows the current RPI.
Column (b) shows the recalculated RPI, relative to a new base period, year 1.

year 1: $250/250 \times 100 = 100$
year 2: $260/250 \times 100 = 104$
⋮
year 5: $315/250 \times 100 = 126$

Column (c) shows Fred's wages over the same five year period.
Column (d) shows the result of deflating Fred's wages, relative to the base period, year 1 – i.e. what Fred's wages are worth, taking price changes (as represented by the RPI) into account. Economists call these deflated wage figures **real wages.**

year 1: £10 000 × 100/100 = £10 000.00
year 2: £10 200 × 100/104 = £9 807.69
⋮
year 5: £10 800 × 100/126 = £8 571.43

Column (e) shows Fred's wages deflated relative to year 5. It indicates that Fred would need to be earning £12 600 in year 5 to enable him to maintain the standard of living he enjoyed in year 1.

year 1: £10 000 × 126/100 = £12 600.00
year 2: £10 200 × 126/104 = £12 357.69
⋮
year 5: £10 800 × 126/126 = £10 800.00

The information shown in table 4.15 can be interpreted in a number of ways. For instance, it is clear that in real terms Fred is earning £8571.43, compared to £10 000 in year 1. For his real wage level to be restored to the year 1 level would require an immediate increase of £1800 per annum (£12 600 – £10 800).

The RPI is an invaluable tool in any negotiating situation, however it is not the only index available. Fred's employers might reasonably draw his attention to table 4.16, which shows the profit index for Slave Labour Co Ltd. Perhaps Fred will have to compromise.

Table 4.16 A comparison of employee remuneration, in real terms, and the profit index of Slave Labour Co Ltd, over a five year period

Year	Profit index base period = year 1	Wages (£)	Real wages Year 1 value (£)	Real wages Year 5 value (£)
1	100	10 000	10 000.00	10 800.00
2	102	10 200	10 000.00	10 800.00
3	103	10 400	10 097.09	10 904.85
4	105	10 600	10 095.24	10 902.86
5	108	10 800	10 000.00	10 800.00

Sample questions

1 (a) Briefly discuss the advantages and disadvantages of the mean, median and mode as measures of central tendency.

(b) Journey times to work of employees of a company:

Times (minutes)	No. of employees
less than 10	14
10 but " " 20	26
20 " " " 30	64
30 " " " 40	46
40 " " " 50	28
50 " " " 60	16
60 " " " 80	8
80 " " " 100	4

(i) Obtain the mean, median and mode of travel time to work.

(ii) Negotiations are under consideration for the payment of travelling time to work. State, giving reasons, which measure of central tendency you would use if you are (1) the employer, and (2) a trade union representing the employees.

CIS; Part 1: Quantitative Studies; June 1987.

Solution/hints

Part (a)

Refer to section 4.2.7.

Part (b)

(i) As the data is given in the form of a frequency distribution, it is only possible to obtain estimates of the various measures of centre (see section 4.2.2).

To estimate the mean it is first necessary to establish the middle values of the intervals of the frequency distribution. It is possible to indicate the cumulative frequency on the same table as is required to estimate the median (to estimate the median using a graphical method refer to section 4.2.6 and fig. 4.4).

Time (min)	Mid value x	Frequency f	$f \times x$	Cumulative frequency
< 10	5	14	70	14
10 but < 20	15	26	390	40
20 but < 30*	25	64	1600	104
30 but < 40	35	46	1610	150
40 but < 50	45	28	1260	178
50 but < 60	55	16	880	194
60 but < 80	70	8	560	202
80 but < 100	90	4	360	206
Totals		$\Sigma f = 206$	$\Sigma fx = 6730$	

The (estimated) mean can now be found:

$$\bar{x} = \Sigma fx/n = 6730/206$$
$$= 32.67 \text{ minutes (2 d.p.)}$$

The median is the middle item and would be found by noting the value of the item with rank $(n+1)/2$, in this case the average of the 103rd and 104th items. We are only given the grouped data so actual values are not known, so we have to be satisfied with an estimate. '20 but < 30' (*) is the median class and, in this case, also the modal class. The appropriate estimate is found as follows:

$$\text{Median} = L + (n_1/n_2) \times l$$
$$= 20 + (63/64) \times 10 \text{ (see section 4.2.5)}$$
$$= 29.84 \text{ minutes (2 d.p.)}$$

An approximation of the mode can be found by referring to the modal class. This can be done graphically (see section 4.2.6 and fig. 4.3). (Left as an exercise.)

(ii) Left as an exercise.

2 In 1988, over a period of ten weeks, a sample of the daily production level was taken for Gearing Engineering Company. The resulting data is as follows:

24.25	11.20	15.70	25.10	20.82	30.19	18.36
29.27	23.25	27.50	27.28	35.68	29.27	28.59
19.36	28.45	20.55	10.40	26.48	28.48	25.52
24.39	30.48	12.47	12.88	16.47	19.05	32.69
33.55	15.25	26.47	34.55	15.01	10.00	28.55
28.44	27.44	14.35	18.29	17.37	26.45	26.45
17.25	19.75	33.80	29.00	28.46	28.45	28.55
26.50						

(information in thousands of units)

(a) Present the data in a format suitable for easier interpretation, then calculate the variance and, from the variance, the standard deviation.

(b) An earlier sample, resulting from a survey of production carried out in 1987, gave the following statistics:

sample mean = 20.5 thousand units
sample standard deviation = 5.950 thousand units

Compare the results for the two years, commenting on your results.

(c) Using your results for 1988, calculate an estimate for the mean and standard deviation for the whole year. Assume a 48-week working year.

The Polytechnic of North London; Accounting Foundation Course.

Solution/hints

Part (a)

The data is presented in a raw state making it difficult to gain an overall impression. A suitable format would be a frequency distribution, as shown below, or a diagram, chart or graph.

To choose the intervals for the frequency distribution, proceed as follows:

Minimum value = 10.00
Maximum value = 34.55
Range = 34.55 – 10.00 = 24.55

It is usual to aim for 8 to 10 intervals, so intervals of width 3 (10–13, 13+–16, etc) will suffice. (13+–16 is read 'over 13 and up to and including 16'.)

Production (thousands of units)	Mid-point x	Frequency (no. of days) f	$f \times x$	$f \times x^2$
10 –13	11.5	5	57.5	661.25
13+–16	14.5	4	58.0	841.00
16+–19	17.5	5	87.5	1 531.25
19+–22	20.5	5	102.5	2 101.25
22+–25	23.5	3	70.5	1 556.75
25+–28	26.5	10	265.0	7 022.25
28+–31	29.5	13	383.5	11 313.25
31+–34	31.5	3	94.5	2 976.75
34+–37	35.5	2	71.0	2 520.50
Totals		$\Sigma f = 50$	$\Sigma fx = 1190$	$\Sigma fx^2 = 30\,624.50$

From this information we can find the mean, variance and standard deviation:

Mean, $\bar{x} = \Sigma fx/n$
$= 23.8$ thousand units

Variance, $\sigma^2 = \dfrac{\Sigma fx^2}{n} - (\bar{x})^2$
$= 30\,624.50/50 - 23.8^2$
$= 612.49 - 566.44$
$= 46.05$

Standard deviation $= \sqrt{46.05} = 6.786$ thousand units

Notes:

1 At this point you must consider whether to use Bessel's correction factor (see section 4.3.5.2). You are told in the question that the data is a sample, consequently we are calculating sample statistics. Using Bessel's correction factor would give the following results:

Sample variance, $\hat{\sigma}^2 = \sigma^2 \times n(n-1) = 46.05 \times 50/49 = 46.99$

Sample standard variation, $\hat{\sigma} = \sigma \times \sqrt{n/(n-1)}$
$= 6.786 \times \sqrt{50/49} = 6.855$

As a general rule, if you are calculating the variance and/or the standard deviation from a sample, then the correction factor should be employed. In practice, if n is large you will notice little difference.

2 Because the statistics were calculated using mid-point values, approximation errors will result. In most cases these will be insignificant. For information, the actual values of the mean, variance and standard deviation are 23.22, 54.69 (with correction, 55.80) and 7.38 (with correction, 7.47).

Part (b) Comparison of data, 1987 and 1988.

It is not clear from the question how large a sample was taken in 1987, so a simple comparison of sample standard deviations, etc, would not make sense. However, if we calculate the coefficient of variation for each sample, then a comparison, though still a crude one, can be made.

1987: coefficient of variation $= 5.950/20.5 \times 100$
$= 29.02$

1988: coefficient of variation $= 6.855/23.8 \times 100$
$= 28.80$

It is apparent that the average production level (represented by the mean values 20.5 and 23.8) has risen slightly, and the variation inherent in the production process (represented by the coefficient of variation values 29.02 and 28.80) has dropped. (Further analysis could be carried out to see if the changes are significant. Appropriate statistical tests are covered in a later chapter.)

Part (c) Estimated mean and standard deviation for the whole year, 1988.

The values used in part (a) are daily production figures for a 10-week period in 1988. If we assume a 5-day working week and that the days and weeks throughout the year are independent from one another, it is possible to calculate the mean and standard deviation as follows:

Average annual production $= 48 \times 5 \times \bar{x}$
$= 48 \times 5 \times 23.8$
$= 5712$ thousand units

Variance (for 48 weeks) $= 48 \times 5 \times 46.99$
$= 11\,277.6$

Standard deviation (for 48 weeks) $= \sqrt{11\,277.6} = 106.20$ thousand units.

Another way of considering the calculation, which is more generally applicable, is as follows:

The mean (for one day) is 23.8 thousand units, and the variance (also for one day) is 46.99. The mean for 48 weeks is equal to mean(day 1) + mean(day 2) + ... + mean(day 240), which in this case is estimated to be 240 × mean(for 1 day) = 240 × 23.8 = 5712 thousand units. Similarly, for variance is the sum of the variances of the 240 days.

Note: It is not possible to get the required standard deviation in the same way.

3 (a) Because of the high level of inflation, it has become common practice in recent years to undertake the indexation of major economic and business variables. Explain what is meant by indexation and how such a practice can help to ease the effect of inflation on business decision-making.

(b) The following information relates to the sales of four commodities over the period 1980 to 1985:

Commodity	Price (£) 1980	Price (£) 1985	Quantity Sold Units ('000) 1980	Quantity Sold Units ('000) 1985
A	2.3	1.93	72.8	103.6
B	2.13	1.83	134.4	145.6
C	1.74	1.86	207.2	176.4
D	0.55	0.59	372.4	410.2

You are required to calculate the relevant price and quantity indices for 1985 (1980 = 100) using both base and current year weightings, and comment on the results obtained.

CIPFA; Diploma in Public Sector Audit and Accounting; Quantitative analysis; March 1986.

Solution/hints

Part (a)

Indexation involves identifying relevant business or economic variables and monitoring the changes in their price or the quantity used. For instance, the steel industry might identify iron ore, gas (as a fuel) and magnesium (as a decarbonating agent), as the main raw material inputs in producing steel. Indexation would involve deciding on the relative importance of these inputs (variables), weighting, and calculating a price (or quantity) index to monitor price (or quantity) change over a period of time. The information so obtained would enable the users of such index numbers to make more informed decisions and take appropriate action. This might mean increasing the price of their products or seeking different suppliers. What is important to notice is that the business decision-maker is less likely to omit the effects of inflation from important decision procedures.

Part (b)
Price indices
(i) base year weights

$$\text{Price index} = \frac{(1.93 \times 72.8) + (1.83 \times 134.4) + (1.86 \times 207.2) + (0.59 \times 372.4)}{(2.30 \times 72.8) + (2.13 \times 134.4) + (1.74 \times 207.2) + (0.55 \times 372.4)} \times 100$$

$$= \frac{991.564}{1019.06} \times 100$$

$$= 97.3$$

(ii) current year weights

$$\text{Price index} = \frac{1036.518}{1080.954} \times 100 = 95.89$$

(The detailed calculations are left as an exercise.)

Both indices indicate an overall drop in prices – 2.7 per cent and 4.11 per cent, respectively.

Quantity indices
(i) base year weights
For a quantity index, it is the relevant prices which are used as weights, in this case the 1980 prices. (The detailed calculations are left as an exercise.)

$$\text{Quantity index} = \frac{1080.954}{1019.06} \times 100 = 106.07$$

(ii) current year weights

$$\text{Quantity index} = \frac{1036.518}{991.654} \times 100 = 104.53$$

Both indices indicate an overall rise in the quantities of the four commodities sold. Considering both sets of indices it is clear that prices have dropped and sales increased. It is possible that the two are connected. This hypothesis could be tested by further lowering the prices.

4 Weekly earnings of females by highest educational qualification:

	1983 percentages		
Range of income (£)	**'A' level**	**'O' level**	**None**
under 60	2	4	8
60 and ,, 80	8	11	23
80 ,, ,, 100	17	29	34
100 ,, ,, 120	16	13	11
120 ,, ,, 140	26	21	16
140 ,, ,, 160	11	9	5
160 ,, ,, 180	11	8	2
180 ,, ,, 200	3	2	0
200 and over	6	3	1

(*Source: General Household Survey*)

(a) Complete the following table:

	'A' level	**'O' level**	**None**
Lower quartile	97.65	–	74.78
Median	125.38	–	91.18
Upper quartile	150.91	–	118.18
Quartile deviation	26.63	–	–
Coefficient of skewness	–	–	0.24

(b) Write a brief report on the earnings of females and highest education qualification.

Formulae:

$$\text{Quartile measure of skewness} = \frac{Q_3 + Q_1 - 2\text{Md}}{Q_3 - Q_1}$$

$$\text{Pearson measure of skewness} = \frac{3(\bar{x} - \text{Md})}{s}$$

CIS; Part 1, Quantitative studies; December 1986.

Solution/hints

Part (a)

To complete the table it is first necessary to calculate the appropriate cumulative percentages:

1983 percentages and cumulative percentages:

Range of income (£)	'A' level	Cumulative %	'O' level	Cumulative %	None	Cumulative %
60–	2	2	2	4	8	8
60 to 80–	8	10	11	15	23	31
80 to 100–	17	27	29	44	34	65
100 to 120–	16	43	13	57	11	76
120 to 140–	26	69	21	78	16	92
140 to 160–	11	80	9	87	5	97
160 to 180–	11	91	8	95	2	99
180 to 200–	3	94	2	97	0	99
200+	6	100	3	100	1	100

(Note: '60–' is read as 'under 60' and '200+' is read as 'over 200', etc)

Lower quartile: 'O' level $Q_1 = 80 + (10/29 \times 20) = £86.90$

Median: 'O' level $\text{Md} = 100 + (6/13 \times 20) = £109.23$

Upper quartile: 'O' level $Q_3 = 120 + (18/21 \times 20) = £137.14$

Quartile deviation: 'O' level $(Q_3 - Q_1)/2 = (£137.14 - £86.90)/2 = £25.12$

Coefficient of skewness: 'O' level

In this case it is appropriate to use the quartile measure as it is only possible to estimate the mean, $\bar{x}$, and the standard deviation, $\hat{\sigma}$, as two classes are 'open' ('60' and '200+').

Therefore, quartile measure of skewness

$$= \frac{137.14 + 86.90 - (2 \times 109.23)}{137.14 - 86.90}$$

$$= 0.111$$

Quartile deviation: no qualifications

$$(Q_3 - Q_1)/2 = (£118.18 - £74.78)/2 = £21.70$$

Coefficient of skewness: 'A' level

This is left as an exercise. (The answer is 0.11.)

Part (b) Report on earnings

The results show that females with better qualifications have, on average, higher levels of income. The difference in income potential is such that none of the middle 50 per cent of unqualified females can expect an income equal to the average income received by those who have 'A' levels, even though there is some overlap between the different categories.

5 (a) Briefly explain the importance of the index of industrial production as an economic indicator.

(b) Index of industrial production, November 1985 (1980 = 100)

Group	Weight	Index
Energy and water supply	264	125
Manufacturing industries:		
Metals	25	116
Mineral products	41	97
Chemicals	68	118
Engineering	325	105
Food, drink and tobacco	99	104
Textiles	52	106
Other	126	103

(*Source: Monthly Digest of Statistics*)

(i) Calculate the index of industrial production for November 1985 with 1980 = 100, for:

(1) all industries; and
(2) all manufacturing industries

(ii) The index of industrial production for all industries for 1980 to 1984 with 1980 = 100 is given below:

1980	1981	1982	1983	1984
100	96.6	98.4	101.9	103.2

Using your result in part (i), comment on the movements of the index of industrial production since 1980.

CIS; Part 1; Quantitative studies; December 1986.

Solution/hints

Part (a)

The index of industrial production measures the changes in the volume of production (hence it is a quantity index) of major sections of the country's industry. It is particularly important because it is used as a measure of the state of the national economy.

Part (b)

(i) In this question the individual indices can be considered as quantity relatives (see price relatives, section 4.5.8): e.g. the index for textiles in November 1985, 106, indicates a 6 per cent rise in volume of production in November 1985, relative to 1980.

(1) Index for all industries = (Σ index × weight)/Σ weights

$$= \frac{(125\times 264)+(116\times 25)+\ldots+(103\times 126)}{(264+25+\ldots+126)}$$

$$= \frac{110\,812}{1000} = 110.812$$

(2) The index for all manufacturing industries can be calculated in the same way or by a short-cut which involves subtracting the contribution of non-manufacturing industries ('Energy and water supply' and 'Other') as follows:

$$\text{Index} = \frac{110\,812-(125\times 264)-(103\times 126)}{1000-264-126}$$

$$= \frac{64\,834}{610} = 106.29$$

(ii) Adding the index for 1985 to the table of indices provided gives the following:

1980	1981	1982	1983	1984	1985
100	96.6	98.4	101.9	103.2	110.8

Apart from a drop in production of 3.4 per cent in 1981 (relative to 1980), production has steadily increased over the period. Using the chain index method of calculation, it is possible to see the increase in production each year relative to the previous year:

1980	1981	1982	1983	1984	1985
100	96.6	101.9	103.5	101.3	107.4

The largest increase (7.4 per cent) occurred in the year to November 1985.

Recommended reading

Owen, F. and Jones, R., *Statistics* (Pitman, 1988), chapters 5, 8 and 9.

Whitehead, P. and Whitehead, G., *Statistics for Business* (Pitman, 1984), chapters 8, 9 and 11.

Thomas, R., *Notes and Problems in Statistics* (Stanley Thornes, 1984), chapter 1.

Curwin, J. and Slater, R., *Quantitative Methods for Business Decisions* (Van Nostrand Reinhold, 1986), chapter 3–5.

5 Time series analysis

5.1 Introduction

A time series is a set of observations of a variable (e.g. employment) measured at successive points in time or over successive periods of time.

You only need to open the pages of one of the better newspapers to see a graph of a time series, illustrating the movement of stocks and shares, the monthly sales figures of a company, or the level of unemployment in a region. The graphs are rarely straightforward to understand and often look like the walk of a drunken insect (see fig. 5.1). Though they may not be obvious at first sight, time series often contain patterns which can be isolated by appropriate analysis. This chapter focuses on the patterns you need to be aware of and explains the methods you will need for successful analysis.

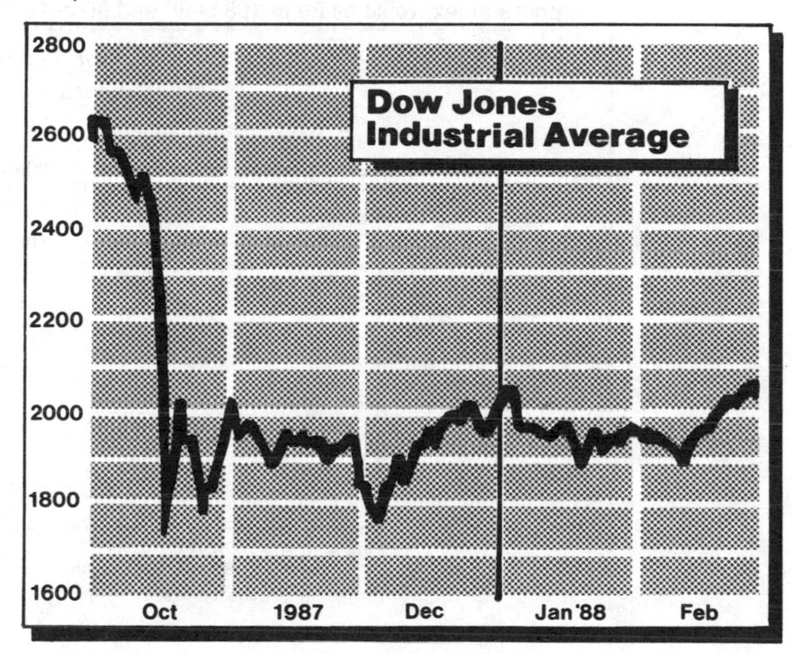

Fig. 5.1 (*Source: Financial Times, 27 February 1988.*)

Examples of time series are company sales (recorded monthly), unemployment (recorded every quarter), the number of cars recorded in a traffic census over a number of weeks and a record of industrial change such as that illustrated in fig. 5.1. The information available from a time series enables the business person or other interested party to monitor historical aspects of data – that is, to observe how the variable has performed over the period of time being monitored. Equally, if not more, important is the potential to use that historical data to forecast future performance, which is a necessary part of business preparation and decision-making.

Using past data to forecast is only possible if certain patterns exist in the data, patterns which are likely to be repeated in the future. Table 5.1 and fig. 5.2 (overleaf) illustrate such a pattern – a **trend**. The ordinary maintenance grant (table 5.1, column 1; fig. 5.2(a)) is increasing year by year – that is, there is an upward trend. However, it is also clear that the grant in real terms (table 5.1, column 2; fig. 5.2(b)) is declining – that is, in real terms, there is a downward trend. The trend is the overall tendency of the data.

Other patterns that might be present in a time series are cyclical variation, seasonal variation and random variation. All of the patterns with which you need to be familiar will be considered in the following section.

Table 5.1 Student awards – real values and parental contributions, England and Wales

	Ordinary maintenance grant[1] (£)	Index (September 1978 = 100) of the real value of the grant deflated by the: retail prices index	average earnings index	Average percentage contribution by parents[2]
1978–79	1100	100	100	16.20
1979–80	1245	97	97	13.41
1980–81	1430	96	97	12.79
1981–82	1535	93	89	14.18
1982–83	1595	90	85	18.98
1983–84	1660	89	82	19.53
1984–85[3]	1775	91	82	25.27
1985–86[3]	1830	88	82	...

[1] Excludes London. Prior to 1982–83, Oxford and Cambridge were also excluded.

[2] Of students receiving parental contributions; in 1984–85 there were 100.3 thousand mandatory award students receiving the maximum grant.

[3] The rate of grant in 1984–85 and 1985–86 includes an additional travel allowance of £50. If this additional allowance were excluded, the index of the real value of the grant deflated 'y thg retcil prices index would be 89 in 1984–85 and 96 in 1985–86.

(Source: Department of Education and Science; Department of Employment; Key Data 87, Table 15.4)

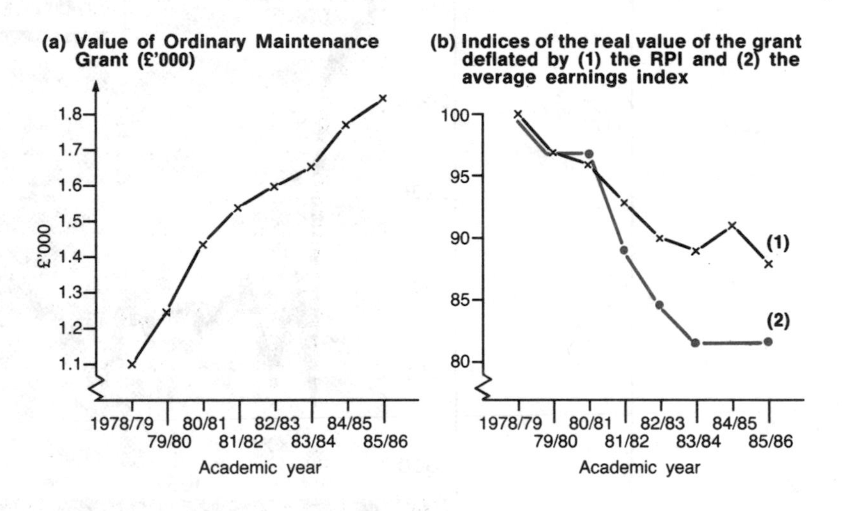

Fig. 5.2 (a) The value of the Ordinary Maintenance Grant 1978–86; (b) Indices of the real value of the grant deflated by the RPI and the average earnings index

5.2 Patterns (or components) of a time series

Table 5.2 exhibits the profit figures, over a period of six years of four departments of Sandton Engineering Company, the graphs of which are displayed in fig. 5.3. The patterns, where they exist, are obvious:

1 **The trend:** The profits of Department A show a marked, regular, upward trend: profits increase by £250 each quarter.

2 **Seasonal variation:** The profits of Department B are affected by seasonal fluctuations or seasonal variation. The 2nd quarter of each year produces a loss of £4000, while the 3rd quarter of each year produces profits of £3000, the highest for the year.

3 **Cyclical variation:** the profits of Department C are cyclical, the cycle repeating itself every five years. The graph illustrates a typical complete cycle – point A (4th quarter of 1983) to point B (4th quarter of 1988).

4 **Random variation:** The profits of Department D show no pattern at all. The data appears to be random and, as such, not predictable.

Table 5.2 Sandton Engineering Company – departmental quarterly record of profits (£'000): 1983 to 1988

Year	quarter	Dept. A	Dept. B	Dept. C	Dept. D
1983	1	2.0	2.0	3.0	0.288
	2	2.25	−4.0	3.3	−0.465
	3	2.5	3.0	3.7	−0.370
	4	2.75	−1.0	3.90	−0.313
1984	1	3.0	2.0	4.1	−0.006
	2	3.25	−4.0	4.25	0.304
	3	3.5	3.0	4.1	0.232
	4	3.75	−1.0	3.9	−0.047
1985	1	4.0	2.0	3.7	−0.079
	2	4.25	−4.0	3.3	0.113
	3	4.5	3.0	3.0	−0.026
	4	4.75	−1.0	2.7	0.224
1986	1	5.0	2.0	2.3	0.119
	2	5.25	−4.0	2.1	0.491
	3	5.5	3.0	1.9	−0.174
	4	5.75	−1.0	1.75	−0.357
1987	1	6.0	2.0	1.9	0.103
	2	6.25	−4.0	2.1	0.029
	3	6.5	3.0	2.3	0.436
	4	6.75	−1.0	2.7	−0.142
1988	1	6.75	2.0	3.0	−0.286
	2	7.0	−4.0	3.3	−0.128
	3	7.25	3.0	3.7	−0.058
	4	7.5	−1.0	3.9	0.047
Type of variation:		Trend	Seasonal	Cyclical	Random

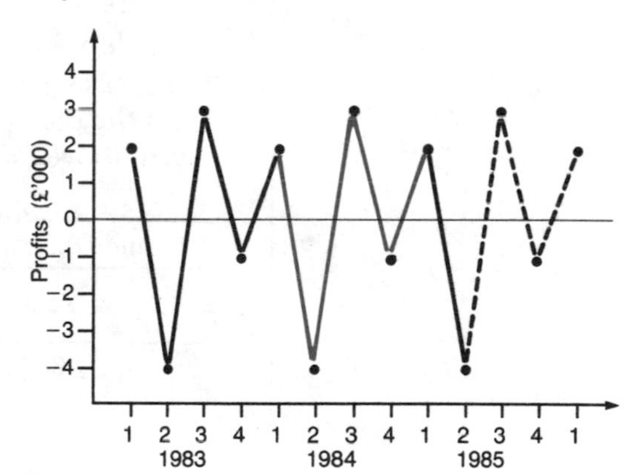

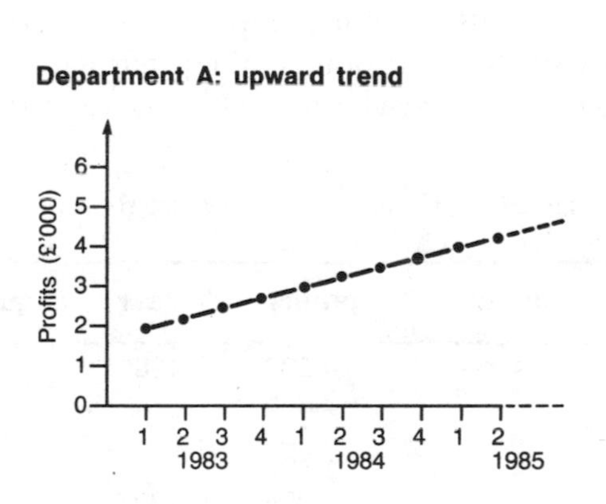

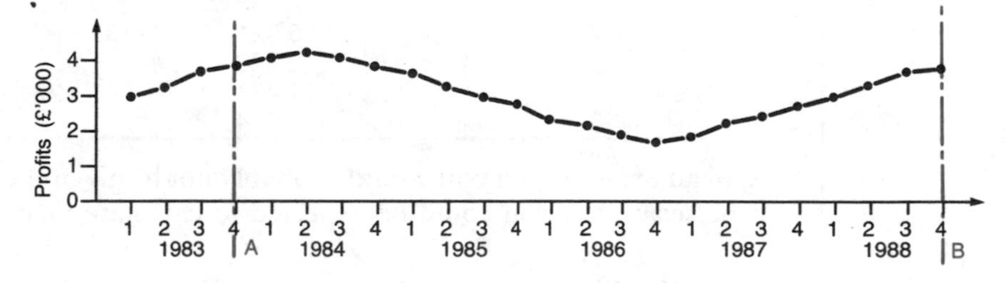

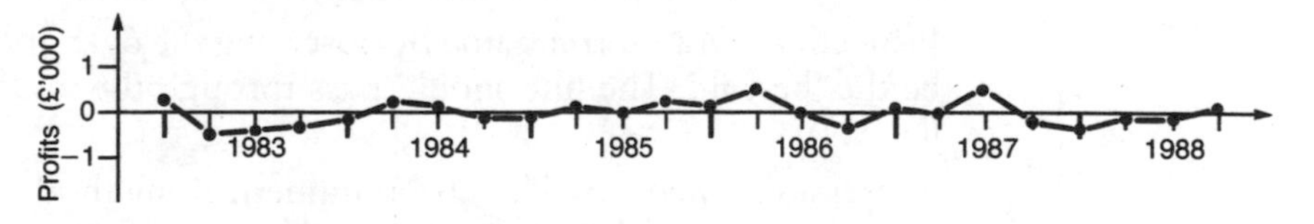

Fig. 5.3 Sandton Engineering Co – departmental quarterly profit records: 1983 to 1988

Clearly the data in this example is artificial, but it illustrates the time series components which you should be aware of. In practice, a data set could have aspects of all four patterns. This is summarized by the following equation:

$$O = C + T + SV + R$$

where O = observed data, such as sales or profits;

C = overall cycle: As the economy expands an upward movement in data such as sales, profits, etc, might be expected. As the economy experiences a slump one would expect to see the opposite – a downward movement in observed data values. This might appear as a wave-like motion, as illustrated by the profits of Department C (fig. 5.3).

T = trend: the overall (average), movement of the data; and

SV = seasonal variation: This refers to short-term fluctuations in data values, which repeat themselves at the same month (or quarter) in each year – e.g. retail sales would be expected to increase just before Christmas and ice-cream sales would be expected to be higher during the summer. In monitoring sales on a day-to-day basis, any regular daily variations would also be called seasonal. For instance, Saturday may be a high sales day (see sample question 2 at the end of this chapter).

R = random variation, sometimes referred to as the **residual**: This is that variation in the data which is not predictable. The effect of such variation can be negligible (e.g. a salesman loses one order because his car breaks down, resulting in a slight, unpredictable, loss in sales) or catastrophic (share prices plummet, resulting in a loss of confidence in the company).

Though you should be aware of cyclical variation, it is unlikely that you will be expected to carry out any calculation connected with isolating it. The mathematical model you need to be aware of excludes any reference to cyclical variation and is:

$$O = T + SV + R \qquad \textbf{(1)}$$

Table 5.3 (below) and the resulting graph, fig. 5.4 (next page), illustrate the form of time series you are likely to encounter: the profit data of Departments A, B and D from the previous example have been combined to give data which reflects the model given above (**1**).

Table 5.3 Sandton Engineering Company. Total quarterly record of profits (£'000) (Departments A, B and D): 1983 to 1988.

Year	quarter	profits	Year	quarter	profits
1983	1	4.288	1986	1	7.119
	2	−2.215		2	1.741
	3	5.130		3	8.326
	4	1.437		4	4.393
1984	1	4.994	1987	1	8.103
	2	−1.554		2	2.279
	3	8.732		3	9.936
	4	2.703		4	5.608
1985	1	5.921	1988	1	8.464
	2	0.363		2	2.872
	3	7.474		3	10.192
	4	3.974		4	6.547

In an examination you would probably not be given the separate components of the time series, but you could be expected to calculate them.

5.3 Calculating the trend

You will be expected to be familiar with three methods:

1 **Inspection:** Draw a trend line by observing the data and judging which line looks to be the 'best fit'. The line should pass through the middle of the recorded data (see fig. 5.4).

2 **Regression analysis:** This is a numerical method of calculating (rather than 'judging') the trend. It assumes that the trend is a straight line. It will be considered in some detail in the next chapter.

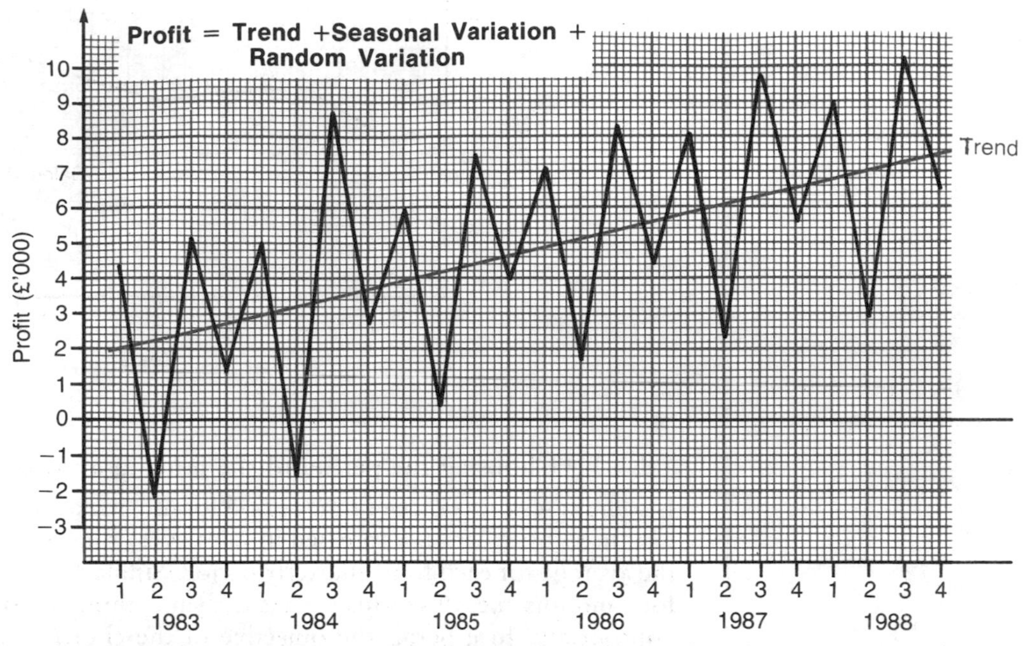

Fig. 5.4 Sandton Engineering Co – total quarterly profit record (Departments A, B and D): 1983 to 1988

3 **Moving averages:** In this method the data are averaged in order to remove certain influences on the data, particularly seasonal variation. This makes it possible to monitor the overall movement of the data, the trend.

Moving averages is the method you are likely to be asked to use to isolate the trend in analysing a time series. We will look at it in some detail now.

5.3.1 Moving averages

In calculating a moving average, it is first necessary to establish the period of time which is being considered. In the following example the period is three years. The average of the first three years is calculated (276.70). The attention is then 'moved' on by one unit of time (one year in this example). The process is repeated for the new period (giving an average of 290.00, in this example). The calculation 'moves on' by one unit of time at each successive calculation producing 'moving' averages.

Year	Sales (£)	Total for the 3-year period	Moving average
1	240		
2	300	(years 1+2+3) 830	276.7
3	290	(years 2+3+4) 870	290.0
4	280	(years 3+4+5) 880	293.3
5	310	930	310.0
6	340	980	326.7
7	330	1040	346.7
8	370		

In this example we have a three-year moving average. The sum of the first three years' sales (240+300+290) is 830 and the average (arithmetic mean) is 276.7. Because this is the average of years 1, 2 and 3, it is related to the mid-point of the overall period – year 2, in this case. As we move on to the end of year 4, the average of years 2, 3 and 4 is calculated and is related to the mid-point of that period, year 3. We continue in this way until the end of year 8 when we find the average of year 6, 7 and 8 (346.7) which relates to year 7.

The moving averages represent the trend of the data (the overall or average tendency). In this example, the sales appear to be increasing (see fig. 5.5, overleaf), a pattern which was not immediately obvious from the original data.

Note: With this method the more recent average is always a number of time periods out-of-date – in this case, one year out-of-date. It is also true that the averages do not start at the same time period as the original data. You should always bear these points in mind when using moving averages to calculate the trend.

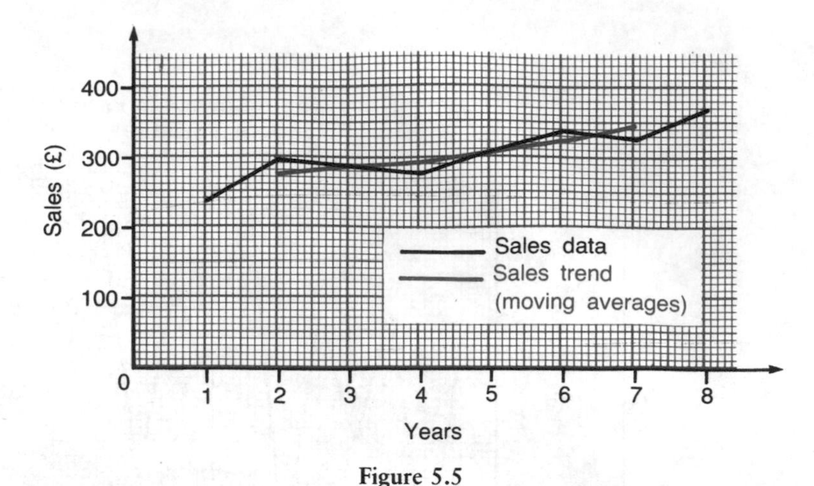

Figure 5.5

5.3.1.1 Choosing the period of the moving average

Is some cases the choice of time period for moving average calculations is obvious. For instance, if the data is expressed in sales every quarter, then it is sensible to calculate the average for each four-quarterly period. Similarly, if sales data were recorded every four months, i.e. three times a year, then a three-period moving average would be appropriate. In general, the objective of the choice is to get as smooth a trend as possible. To help you decide, you should inspect the data to see if there are any obvious cycles. The data relating to the profits of Department C, shown in fig. 5.3, has a five-year cycle, so the smoothest trend would be achieved by calculating a five-year moving average. This, however, would result in the trend always being two and a half years out-of-date. In contrast, the data of Department B has a cycle of four quarters, indicating a four-point average.

The method is best explained by considering another example: Table 5.4 and fig. 5.6 show the record of live births in the UK (1968 – 85). Initially the trend was calculated using a three-year moving average but, as can be seen from fig. 5.6, the trend obtained is no more informative than the original data. By inspecting the graph of the original data, it can be seen that there is a peak in year 1971 and again in 1980, nine years later. When the trend is recalculated using a nine-year moving average, a clear downward trend emerges. However, unlike the three-year moving average which is only one year out-of-date by the end of the period (1985), the nine-year moving average is four years out-of-date in 1985.

Table 5.4 United Kingdom births annual averages 1968 – 85

Year	Total live births male and female ('000)	Sum of 3s	Trend	Sum of 9s	Trend
1968	947				
1969	920	2771	924		
1970	904	2726	909		
1971	902	2640	880		
1972	834	2516	839	7398	822
1973	780	2351	784	7108	790
1974	737	2215	738	6875	764
1975	698	2111	704	6706	745
1976	676	2031	677	6558	729
1977	657	2020	673	6455	717
1978	687	2079	693	6394	710
1979	735	2176	725	6378	709
1980	754	2220	740	6410	712
1981	731	2204	735	6484	720
1982	719	2171	724		
1983	721	2170	723		
1984	730	2202	734		
1985	751				

(Source: Office of Population Censuses and Surveys: Key Data 1987)

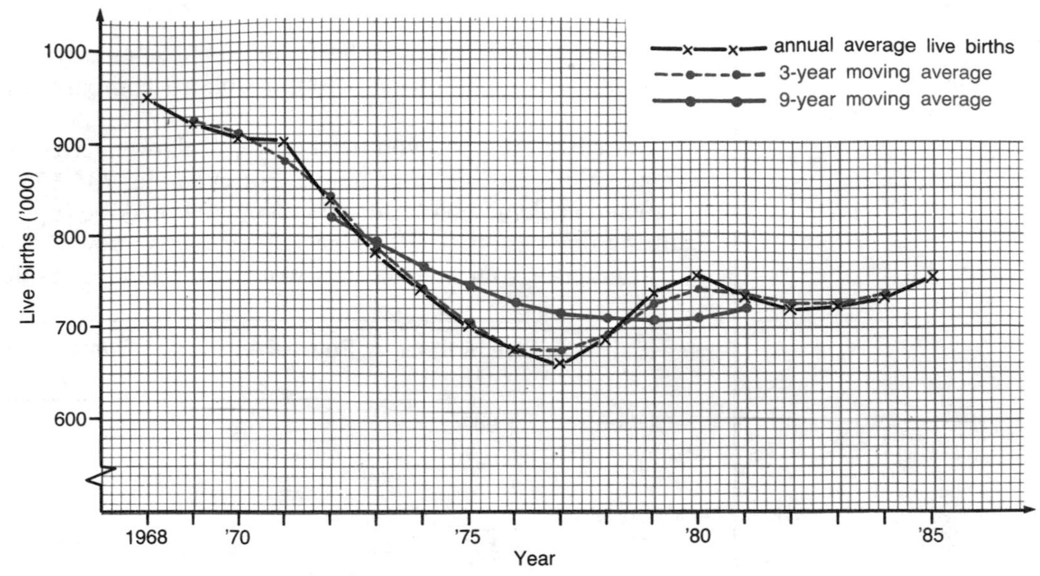

Fig. 5.6 United Kingdom birth – annual averages 1968 – 85 and three- and nine-year moving averages

If the time period chosen is too short the trend obtained will be unstable and very sensitive – it reacts to every change in the data. If the time period is too long the trend will be very stable but not sufficiently sensitive to changes in the data. The aim is to get a balance between the two. In the example just considered the nine-year moving average produces a stable (smooth) trend which is still sensitive to change. For instance, the rise in live births which seems to be occurring from 1977/78 is reflected by the trend.

5.3.1.2 Centering the trend

In all the examples looked at so far, the time period chosen has been an odd number of years. This has resulted in the average for any period being related to the middle term of the chosen period. For instance, consider again the UK births data (table 5.5), part of which is reproduced here:

Live births	Trend
947	—
920	924
904	909
902	880
834	839
⋮	⋮

924 is the average of 947, 920 and 904 and is positioned in the centre of that period (as an average is a measure of centre). If the chosen period had been four years, the table would have looked like this:

Live births	Sum (of 4s)	Trend	Centred trend
947			
920			
	3673	(÷4 =) 918.25	
904			(918.25 + 890)/2 = 904.125
	3560	(÷4 =) 890	
902			(890 + 855)/2 = 872.5
	3420	(÷4 =) 855	
834			
780	:	:	
:	:	:	

In this case the trend still relates to the centre of the data, but the centre lies between two values. For instance, the trend value 918.25 (the average of 947, 920, 904 and

902) is positioned between 920 and 904. The next trend value is positioned between 904 and 902. This makes it impossible to compare the trend with an actual data value.

To centre the trend you need only calculate the average of the appropriate pairs of trend values, as illustrated above. For example, 904.125 is the average of 918.25 and 890 and is aligned with the data value 904, making it possible to note that 'live births during this period, 1970, were 125 below average'.

In practice your calculations would look like the table below:

Live births (1)	Sum (of 4s) (2)	Sum (of 8s) (3)	Trend (4)
947			
920			
	3673		
904		7233	(÷ 8 =) 904.125
	3560		
902		9680	(÷ 8 =) 872.5
	3420		
834		6673	(÷ 8 =) 834.125
	3253		
780			

You will notice that it is unnecessary to find the four-point moving average and then average again. A small amount of work is saved by calculating the sum of fours (column 2), calculating the appropriate sum of eights (column 3) and then finding the (centred) trend.

5.4 Calculating the seasonal variation

Having found the trend, the seasonal variation is the next component to isolate– assuming that there is evidence of seasonal variation, which would usually be obvious from a graph of the data.

As stated in section 5.2, the methods and results being considered are based on the mathematical model: $O = T + SV + R$. If the model is a good representation of the data and the random variation is relatively small (and therefore negligible), the seasonal variation can be found as follows:

$$SV = O - T$$

Consider the example illustrated in table 5.5 (opposite): note the way in which the data and the resulting calculations are set out. (It is a format which could prove useful for examination questions.)

A four-point moving average has been calculated as the data in quarterly sales. It is therefore necessary to centre the data by combining the sums of four, as shown. The trend can then be calculated from the sums of eight. The final column in the table is the difference between the observed data, O, and the trend, T. The figures in this column represent seasonal (and random) variation. We can proceed to isolate the seasonal variation. Table 5.6 summarizes this information and makes it possible to estimate the seasonal variation.

You will notice that the deviations from the trend for a particular quarter vary. This is because $O - T$ contains not only seasonal variation but also random variations. In calculating the averages of the deviations, $O - T$, for each quarter, the random elements are expected to cancel out, or at least to be reduced to a negligible level. The resulting averages are estimates for the seasonal variations in the data. In the case of Holden Enterprises it is clear from the graph (fig. 5.7) that the model is little affected by random variation, so we should be able to feel confident about the estimates for both the trend and seasonal variation. This is not always the case as will be seen in section 5.7.

Before looking at how the results obtained can be used for forecasting we will look at the calculation of correction factors and the effects of rounding of data; see sections 5.4.1 and 5.4.2 on page 86.

Table 5.5 Profits of Holden Enterprises (£'000)

Year	Quarter	Profits (O)	Sum of 4s	Sum of 8s	Trend (T)	Deviation from trend (O − T)	Quarter
1983	1	4.3			*		1
	2	9.8			*		2
			32.9				
	3	11.1		66.5	(÷8 =) 8.3125	2.7875	3
			33.6				
	4	7.7		68.4	(÷8 =) 8.55	−0.85	4
			34.8				
1984	1	5.0		71.4	(÷8 =) 8.925	−3.925	1
			36.4				
	2	11.0		73.8	(÷8 =) 9.225	1.775	2
			37.4				
	3	12.7		75.7	(÷8 =) 9.4625	3.2375	3
			38.3				
	4	8.7		77.9	(÷8 =) 9.7375	−1.0375	4
			39.6				
1985	1	5.9		79.9	(÷8 =) 9.9875	−4.0875	1
			40.3				
	2	12.3		81.6	(÷8 =) 10.2	2.1	2
			41.3				
	3	13.4		83.6	(÷8 =) 10.45	2.95	3
			42.3				
	4	9.7		85.1	(÷8 =) 10.6375	−0.9375	4
			42.8				
1986	1	6.9		86.5	(÷8 =) 10.8125	−3.9125	1
			43.7				
	2	12.8		88.4	(÷8 =) 11.05	1.75	2
			44.7				
	3	14.3		90.4	(÷8 =) 11.3	3.0	3
			45.7				
	4	10.7		92.8	(÷8 =) 11.6	−0.9	4
			47.1				
1987	1	7.9		95.0	(÷8 =) 11.875	−3.975	1
			47.9				
	2	14.2		96.7	(÷8 =) 12.0875	2.1125	2
			48.8				
	3	15.1		98.1	(÷8 =) 12.2625	2.8375	3
			49.3				
	4	11.6		99.4	(÷8 =) 12.425	−0.825	4
			50.1				
1988	1	8.4		101.3	(÷8 =) 12.6625	−4.2625	1
			51.2				
	2	15.0			*		
	3	16.2			*		

Table 5.6 Summary of seasonal variation: O − T

Year	Quarter 1	Quarter 2	Quarter 3	Quarter 4
1983			2.7875	−0.85
1984	−3.925	1.775	3.2375	−1.0375
1985	−4.0875	2.1	2.95	−0.9375
1986	−3.9125	1.75	3.0	−0.9
1987	−3.975	2.1125	2.8375	−0.825
1988	−4.2625			
Totals	−20.1625	7.7375	14.8125	−4.55
Average = estimated seasonal variation	−4.0325	1.9344	2.9625	−0.91

Total variation = −0.0456

The correction to reduce the total variation to zero is +0.0114 (see section 5.4.1)

	Quarter 1	Quarter 2	Quarter 3	Quarter 4
Correction:	0.0114	0.0114	0.0114	0.0114
	−4.0211	1.9458	2.9739	−0.8986
Final estimate of seasonal variation (these might be rounded up or down):	−4.0	1.9	3.0	−0.9

5.4.1 Correction factors

The trend is an average and the seasonal variations are deviations from that average. Therefore the seasonal variations should balance out over the year, summing to zero. In practice this is rarely the case because of random variation. In table 5.6 (previous page) the total variation is −0.0456 (−4.0325 + 1.9344 + 2.9625 + −0.91), which is between 1 and 5 per cent of the estimates for seasonal variation. The seasonal estimates must therefore be corrected so that they add up to zero. In this example this means adding a quarter of the amount (0.0456 ÷ 4 = 0.0114) to each of the four estimates. The corrected estimates (−4.0211, 1.9458, 2.9739 and −0.8986) now sum to zero.

5.4.2 Rounding data

Having got as far as estimating the seasonal variation, including any necessary corrections, you must decide whether to 'round' the seasonal estimates. In table 5.6 the estimates for the seasonal variations were rounded to the nearest first decimal place.

One must bear in mind the accuracy of the original data and how confident one can be about the accuracy of the seasonal estimates. The values of the seasonal estimates should not imply a level of accuracy that did not exist in the original data. For instance, if we knew that the profits data of Holden Enterprises were only accurate to the nearest £1000, then it would be misleading to quote seasonal estimates of −4.0211 (−£4021.10), etc, which imply accuracy to the nearest 10 pence. Further uncertainty is introduced because of random variation.

As a guide, you should round the seasonal estimates to match the level of accuracy of the original data.

In an examination you would be expected to comment on the trend and/or the seasonal variation. For the Holden Enterprises data you could say:

1 the trend is increasing steadily, reflecting the improvement in (average) profits performance.

2 there is a marked seasonal effect. The first quarter is consistently the worst and the third quarter the best.

3 the data seems to be well represented by the additive model:

$$\text{Sales} = \text{Trend} + \text{Quarterly Variation} + \text{Random Variation}$$

Having calculated and commented on the historical aspects of the data you might be expected to use the data to forecast future sales performance.

5.5 Forecasting by extrapolation

One interpretation of strong patterns in a time series is that there is a good chance of them continuing in the near future. We can assume from this that it is acceptable to forecast future performance. For instance, if the trend appears to be increasing by, on average, 2.5 points per quarter, then we assume it will continue to do so. Further, if the first quarter of the financial year consistently shows an above average sales record, then we can assume that future first quarters will also show an above average sales record. We will continue to refer to the example of the previous section, Holden Enterprises, to illustrate the methods used.

5.5.1 Future trends

The trend increases from a value of 8.3125 (£'000) during the third quarter of 1983, to 12.6625 during the first quarter of 1988 (table 5.5).

Notice that the trend does not start at the same period as the original data; nor does it finish at the same period. The trend figures cover 19 quarters giving 18 quarters' increase. That is, an overall increase of 4.35 (12.6625 − 8.3125), which gives an average increase of 0.24 (4.35 ÷ 18 = $0.241\dot{6}$).

Forecast estimates for the missing trend values can now be calculated using the average trend increase figure.

Year	Quarter	Profits	Trend	
⋮	⋮	⋮	⋮	
	4	11.6	12.425	
1988	1	8.4	12.6625	
	2	15.0	(12.6625 + 0.24 =) 12.9025	Forecast trends
	3	16.2	(12.9025 + 0.24 =) 13.1425	
	4		(13.1425 + 0.24 =) 13.3825	
1989	1		(13.3825 + 0.24 =) 13.6225	
	2		(13.6225 + 0.24 =) 13.8625	
	3		(13.8625 + 0.24 =) 14.1025	

5.5.2 Future seasonal variation

We can also take the estimates of the seasonal variation to be representative of future occurrences, i.e. (from table 5.6):

Quarter	Variation
1	−4.0
2	+1.9
3	+3.0
4	−0.7

We are now in a position to estimate likely future profits for Holden Enterprises, as follows:

Year	Quarter	Profits	Trend	Seasonal variation
⋮	⋮	⋮	⋮	
	4	11.6	12.425	
1988	1	8.4	12.6625	
	2	15.0	12.9025	
	3	16.2	13.1425	
		Forecasts (O = T + SV)		
	4	12.5 (13.3825 − 0.9)	13.3825	−0.9
1989	1	9.6 (13.6225 − 4.0)	13.6225	−4.0
	2	15.8 (13.8625 + 1.9)	13.8625	+1.9
	3	17.1 (14.1025 + 3.0)	14.1025	+3.0

When forecasting, the steps to follow are as follows (taking as an example, the calculation of the forecast for quarter 3, 1989):

1 Having corrected for random variation assume that the mathematical model representative of the data is O = T + SV.

2 Assume the trend will continue in the future as it has done in the past. In this case the trend is increasing (on average) by 0.24 (£'000) per quarter, so by the third quarter of 1989 it is predicted (or forecast) that it will have a value of 14.1025 ($12.6625 + 6 \times 0.24$: 12.6625 is the last trend value calculated from the existing data, quarter 1, 1988).

3 The seasonal variation is similarly assumed to continue affecting the data as it has done in the past. The relevant seasonal variation (for the third quarter) is +3.0 (£'000).

4 At this point the forecast can be made:

We assume O = T + SV,
and have found T = 14.1025 and SV = +3.0,
so forecast profits for 1989 will be

$$14.1025 + 3 = 17.1025$$
$$= 17.1 \text{ (to a comparable level of accuracy)}$$

A graph of the data for Holden Enterprises, including forecast profits, is shown in fig. 5.7, overleaf.

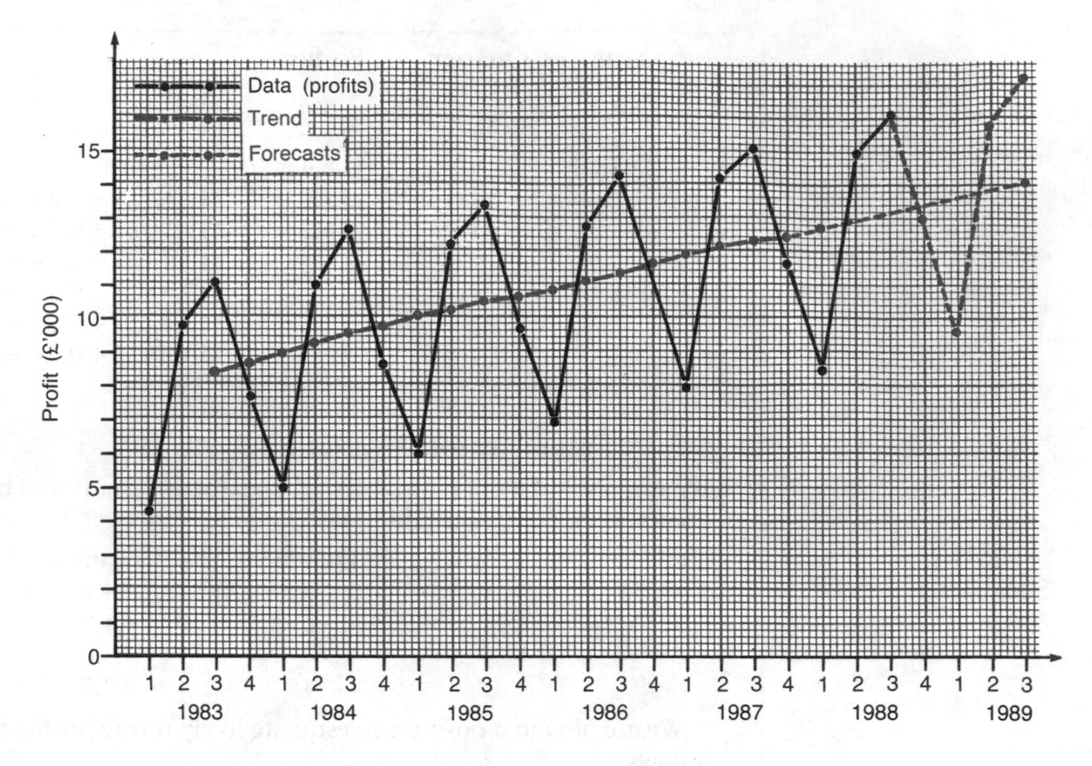

Fig. 5.7 Holden Enterprises: quarterly profit record 1983–88, with forecast for 1989

5.6 Time series: multiplicative model

So far, we have been considering the so-called additive model; that is, O = T + SV(+R). Mathematically, this demands that the different types of variation are independent of one another, for example, that an increasing trend would not influence the value of the seasonal variation. Many statisticians believe this to be an unacceptable assumption and encourage the use of a multiplicative model:

Observed data = Trend × Seasonal Ratio (× **Random Variation**)

or, symbolically, **O = T × SR** (× **RV**).

Reworking the Holden Enterprises example will illustrate the main differences between the two models.

Table 5.7 Profits of Holden Enterprises (£'000): multiplicative model

Year	Quarter	Profits (O)	Trend (T)	Sales/trend ratio (O/T)	Quarter
1983	1	4.3	*	*	1
	2	9.8	*	*	2
	3	11.1	8.3125	1.335	3
	4	7.7	8.55	0.901	4
1984	1	5.0	8.925	0.560	1
	2	11.0	9.225	1.192	2
	3	12.7	9.4625	1.342	3
	4	8.7	9.7375	0.893	4
1985	1	5.9	9.9875	0.591	1
	2	12.3	10.2	1.206	2
	3	13.4	10.45	1.282	3
	4	9.7	10.6375	0.912	4
1986	1	6.9	10.8125	0.638	1
	2	12.8	11.05	1.158	2
	3	14.3	11.3	1.265	3
	4	10.7	11.6	0.922	4
1987	1	7.9	11.875	0.665	1
	2	14.2	12.0875	1.175	2
	3	15.1	12.2625	1.231	3
	4	11.6	12.425	0.934	4
1988	1	8.4	12.6625	0.663	1
	2	15.0	*		
	3	16.2	*		

The trend is calculated in exactly the same way as before. It is only the seasonal variation that is treated differently. You will notice, from table 5.7, that in this case the seasonal variation is the ratio of the original data to the trend, unlike the additive model where the seasonal variation was the difference. Table 5.8 illustrates the calculation of the seasonal ratios. Particularly compare the correction calculations with the additive model method (table 5.6, p.85).

Table 5.8 Summary of seasonal ratio: O/T

Year	Quarter 1	Quarter 2	Quarter 3	Quarter 4
1983			1.335	0.901
1984	0.560	1.192	1.342	0.893
1985	0.591	1.206	1.282	0.912
1986	0.638	1.158	1.265	0.922
1987	0.665	1.175	1.231	0.934
1988	0.663			
Totals	3.117	4.731	6.455	4.562
Average	0.6234	1.18275	1.291	0.9124

= estimated seasonal variation
Total average ratio = 4.00955.
The expected total is 4. Each ratio must be multiplied by 4/4.00955 to obtain the correct seasonal ratio:
Final estimate of seasonal ratio:

	Quarter 1	Quarter 2	Quarter 3	Quarter 4
	0.622	1.180	1.288	0.910

Forecasting follows a similar procedure to that used with the additive model. Compare the calculations below with those for the additive model example on page 87.

Year	Quarter	Sales	Trend	Seasonal ratio
⋮	⋮	⋮	⋮	
	4	11.6	12.425	
1988	1	8.4	12.6625	
	2	15.0		
	3	16.2		
		Forecasts (O = T × SR)		
	4	12.2 (13.3825 × 0.910)	13.3825	0.910
1989	1	8.5 (13.6225 × 0.622)	13.6225	0.622
	2	16.4 (13.8625 × 1.180)	13.8625	1.180
	3	18.2 (14.1025 × 1.288)	14.1025	1.288

5.7 Forecasting: caution needed!

The examples considered so far have been very stable. That is, the trend and seasonal variation (or ratio) have been consistent throughout the data. In such cases one can feel confident about forecasting, though to forecast too far into the future can be very risky: the circumstances which make the data behave in a particular fashion could change, reflecting changes in company policy or the economic climate. One important aspect of a time series, which affects the confidence we are able to have in any forecasting procedure, is how much of the variation is predictable and how much is unexplainable and therefore random.

5.7.1 Random variation or residuals

Random variation or residual values cannot be included in any forecasting, but they are always present. It is these unknown, unpredictable factors which upset our forecasts. As long as the random element is small this need not bother the forecaster, as the error introduced will also be small. If, however, the random variation is relatively large then the effect on forecasts can make associated planning useless. An example (table 5.9, fig. 5.8; next page), will illustrate the point.

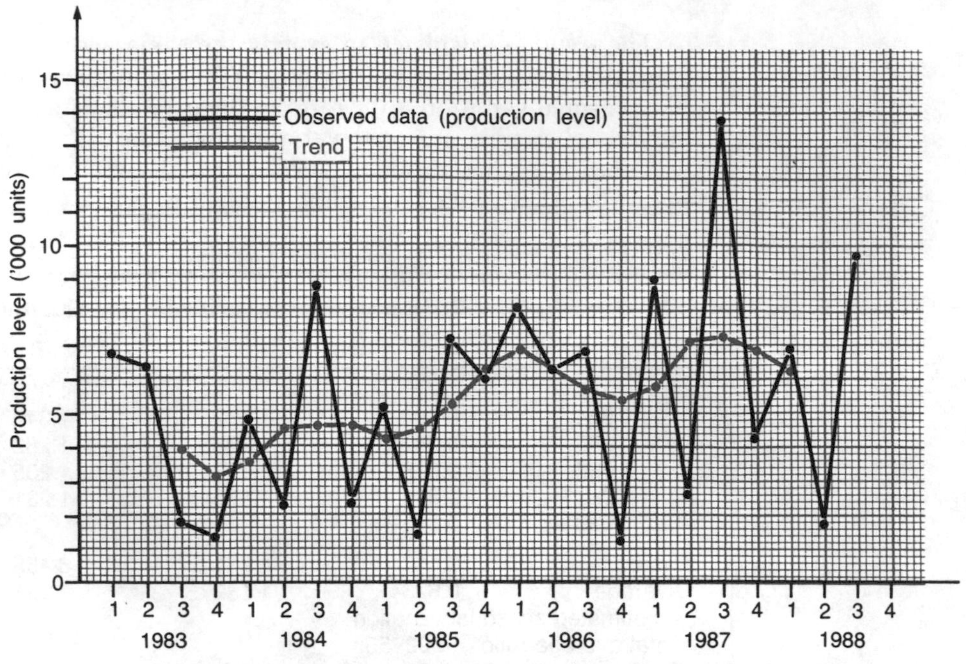

Fig. 5.8 Breeze Boats Company: quarterly production figures 1983–88

Table 5.9 Production level of Breeze Boats Company ('000 units)

Year	Quarter	Production (O)	Trend (T)	Seasonal variation (SV)	Random variation (O – T – SV) absolute	percentage of production
1983	1	6.8	*	*		
	2	6.4	*	*		
	3	1.8	3.9	2.5	−4.6	256.0
	4	1.4	3.1	−2.1	0.4	28.0
1984	1	4.9	3.5	1.7	−0.3	6.1
	2	2.3	4.5	−2.3	0.1	4.3
	3	8.8	4.6	2.5	1.7	19.3
	4	2.3	4.6	−2.1	−0.2	8.7
1985	1	5.2	4.2	1.7	−0.7	13.5
	2	1.4	4.5	−2.3	−0.8	57.1
	3	7.2	5.3	2.5	−0.6	8.3
	4	6.0	6.3	−2.1	1.8	30.0
1986	1	8.2	6.9	1.7	−0.4	4.9
	2	6.2	6.2	−2.3	2.3	37.1
	3	6.8	5.7	2.5	−1.4	20.6
	4	1.2	5.4	−2.1	−2.1	175.0
1987	1	9.0	5.8	1.7	1.5	16.7
	2	2.6	7.1	−2.3	−2.2	84.6
	3	13.9	7.2	2.5	14.9	107.2
	4	4.3	6.8	−2.1	−0.4	9.3
1988	1	6.9	6.1	1.7	−0.9	13.0
	2	1.7				
	3	9.7	*			

You will notice from the graph how eratic Breeze Boats' production seems to be. The trend line is not as smooth as in the previous examples. With such a picture you would expect the random element to be large and this is confirmed by reference to the last two columns in table 5.9.

Random variation is almost three times the level of the data at some points. One could not feel confident about using such a model for forecasting. As an exercise, calculate the random variation of the Holden Enterprises data. You should find the random element is very small.

5.8 Deseasonalizing data

You should be aware by now that a drop in sales in a particular quarter does not necessarily mean economic disaster, nor even that the salesman should be fired. As long as the overall trend is acceptable, which probably means it is increasing, then the salesman's job should be safe. However, the overall trend is not always clear. In some situations it is obscured by seasonal and other variation. In such cases it is useful to remove the seasonal variation altogether. This process is called deseasonalizing. Deseasonalized data will, however, still contain trend and random variation.

The method used to deseasonalize data depends on how the seasonal effect has been calculated. In the case we will consider here, it is assumed that the multiplicative model best describes the variability of the data (i.e. $O = T \times SR \times RV$). For the multiplicative model, the deseasonalized data is found by dividing the observed data by the seasonal ratio:

$$\textbf{Deseasonalized data} = O/SR = T \times RV$$

The necessary calculations to deseasonalize the Holden Enterprises data (multiplicative model) are shown in table 5.10 and the results are illustrated in fig. 5.9.

Table 5.10. Profits of Holden Enterprises (£'000): multiplicative model – deseasonalized profits

Year	Quarter	Profit (O)	Seasonal ratio (SR)[1]	Deseasonalized profits (O/SR)
1983	1	4.3	0.622	6.9
	2	9.8	1.180	8.3
	3	11.1	1.288	8.6
	4	7.7	0.910	8.5
1984	1	5.0	0.622	8.0
	2	11.0	1.180	9.3
	3	12.7	1.288	9.9
	4	8.7	0.910	9.6
1985	1	5.9	0.622	9.5
	2	12.3	1.180	10.4
	3	13.4	1.280	10.4
	4	9.7	0.910	10.7
1986	1	6.9	0.622	11.1
	2	12.8	1.180	10.8
	3	14.3	1.288	11.1
	4	10.7	0.910	11.8
1987	1	7.9	0.622	12.7
	2	14.2	1.180	12.0
	3	15.1	1.288	11.7
	4	11.6	0.910	12.7
1988	1	8.4	0.622	13.5
	2	15.0	1.180	12.7
	3	16.2	1.288	12.6

[1] Using predicted SR values calculated in table 5.8.

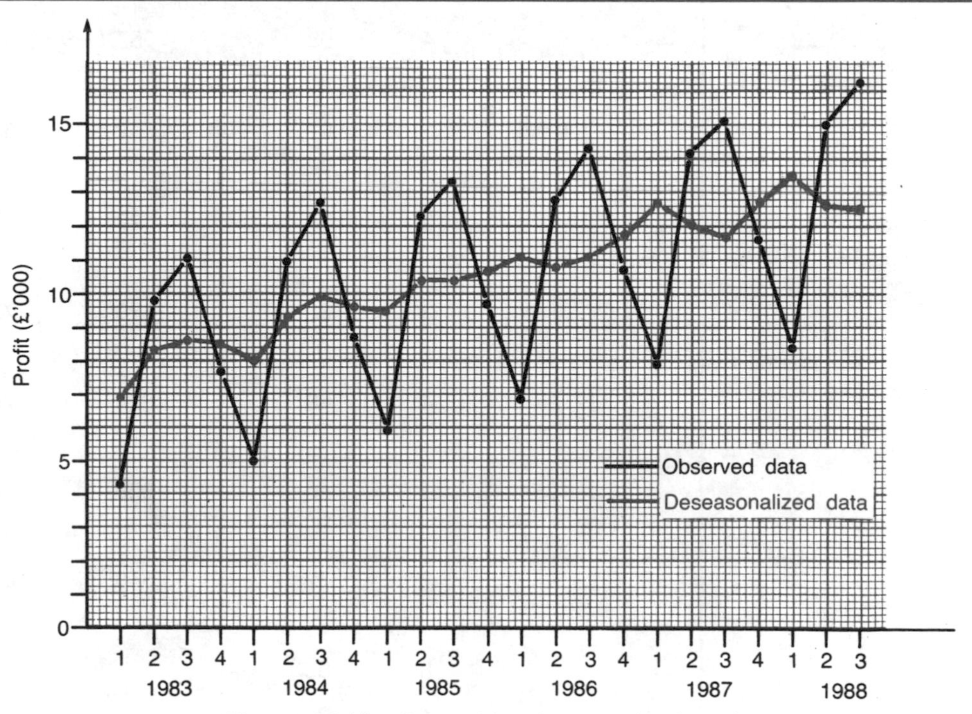

Fig. 5.9 Holden Enterprises: deseasonalized profits

To deseasonalize data in which the additive model is assumed the seasonal variation is simply subtracted from the observed data.

Sample questions

1 (a) Briefly outline the components which make up a typical time series.

(b)

Road Casualties 1982–84 ('000)

Quarter	1	2	3	4
1982	70	82	90	92
1983	67	77	82	83
1984	69	79	85	91

(*Source: Monthly Digest of Statistics*)

(i) By means of a moving average, find the trend and the seasonal adjustments.

(ii) Give the data for 1984 seasonally adjusted. Explain why it is necessary to seasonally adjust data.

CIS; Part 1; Quantitative Studies; December 1986.

Solution/hints

Part (a)

See sections 5.1 and 5.2.

Part (b)

(i) Table 1 below shows the necessary analysis to find the trend and the first stages in establishing the variation (O – T). The variation from the trend contains seasonal and random variation. The calculations needed to establish the (average) seasonal variation are shown in table 2.

Table 1

Year	Qtr	Road casualties (O)	Totals 4s	Totals 8s	Trend (T)	Variation from the trend (O – T)
1982	1	70				
	2	82				
			334			
	3	90		665 (÷8 =)	83.125	6.875
			331			
	4	92		657 (÷8 =)	82.125	9.875
			326			
1983	1	67		642	80.25	–13.25
			318			
	2	77		627	78.375	–1.375
			309			
	3	82		620	77.5	4.5
			311			
	4	83		624	78.0	5.0
			313			
1984	1	69		629	78.625	–9.625
			316			
	2	79		640	80.0	–1.0
			324			
	3	85				
	4	91				

Table 2 Seasonal variation: calculations

Year	Quarter 1	Quarter 2	Quarter 3	Quarter 4
1982			6.875	9.875
1983	−13.25	−1.375	4.5	5.0
1984	−9.625	−1.0		
Total	−22.875	−2.375	11.375	14.875
Average	−11.4375	−1.1875	5.6875	7.4375
Sum of averages = 0.5				
Correction factor = 0.5 ÷ 4 = 0.125				
Seasonal variation	−11.6	−1.3	5.6	7.3

Note: As there is no indication of the level of accuracy in the original data it is unwise to retain a large number of decimal places – to do so implies a level of precision which may not be present.

(ii) Seasonally adjusted data for 1984 ('000)

	Quarter 1	2	3	4
1984	57.4	77.7	90.6	98.3

e.g. Quarter 1: $O - SV = 69 + (-11.6) = 57.4$

It is necessary to seasonally adjust data so that the overall movement of the data (the trend) can be more easily observed. If seasonal influences are left in the data, then it is difficult to establish if changes are due to the effects of seasonal influences or other reasons (trend or random effects).

2 The following figures represent cash sales of a store which does not open for business on a Monday.

Week Number	**1**	**2**	**3**	**4**
	£	£	£	£
Day of week				
Tuesday	360	350	380	390
Wednesday	400	430	440	450
Thursday	480	490	490	500
Friday	660	680	690	690
Saturday	660	680	690	690

From the data supplied, calculate:

(a) the trend using a 5-point moving average technique;

(b) the seasonal, or in this case, average daily variations;

(c) predict the cash sales of the store on Tuesday of week 5 and comment on the result obtained.

CIPFA; Diploma in Public Sector Audit and Accounting, Quantitative analysis; May 1987.

Solution/hints

Part (a)

To calculate the trend using a five-point moving average is easier than for instance, a four-point moving average. The trend can immediately be aligned with the data. This is true for all odd-number moving averages. Some of the necessary calculations are shown here; the rest have been left as an exercise (marked * in the tables).

Week no.	Day	Sales (£) (a)	Total 5s	Trend (b)	Variation from the trend (a − b)
1	Tues.	360			
	Weds.	400			
	Thur.	480	2560 (÷ 5 =)	512	−32
	Fri.	660	2550	510	+150
	Sat.	660	2580	516	+144
2	Tues.	350	2590	518	−168
	Weds.	430	2610	522	−92
	Thur.	490	2630	526	−36
	Fri.	680	2660	532	+148
	Sat.	680	⋮	⋮	⋮
3	Tues.	380			

Part (b)
Seasonal (average daily) variations:

Week no.	Tues.	Weds.	Thur.	Fri.	Sat.
1			−32	+150	+144
2	−168	−92	−36	+148	*
3	−92	*	−48	*	*
4	*	*	−44		
Total	−476	*	−160	*	*
Average	−158.67	*	−40	*	*

Part (c)
Left as an exercise.

3 You are the personal assistant of the sales manager for Newsports, a retailer of winter sports equipment. The sales of skis over the period 1983 to 1988 are given in table 1 (opposite page). She wishes you to prepare a report which will focus on the past sales of skis and future sales prospects.

A part analysis has been provided for your use in table 2 (opposite page).

The sales manager needs the report in a suitable format for tomorrow's board meeting. She would like her attention drawn to any important factors in your analysis.

Solution/hints

BTEC questions are often placed in some relevant context. This one requires the solution to be presented in the form of a report to be used in the context of a board meeting. Before starting the calculations, which should be straightforward, the reader should think carefully about what might be required. A chart, graph or diagram would be a useful form of presentation to have at a meeting. For any time series analysis problem a graph will give most of the information required. Generally the information should be brief and easy to use in the context of a business meeting.

4 The number of visitors to a local museum has been recorded on a quarterly basis, as follows, for the years 1982–85 inclusive:

	1982	1983	1984	1985
Quarter 1	29.5	30.1	35.0	34.1
,, 2	32.9	31.5	38.6	36.3
,, 3	34.4	26.5	29.2	35.8
,, 4	32.5	26.8	38.5	40.2

All the figures are in thousands.

YOU ARE REQUIRED TO:

(a) calculate the trend for the period 1982–85 by the method of moving averages;
(b) determine the seasonal component for each quarter;
(c) forecast the likely number of visitors in the first quarter of 1986;
(d) indicate why the forecast figure may differ from the actual number of visitors.

CIPFA; Diploma of Public Sector Audit and Accounting; Quantitative analysis; March 1986.

The solution to this question is left as an exercise.

Recommended reading

Owen, F. and Jones, R., *Statistics* (Pitman, 1988), chapters 6 and 7.

Whitehead, P. and Whitehead G., *Statistics for Business* (Pitman, 1984), chapter 10.

Key Data 87, Central Statistical Office (HMSO, 1987).

Curwin, J. and Slater, R., *Quantitative Methods for Business Decisions* (Van Nostrand Reinhold, 1986), chapter 19.

Lucey, T., *Quantitative Techniques* (D.P. Publications, 1980), chapter 5.

Table 1 Sales of skis (all UK branches)

	('000s of units)			
		Quarter ended		
Year	March	June	September	December
1983	295	329	344	325
1984	301	315	265	368
1985	350	386	262	405
1986	383	419	281	432
1987	393	436	302	449
1988	425	460		

Table 2

Year	Qtr	Sales ('000s)	Totals 4s	Totals 8s	Trend	Sales – Trend
1983	1	295				
	2	329				
			1293			
	3	344		2592 (÷8 =)	324	
			1299			
	4	325		2584	323	
			1285			
1984	1	301		2491	311.375	
			1206			
	2	315		2455	306.875	+8.125
			1249			
	3	265		2547	318.375	−53.375
			1298			
	4	368		2667	333.375	+34.625
			1369			
1985	1	350		2735		
			1366			
	2	386		2769		
			1403			
	3	262		2839		
			1436			
	4	405		2905		
			1469			
1986	1	383		2957		
			1488			
	2	419		3003		
			1515			
	3	281		3040		
			1525			
	4	432				
1987	1	393				
	2	436				
	3	302				
	4	449				
1988	1	425				
	2	460				

The Polytechnic of North London; BTEC HND, Business and Finance (part question); June 1988.

6 Linear regression and correlation

6.1 Introduction

It is often the case that two (or more) variables are associated or related in some way. For instance, it is generally true that a person's weight is related to their height and/or to a number of other factors, such as age and sex.

In the last chapter it was seen that a company's sales, profits or level of production can, in certain circumstances, be related to the passage of time. In particular, the trend was often seen to be related linearly to time. That is, for each unit increase in time (week, month, quarter, etc) the trend seemed to increase by a fixed amount, resulting in a straight-line (linear) relationship.

It is important for the business person or economist to be able to identify related variables. For instance, if a company spends a great deal on advertising it is probably because they believe there is a relationship between advertising and the level of sales. Once a likely relationship has been identified, such as a relationship between advertising expenditure and sales revenue, it is necessary to determine:

1 The nature of the relationship – you will be mainly concerned with linear relationships and finding equations which describe these relationships.

2 The extent to which the calculated equations reflect the relationship under investigation. An increase in advertising expenditure may explain some of the observed increase in sales but other factors may have contributed to the increase, such as a more aggressive marketing policy or change in selling price.

This chapter is concerned with the two areas just outlined, namely, the identification of a relationship between two variables, usually in the form of an equation, and the determination of how well that equation explains the perceived relationship.

You will be expected to be familiar with a number of ways of finding the appropriate equations, ranging from simple graphical methods to complex regression analysis. You will also need to understand the principles behind measuring a relationship using correlation techniques.

At this point it is appropriate to consider some of the factors with which you will need to be familiar:

6.2 The scatter diagram

You have already encountered scatter diagrams, or scatter graphs, in chapter 3. They provide a means of ascertaining whether a relationship exists between two variables. For example, consider the data shown in table 6.1 and represented in a scatter diagram in fig. 6.1 (opposite). Though it is difficult to see from the tabulated data whether a relationship exists between sales revenue and advertising expenditure, it is clear from the scatter diagram that increased spending on advertising produces a positive response in sales. The two variables, advertising expenditure and sales revenue are said to be positively correlated.

It is important that you understand the vocabulary you are likely to meet in this topic area. If you are familiar with the terms independent and dependent variable, line of best fit, interpolation and extrapolation, then move on to section 6.6.

6.3 Independent and dependent variables

In the example cited above, advertising expenditure would be thought of as the independent variable. In contrast, sales revenue is thought to 'depend' on the amount spent on advertising – consequently, it is called the dependent variable.

Table 6.1 Stevenson's Crafts: Record of advertising expenditure (£'000) and sales (£ million) for the period 1970–89

	Advertising (X) (£'000)	Sales (Y) (£m)
1970	70	13.0
1971	75	16.0
1972	85	16.0
1973	105	22.0
1974	90	19.8
1975	100	23.2
1976	95	20.8
1977	90	20.2
1978	75	14.5
1979	85	17.2
1980	85	17.0
1981	90	18.0
1982	90	17.8
1983	80	16.0
1984	110	25.0
1985	100	24.0
1986	90	19.0
1987	80	18.0
1988	95	20.5
1989	70	14.1

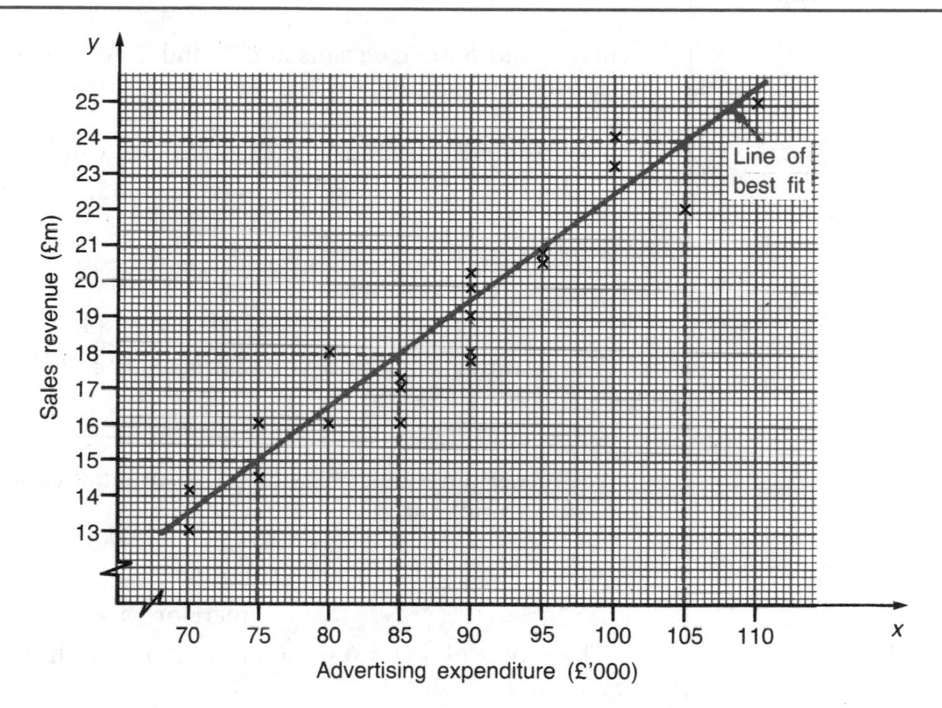

Fig. 6.1 Stevenson's Crafts: sales revenue *versus* advertising expenditure

It is not always obvious which variable is independent and which dependent. For instance, suppose we investigate the existence of a relationship between the supply of money in the economy, as represented by M3*, and the level of unemployment in the UK. There is uncertainty as to which factor is influencing which, or even if it makes sense to relate them at all. It could be that some other economic feature, such as Government legislation, is influencing both.

Once the variables have been classified it is conventional to label the independent variable X and the dependent variable Y. The example illustrated in table 6.1 and fig. 6.1 has been labelled accordingly. In this example there is a very definite relationship between advertising and sales. Consequently it makes sense to add a 'line of best fit' which reflects this relationship. It is impossible to be absolutely sure that the relationship is linear, but a linear relationship is strongly suggested by the scatter diagram. We will look briefly at the underlying ideas of lines of best fit.

*M3 includes notes and coins in circulation, various bank and building society accounts and UK residents' foreign currency deposits. For a more detailed explanation see the first reference in the recommended reading list at the end of this chapter.

6.4 The line of best fit (I)

The line of best fit represents the best linear relationship between two variables. The line in fig. 6.1 was drawn by eye. That is, it was drawn to pass through the middle of the data points, thereby having as many data points below the line as above it. Later on we will see more accurate, though complex, ways of ensuring best fit.

6.4.1 Forecasting from the graph

Once the line of best fit is drawn, it can be used for forecasting. For instance, if Stevenson's Crafts agree on a budget of £75 000 for advertising in 1990, then (from the graph) the forecast sales will be £15m. There were two previous occasions when there was this level of advertising expenditure: 1971 and 1978. The resulting sales figures were £16m and £14.5m. The forecast is, at worst, just over 6 per cent different from these figures.

6.4.2 The line of best fit: calculating the equation

It is often useful to know the equation of the line of best fit. Later sections will explain how this is done using the method of least squares, but it is also useful to be able to do it from a line drawn by eye.

As the line of best fit is a straight line, it must have an equation of the form

$$Y = a + bX$$

where a and b are constants and X and Y represent the independent and dependent variables, respectively. To find the constants we need only consider two points on the line.

For example, from fig. 6.1, the following can be seen:

$$\text{when } x = 85, \quad y = 18$$
$$\text{and when } x = 105, \quad y = 24.$$

These values can be substituted into the equation, $Y = a + bX$, to give:

$$24 = a + 105b \qquad (1)$$
$$\text{and } 18 = a + \ 85b \qquad (2)$$

The resulting simultaneous equations, (1) and (2), can be solved to find the values of a and b, as follows:

Subtract (2) from (1) to give $6 = 20b$; therefore, $b = 0.3$.

Substitute this value of b into either original equation to find the value of a. e.g. substituting into (1):

$$24 = a + (105 \times 0.3)$$
$$24 - 31.5 = a$$
$$\text{therefore } a = -7.5$$

The values of a and b so obtained allow our line of best fit to be described by the equation

$$Y = -7.5 + 0.3X$$

If you had any difficulty following these calculations, see sections 9.3 and 9.4 which deal with linear and simultaneous equations.

6.4.3 Forecasting from the equation

The equation can be used to forecast sales revenue (Y) for any given level of advertising expenditure (X). For example, suppose the proposed advertising budget (X) is £150 000; using the equation, the forecast sales (Y) could be found as follows:

$$y = -7.5 + (0.3 \times 150^*)$$
$$y = -7.5 + 45.0$$
$$y = 37.5$$

That is, the forecast sales value is £37.5m.

*Recall that for the purposes of the graph, and therefore the equation, advertising expenditure is in £'000, so the value of x is 150, not 150 000.

The two forecasts are different and not only because they involve different numbers or one was done directly from the diagram, while the other used the equation of the line of best fit. The important difference is that forecast 1 was made within the limits of the available data (advertising expenditure ranged from £70 000 to £110 000) and forecast 2 was made outside of these limits. The following section explains the significance of this distinction.

6.5 Interpolation and Extrapolation

An **interpolation** is a prediction made within the limits of the collected data, such as the forecast in section 6.4.1. An **extrapolation** is one made outside the limits of the available data, such as the forecast in section 6.4.3. The forecasts made in the previous chapter (time series analysis) were extrapolations.

It must be remembered that the line of best fit and the associated equation represent the best possible linear relationship for a given data set and then, only within the limits of the data. There is no guarantee that the relationship will be valid outside of the limits of the available data. For instance, is it reasonable to expect sales to increase indefinitely and at the same rate as advertising revenue increases?

Suppose a wider range of values were known. Figure 6.2 illustrates what this extended range of values might look like. It can be seen that the effect of increasing advertising expenditure on sales gradually decreases. It appears as though a market saturation point is reached. That is, the demand for the goods reaches a ceiling beyond which sales will increase very little, in spite of increased advertising. The extrapolated forecast for advertising expenditure of £150 000 is almost 40 per cent too large.

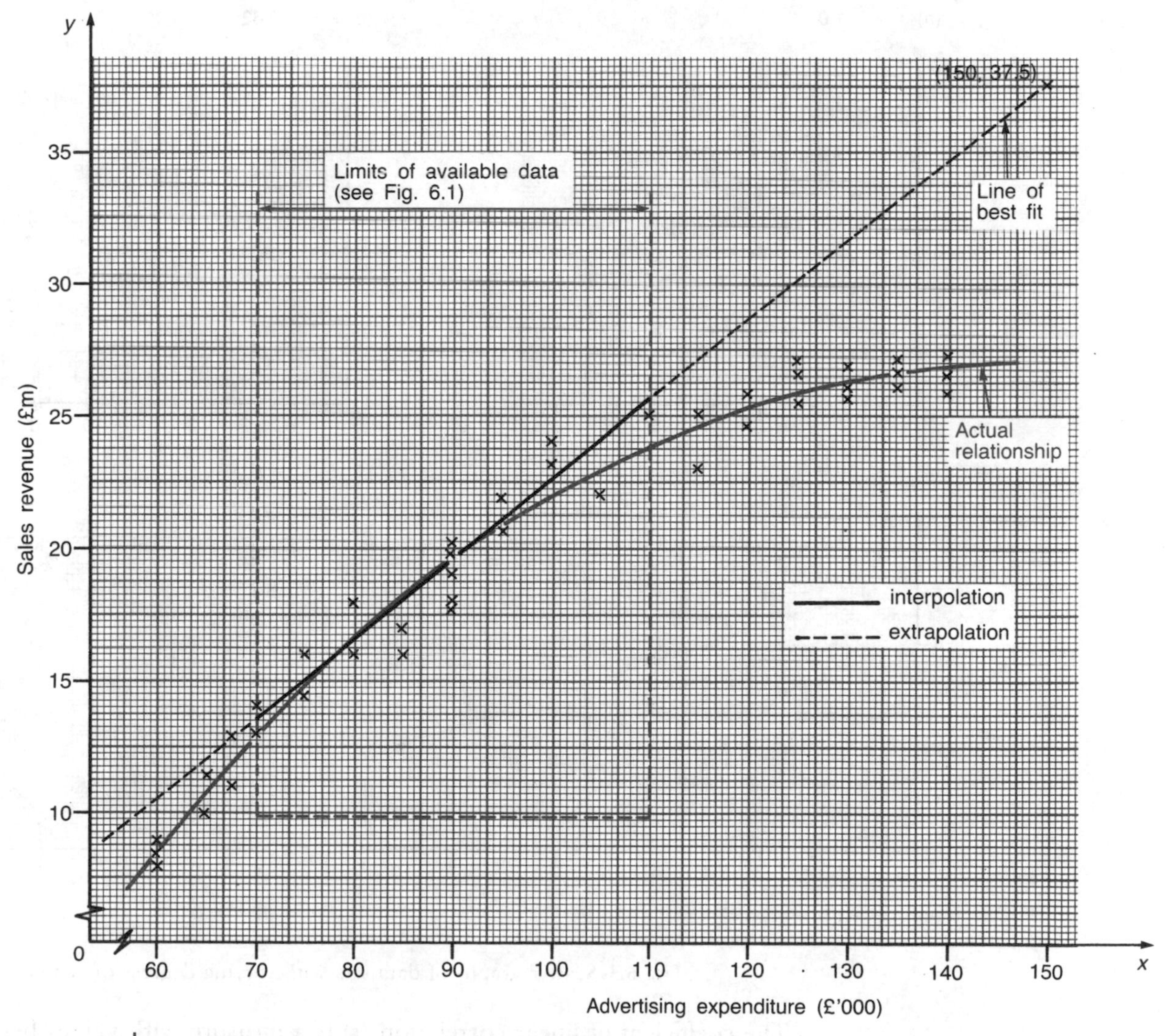

Fig. 6.2 Stevenson's Crafts: sales revenue *versus* advertising expenditure – extended data

In examinations you are likely to encounter data which has a linear, or near linear, relationship and it is probable that you would be expected to use the data for forecasting.

6.6 Correlation

Strictly speaking, it would not be sensible to consider using data to make a forecast unless the relationship between the variables is a strong one. One way to measure the strength of relationship between two variables (or degree of association between two variables) is to calculate a **coefficient of correlation.**

Correlation is an indication of the degree to which one variable is linearly related to, or varies with, another variable. Figure 6.3 shows examples of varying degrees of correlation.

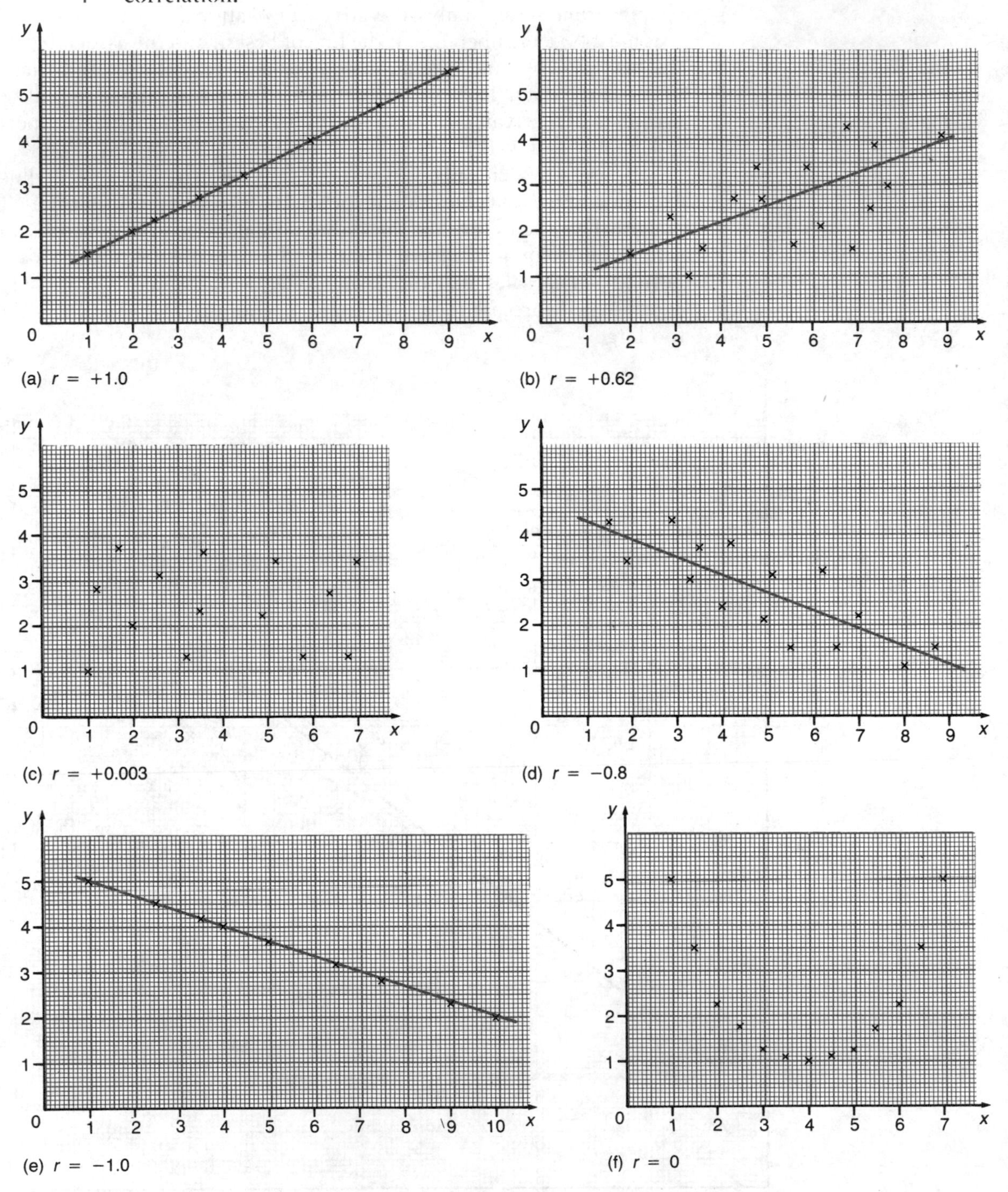

Fig. 6.3 Scatter graphs of data sets with varying degrees of correlation

The coefficient of linear correlation (r) is a measure with values between −1 and +1, which can be interpreted as follows:

1 $r = +1.0$ Perfect correlation: Y is linearly and positively related to X.

2 $r = +0.5$ Partial correlation: moderate positive relationship between X and Y. This level of correlation would not allow confident forecasting.

3 $r = 0$ No correlation: no evidence of a linear relationship. *Note:* It does not mean that there is no evidence of *any* relationship (see fig. 6.3, diagram f).

4 $r = -0.5$ As 2 but Y is decreasing as X increases – i.e. there is a moderate negative correlation.

5 $r = -1.0$ As 1 but the linear relationship is a negative linear relationship (see fig. 6.1, diagram e).

In an examination you could be asked to interpret or calculate a coefficient or correlation. In most cases it is impossible to judge with any accuracy the correlation value of a paired set of data by merely observing a scatter diagram. At this stage you should be aware that the range of possible values for a measure of correlation is from −1 to +1, that is

$$-1 \leqslant r \leqslant +1$$

and that the nearer to +1 (or −1) the coefficient is, the more confident one can be about using the related line of best fit for forecasting purposes within the range of variation of the data. The coefficient of correlation has no unit of measurement. A consequence of this is that it is not true that $r = 0.8$ represents twice as much correlation as $r = 0.4$. It is true, though, that $r = -0.8$ represents as close a relationship between two sets of data as $r = +0.8$.

There are two important coefficients you need to be aware of:

1 Pearson's product moment coefficient of correlation, r.

2 Spearman's coefficient of rank correlation, R.

(The values of r indicated on fig. 6.3 were calculated using Pearson's coefficient.)

6.6.1 Pearson's product moment correlation coefficient

Pearson's product moment correlation coefficient, commonly referred to just as the **correlation coefficient**, is based on the amount of variability in a set of paired data (often referred to as **bivariate data**). You will recall from chapter 4 that variability can be measured by considering the differences between each data value and the mean – these differences were used to calculate the variance and the standard deviation.

The formula for calculating the correlation coefficient involves terms which look very like the standard deviation. This is because the two measures are related, both being measures of variability. Unlike the standard deviation though, the coefficient does not have units. There are a number of texts you could consult for detailed explanation of the derivation of the correlation coefficient (see the 'Recommended reading' list at the end of this chapter). For the purposes of this text it is only necessary to explain what it is and how to use it. You will encounter many different versions of the formula but the following is one which is relatively easy to use:

$$r = \frac{\Sigma XY - n\bar{X}\bar{Y}}{\sqrt{(\Sigma X^2 - n\bar{X}^2)(\Sigma Y^2 - n\bar{Y}^2)}}$$

Consider table 6.2 and fig. 6.4 (overleaf). They show data relating to transit times of Oldham Transport Company. There is a relationship between the distance travelled by the company vehicles and the time taken to travel those distances. A line of best fit has been added, which reflects the positive linear relationship between the two sets of data.

Table 6.2 Distances travelled and related transit times of vehicles belonging to Oldham Transport Company

Distance (km) (X)	Transit time (hours) (Y)
200	3.2
120	2.0
175	3.0
150	2.0
300	4.7
320	5.5
240	3.8
180	2.8
210	3.4
260	4.5

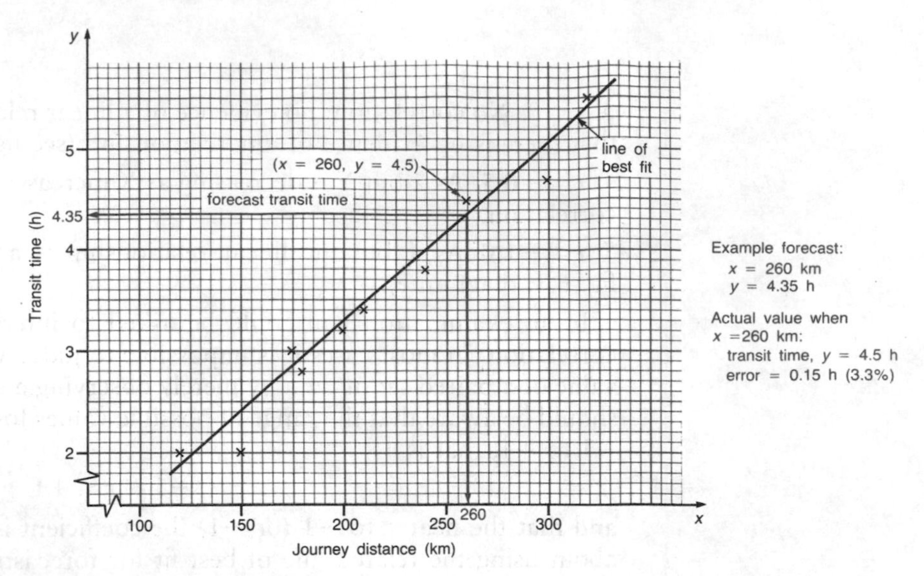

Fig. 6.4 Oldham Transport Company: transit time *versus* journey distance

If the variability in transit time could be completely explained by the distance travelled, then the relationship would be perfectly correlated ($r = 1$), giving a linear relationship. That is, all the observations would be on an upward-sloping straight line. In this example the observations are close to the line of best fit, but not on it. We can conclude that most, but not all, of the variability in transit time is explained by the distance travelled. The correlation coefficient allows one to quantify this variability.

Calculation of the correlation coefficient is complicated and it is very important that you become familiar with the steps involved. A suitable method for organizing the necessary preliminary calculations is shown in table 6.3. They relate to the example illustrated in table 6.2 and fig. 6.4.

Table 6.3 Calculation of Pearson's product-moment correlation coefficient, for the distances travelled (X) and the corresponding transit times (Y) for Oldham Transport Company

$$r = \frac{\Sigma XY - n\bar{X}\bar{Y}}{\sqrt{(\Sigma X^2 - n\bar{X}^2)(\Sigma Y^2 - n\bar{Y}^2)}}$$

X (km)	Y (hours)	X^2	Y^2	XY
200	3.2	40 000	10.24	640
120	2.0	14 400	4.00	240
175	3.0	30 625	9.00	525
150	2.0	22 500	4.00	300
300	4.7	90 000	22.09	1410
320	5.5	102 400	30.25	1760
240	3.8	57 600	14.44	912
180	2.8	32 400	7.84	504
210	3.4	44 100	11.56	714
260	4.5	67 600	20.25	1170
$\Sigma X = 2155$ $\bar{X} = 215.5$	$\Sigma Y = 34.9$ $\bar{Y} = 3.49$	$\Sigma X^2 = 501\,625$	$\Sigma Y^2 = 133.67$	$\Sigma XY = 8175$

$$r = \frac{8175 - (10 \times 215.5 \times 3.49)}{\sqrt{[501\,625 - 10 \times (215.5)^2][133.67 - 10 \times (3.49)^2]}}$$

$$= \frac{8175 - 7520.95}{\sqrt{(501\,625 - 464\,402.5)(133.67 - 121.801)}}$$

$$= \frac{654.05}{\sqrt{37\,225.5 \times 11.869}} = \frac{654.05}{\sqrt{441\,793.85}} = \frac{654.05}{664.68}$$

$$r = 0.98$$

The value of the correlation coefficient, 0.98, is very close to 1.0. This indicates a high level of positive, or direct, correlation.

Table 6.4 shows the values of the data illustrated in fig. 6.3. As an exercise, calculate the correlation coefficients to confirm the values indicated in the figure. Your answers may differ from those given: minor differences will be due to rounding error. It is advisable to retain as much detail (number of decimal places) as possible for as long as possible, only rounding up/down when it is unavoidable (because of calculator capacity) or at the end of the calculation.

Table 6.4 Data values for the scatter diagrams of fig. 6.3

(a)		(b)		(c)		(d)		(e)		(f)	
x	y	x	y	x	y	x	y	x	y	x	y
1.0	1.5	2.0	1.5	1.0	1.0	1.5	4.25	1.0	5.0	1.0	5.0
2.0	2.0	2.9	2.3	1.2	2.8	1.9	3.4	2.5	4.5	1.5	3.5
2.5	2.25	3.3	1.0	1.7	3.7	2.9	4.3	3.5	4.15	2.0	2.25
3.5	2.75	3.6	1.6	2.0	2.0	3.3	3.0	4.0	4.0	2.5	1.75
4.5	3.25	4.3	2.7	2.6	3.1	3.5	3.7	5.0	3.65	3.0	1.25
6.0	4.0	4.8	3.4	3.2	1.3	4.0	2.4	6.5	3.15	3.5	1.1
7.5	4.75	4.9	2.7	3.5	2.3	4.2	3.8	7.5	2.8	4.0	1.0
9.0	5.5	5.6	1.7	3.6	3.6	4.9	2.1	9.0	2.3	4.5	1.1
		5.9	3.4	4.9	2.2	5.1	3.1	10.0	2.0	5.0	1.25
		6.2	2.1	5.2	3.4	5.5	1.5			5.5	1.75
		6.8	4.3	5.8	1.3	6.2	3.2			6.0	2.25
		6.9	1.6	6.4	2.7	6.5	1.5			6.5	3.5
		7.3	2.5	6.8	1.3	7.0	2.2			7.0	5.0
		7.4	3.9	7.0	3.4	8.0	1.1				
		7.7	3.0			8.7	1.5				
		8.9	4.1								

6.6.2 Interpretation of the correlation coefficient

In general, values of r near to -1.0 or $+1.0$ indicate a strong linear relationship between the variables (X and Y) and values near to zero indicate a lack of evidence of a linear relationship. Remember that $r = 0$ does not necessarily imply that there is no relationship. The difficulties arise in interpreting intermediate values of r, such as 0.8 or -0.35.

6.6.3 The coefficient of determination

The coefficient of determination provides a more meaningful measure. It is usually expressed as a percentage and measures the extent to which the change in one variable can be explained by a linear relationship with the other variable. It is acquired by squaring the correlation coefficient.

In the example considered above (section 6.6.1), we have the correlation coefficient $r = 0.98$. Therefore the coefficient of determination, $r^2 = 0.96 = 96\%$ – i.e. 96 per cent of the variation in the transit time is explained by the distance travelled. The other 4 per cent of the variation is due to other factors – perhaps traffic or weather conditions.

A correlation coefficient of 0.40 would result in a coefficient of determination of 0.16 or 16 per cent. That is, 16 per cent of the variation in Y could be explained by the variation in X. The other 84 per cent of the variation in Y is unexplained.

6.6.4 Correlation: caution needed!

There are, however, other considerations to be taken into account.

1 A correlation coefficient is a measure of association only. It does not indicate cause and effect. For instance, prices, wages, sales values and many other sets of data are influenced by inflation. It is possible to get a high correlation coefficient between pairs of these data sets, such as wages and prices. This might appear to indicate a cause and effect link between the two sets of data (that one is influencing the other), but it might simply mean that wages and prices are *both* influenced by inflation.

   We could be forgiven for thinking that a relationship exists between wages and prices. There are cases though, where the apparent correlation is obviously spurious. The correlation coefficient between the reading ability of children and their shoe sizes is high. It hardly means that improving a child's reading will cause their feet to grow. Both of these variables are related to a third variable, age.

2 A high correlation coefficient does not necessarily indicate that the data has a significant correlation. The reason is that the number of data pairs used in the calculation affects the resulting correlation coefficient. The 10 sets of data for the Oldham Transport Company shown in table 6.2 and fig. 6.4 are a small sample. The correlation coefficient ($r = 0.98$) suggests a high degree of association between X and Y but it is not as difficult to fit a straight line to 10 points as to, for example, 50 points. Caution must be exercised in interpreting the correlation coefficient of a small set of data.

There are sophisticated methods one can employ to test the significance of a correlation coefficient but, for the purposes of examination questions you are likely to encounter, it is probably sufficient to note the following:

For small sample size ($n \leq 9$), values of r greater than 0.9 can be taken as significant.

In general, for $n > 9$, the correlation coefficient, r, will be significant if the absolute value of r ($|r|$) is greater than $3/\sqrt{(n-1)}$.

3 Pearson's correlation coefficient is only applicable if the data under consideration is quantitative. That is, if numerical values are available. If we are dealing with qualitative data, such as taste preference or order of merit, then we need another measure of correlation.

Suppose a company wants objective tests for recruitment purposes. The company may set two types of test, an aptitude test and a training test. Suppose, further, that it is not suitable to give overall marks and the candidates are ranked in order of merit. The company may wish to know if the results of the two tests are correlated. To do this an alternative to Pearson's correlation coefficient must be used.

6.6.5 Spearman's rank correlation coefficient

Spearman's rank correlation coefficient is calculated by referring to two sets of ranks (rather than two sets of numerical data values). The formula for Spearman's coefficient is

$$R = 1 - \frac{6\Sigma d^2}{n(n^2-1)}$$

where d is the difference between the corresponding ranks.

Table 6.5 The ranks of 15 candidates' aptitude and training test results

Candidate	Aptitude test rank	Training test rank	Difference in rank, d	d^2
E. Holden	4	3	1	1
J. Oldham	1	4	−3	9
H. Stevenson	12	15	−3	9
I. Strange	13	10	3	9
S. Patel	3	2	1	1
E. Werner	2	1	1	1
T. Singh	15	13	2	4
K. Childs	8	9	−1	1
J. Sedgewick	10	12	−2	4
S. Archbold	6	5	1	1
A. Clarke	5	6	−1	1
M. Topple	14	14	0	0
C. Holland	9	11	−2	2
L. Breeze	7	8	−1	1
J. Brown	11	7	4	16
Total				$\Sigma d^2 = 59$

For the example above (table 6.5), Spearman's rank correlation coefficient is calculated as follows:

$$\begin{aligned} R &= 1 - [6 \times 59/15(225-1)] \\ &= 1 - (354/3360) = 1 - 0.105 \\ &= 0.90 \text{ (to 2 decimal places)} \end{aligned}$$

Spearman's rank correlation coefficient can be used as a preliminary to calculating Pearson's product moment correlation coefficient. It is easy to use and gives an early indication of the level of correlation that might exist. The calculation for the Oldham Transport Company data is given as an example (table 6.6). The value of R is 0.98, the same as the value of the Pearson correlation coefficient (table 6.3). It should be noted that this is a coincidence; the values of R and r are usually different.

The data are ranked from the largest (rank 1) to the least (rank n). Note the way of dealing with values which have tied ranks: Y has two values equal to 2.0, which would both be ranked 9, leaving rank 10 unallocated. The solution is to give the two values the mean of the two ranks, which in this case is 9.5 [(9 + 10)/2]. If there had been three equal values, which if not tied would have been allocated ranks 8, 9 and 10, then each value would have been given rank 9 [(8 + 9 + 10)/3].

Table 6.6 Oldham Transport Company: rank order of the distances travelled and the related transit times

Distance (km) X	Transit time (h) Y	X and Y ranked		d	d^2
200	3.2	6	6	0.0	0.0
120	2.0	10	9.5	0.5	0.25
175	3.0	8	7	1.0	1.0
150	2.0	9	9.5	−0.5	0.25
300	4.7	2	2	0.0	0.0
320	5.5	1	1	0.0	0.0
240	3.8	4	4	0.0	0.0
180	2.8	7	8	−1.0	1.0
210	3.4	5	5	0.0	0.0
260	4.5	3	3	0.0	0.0
					$\Sigma d^2 = 2.5$

$R = 1 - [(6 \times 2.5)/(10 \times 99)]$
$= 1 - 0.015$
$= 0.98$ (to 2 decimal places)

6.7 The line of best fit (II) – regression

A major consequence of establishing a relationship between two sets of data is the capacity for that relationship to be used for forecasting. If the (linear) relationship proves to be a 'good fit' then any forecasts made should be relatively reliable. Figure 6.4 illustrated such an example. It is important to be able to find the equation of the line of best fit. The method used in section 6.4.2 relied on fitting the line by eye – the best line was the one that 'looked right'. This is neither easily done nor very accurate. Consequently, this method cannot be relied on as the basis for important decision-making.

We need a criterion for judging which line is the 'best' and a method for finding the equation of that line. It would seem reasonable to choose the line which 'minimizes the errors'.

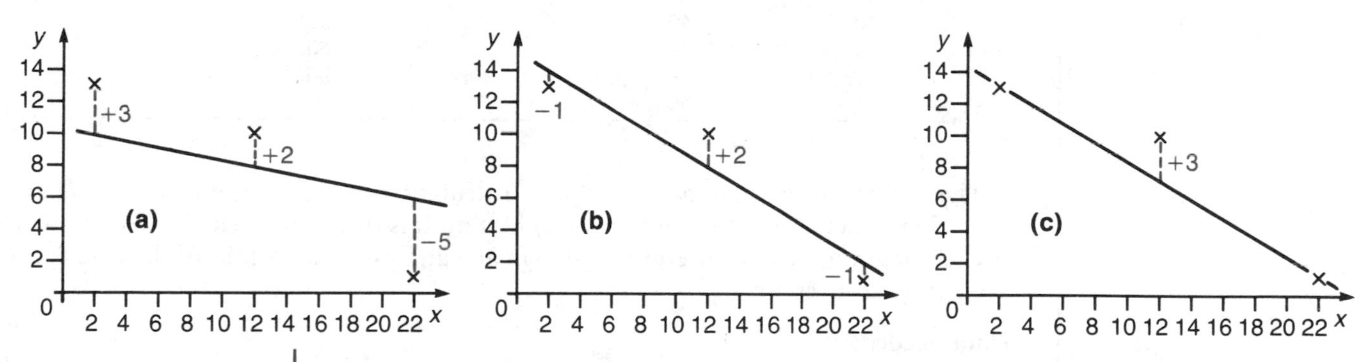

Fig. 6.5

An example of fitting a 'best line' by eye, with the intention of minimizing the **net error**, that is the sum of all the errors, is shown in fig. 6.5, diagrams (**a**) and (**b**). Both lines give a total net error of zero, yet the line in (**b**) is clearly the better fit. The result of minimizing the **absolute error** (i.e. ignore the signs of the error values) would make the line in diagram (**c**) the choice. But, intuitively, (**b**) is still the better choice.

6.7.1 The least squares method

The criterion developed by statisticians defines the line of best fit as 'that line which minimizes the sum of the **squares of the errors** measured with respect of the dependent variable'. This is called the **least squares regression line.** You will see that according to this criterion, the line in (**b**) (fig. 6.5) is better than those in (**a**) or (**c**). It is in fact the best line of fit.

The following formulae may appear complicated, but they use terms you should already be familiar with from correlation analysis. As we are interested in finding the equation of a straight line, the equation will be of the form

$$Y = a + bX$$

so the aim is to find the values of the constants, a and b. The method of least squares enables us to find a and b using the following formulae:

$$b = \frac{\Sigma XY - n\bar{X}\bar{Y}}{\Sigma(X^2) - n\bar{X}^2}$$

$$a = \bar{Y} - b\bar{X}$$

The data shown in table 6.7 will be used as a basis for finding the correlation coefficient and the equation of the line of best fit, 'the regression line'.

Table 6.7 Annual advertising expenditure and annual turnover of 25 small firms

Annual advertising expenditure (£'000) X	Annual Turnover (£'000) Y
17	220
18	240
20	250
20	240
21	260
24	300
25	310
27	330
28	350
28	360
29	350
32	390
35	430
35	410
36	440
37	450
40	500
42	510
43	525
48	580
48	575
49	600
49	605
49	605
50	610

The following is a summary of the calculations used to find the correlation coefficient and the equation of the line of best fit. It is recommended that you draw the scatter diagram, as it is a useful first step in any analysis. The details of the calculation are left as an exercise.

Data needed:

$n = 25$

$\Sigma X = 850$ $\qquad \Sigma Y = 10\,440$

$\bar{X} = 34$ $\qquad \bar{Y} = 417.6$

$\Sigma X^2 = 31\,856$ $\qquad \Sigma Y^2 = 4\,781\,400$

$\Sigma XY = 390\,205$

Correlation coefficient:

$$r = \frac{\Sigma XY - n\bar{X}\bar{Y}}{\sqrt{(\Sigma X^2 - n\bar{X}^2)(\Sigma Y^2 - n\bar{Y}^2)}}$$

$$= \frac{390\,205 - 25 \times 34 \times 417.6}{\sqrt{[31\,856 - 25 \times (34)^2][4\,781\,400 - 25 \times (417.6)^2]}}$$

$= 0.99$ (to 2 d.p.)

Line of best fit:

$Y = a + bX$

$$b = \frac{n\Sigma XY - \Sigma X\Sigma Y}{n\Sigma X^2 - (\Sigma X)^2} = \frac{\Sigma XY - n\bar{X}\bar{Y}}{\Sigma X^2 - n\bar{X}^2}$$

$$= \frac{390\,205 - 25 \times 34 \times 417.6}{31\,856 - 25 \times (34)^2}$$

$$= 11.9$$

$$a = \bar{Y} - b\bar{X}$$

$$= 417.6 - (11.9 \times 34)$$

$$= 13.0$$

Therefore the line of best fit is given by the equation **$Y = 13.0 + 11.9X$.**

6.8 Correlation and regression: summary

1 Correlation and regression are techniques used to investigate the relationship between two variables. **Regression analysis,** finding the equation of the line of best fit, is concerned with establishing the equation of a linear relationship between two sets of data. **Correlation analysis** is concerned with finding the degree or strength of that relationship.

2 Drawing a **scatter diagram** is recommended as a preliminary to any further analysis. It is usually possible to determine from this whether a linear relationship exists between the two variables. The strength of the relationship, the correlation, can also be estimated from a scatter diagram.

3 If a linear relationship is apparent, the equation of the **line of best fit** can be determined. A rough idea can be obtained **'by eye'**, drawing the line that appears to give the best fit. The method of **least squares** gives more accurate results.

4 The regression equation (equation of the line of best fit) can be used for **forecasting,** i.e. to predict the dependent variable for given values of the independent variable. Care must be exercised if the forecast is an extrapolation.

5 The **correlation coefficient,** *r*, is calculated to determine the significance of a linear relationship. The nearer *r* is to +1 (or to −1), the stronger the relationship. A value of 0 (zero) means that there is no evidence of a linear relationship, but it does not mean that there is *no* relationship between the variables – it could be non-linear.

6 The formula for **Pearson's correlation coefficient,** *r*, is

$$r = \frac{\Sigma XY - n\bar{X}\bar{Y}}{\sqrt{(\Sigma X^2 - n\bar{X}^2)(\Sigma Y^2 - n\bar{Y}^2)}}$$

7 By squaring the above coefficient we obtain the **coefficient of determination.** For instance, if $r = 0.8$, then the coefficient of correlation is $r^2 = 0.64 = 64$ per cent. This means that 64 per cent of the variation in variable Y can be explained by the association with variable X.

8 If the data is not quantitative (not numerical or in different units e.g. X in km and Y in hours) then an alternative coefficient, **Spearman's rank correlation coefficient,** is used. The formula for this is

$$R = 1 - \frac{6\Sigma d^2}{n(n-1)}$$

Like Pearson's correlation coefficient, *R* takes values between −1 and +1.

9 Regression and correlation analysis do not indicate cause and effect. Even if $r = 1.0$, the correlation could still be spurious – both variables could be being influenced by a third.

Sample questions

1 **(a)** Explain the assumptions behind, and the differences between, the techniques of regression and correlation.

(b) The following information has been extracted from the 1985/86 annual report of the Metroshire Metropolitan County Council:

	Number of passenger journeys by public transport (road only) ('000)	Number of serious or fatal road accidents
1976/7	866	112
1977/8	845	105
1978/9	840	103
1979/80	832	106
1980/1	830	100
1981/2	801	99
1982/3	803	103
1983/4	799	99
1984/5	798	97

Compute the correlation coefficient between the number of passenger journeys by public transport and the number of serious or fatal accidents and comment on your findings.

CIPFA; Diploma in Public Sector Audit and Accounting; Quantitative analysis; March 1986.

Solution/hints

Part (a)

See main text, particularly section 6.8.

Part (b)

The detailed calculations are left as an exercise. You should get the result $r = 0.85$ (to 2 decimal places).

Comment: The correlation suggests a fairly strong relationship between the numbers of journeys on public transport and fatal accidents. However, the sample is a very small one so care should be taken in drawing any conclusion. Also, are we to assume the fatal accidents all involved public transport? One would require further information before being committed to any conclusions. Having said that, the available information does suggest some connection between the number of passenger journeys on public transport and the number of fatal accidents, an example of indirect correlation via a link to total journeys of all type, private and public.

2 (**a**) Give an Accounting example of a pair of variables you would expect to be positively correlated and a pair of variables you would expect to be negatively correlated.

(**b**) A large manufacturing company is investigating the cost of sickness amongst production workers who have been employed by the company for more than one year. The following regression equation, based on a random sample of 50 such production workers, was derived for 1985:

$$Y = 15.6 - 1.2X,$$

where Y represents the number of days absent because of sickness and X represents the number of years employment with the company. The coefficient of determination, r^2, was 0.9.

You are required to:

(**i**) state the value of the correlation coefficient;

(**ii**) state the meaning of the coefficients in the equation;

(**iii**) predict the number of days absence through sickness to be expected of an employee who has been with the company for eight years;

(**iv**) draw a scatter diagram, including the regression line, to illustrate approximately the situation in this manufacturing company;

(**v**) list any limitations or problems of using this equation in practice.

CIMA; Foundation Stage, Section B; Mathematics and statistics; May 1986.

Solutions/hints

Part (a)

Left as an exercise.

Part (b)

(i) $r = -0.95$ $(\sqrt{0.9})$

Note r is negative: the equation tells us that Y decreases as X increases, that is, X and Y are negatively or inversely correlated.

(ii) The coefficient -1.2 is the rate of change of the variable Y with respect to unit change in X. That is, every time X increases by 1 (year), Y *decreases* by 1.2 (days). It appears that longer-serving workers are likely to have less time off work because of sickness than do more recently employed workers.

The constant 15.6 is the y-intercept of the equation (i.e. where the line meets the y-axis). If we were to interpret this literally we would have to conclude that those workers not employed (employed for 0 years) would be forecast as having 15.6 days off work because of sickness. This is obviously nonsense. The important point to bear in mind is that, in general, the equation may only be valid for a limit range of values (see section 6.5). In this case, 15.6 is a function of the equation-estimating process and suggests its range of valid use is limited.

(iii) Substituting $x = 8$ into the regression equation, $y = 15.6 - (1.2 \times 8) = 6$.

(iv) Left as an exercise.

(v) See (ii) above.

3 In a regression analysis the regression of Y on X has been calculated to be $Y = 100 - 5X$ and the regression of X and Y has been calculated to be $X = 60 - 0.162Y$. The correlation coefficient, r, between X and Y is

A -0.90 **B** -0.60 **C** 0.81 **D** 0.90

E impossible to calculate without further information.

CIMA; Foundation Stage, section B; Mathematics and statistics; November 1986.

Solution/hints

It has been assumed throughout this text that the independent variable is X and the dependent variable is Y, giving an equation for the line of best fit in the form $Y = a + bX$. It is, of course, possible to find the line of best fit such that Y is the independent variable and X is the dependent variable: forecasts of X could be made from given values of Y. The resulting equation in that case would be in the form: $X = c + dY$. The appropriate least squares formulae are:

$$b = \frac{\Sigma XY - n\bar{X}\bar{Y}}{\Sigma X^2 - n\bar{X}^2} \qquad d = \frac{\Sigma XY - n\bar{X}\bar{Y}}{\Sigma Y^2 - n\bar{Y}^2}$$

$$a = \bar{Y} - b\bar{X} \qquad c = \bar{X} - d\bar{Y}$$

$b \times d$ gives the coefficient of determination, r^2, and the correlation coefficient, r, is therefore $\sqrt{d \times b}$.

$b = -5$ and $d = -0.162$

Therefore $r^2 = -5 \times -0.162 = 0.81$

and $r = \sqrt{0.81} = \pm 0.90$

$r = -0.90$(**A**) since both regression equations show the correlation is negative.

4 The management accountant of Atlas Stores plc is analysing the profitability of one of the company's larger shops. From the figures for the latest quarter, he extracts the following data which give the number of customers and the shop turnover for each week during the quarter.

Week	No. of customers ('000)	Turnover (£'000)
1	5.2	10.7
2	6.1	11.9
3	4.8	9.6
4	5.6	11.1
5	5.0	10.4
6	6.2	12.1
7	4.7	9.6
8	5.8	11.4
9	4.9	9.7
10	4.6	9.5
11	5.8	11.6
12	5.7	11.2
13	5.8	11.8

Using a linear regression analysis to examine the relationship between the number of customers and the total turnover gives rise to the following equation:

$$\text{Turnover} = 1.60 + 1.71 \times \text{Customers}$$

In this situation, however, it appears that the intercept (1.60) should really be zero since, with no customers, there would be no turnover. Instead of the above regression equation, therefore, he decides to use a zero-intercept regression of the form Y = bX, where, with the least squares estimate of the gradient b is given by

$$b = \Sigma XY / \Sigma X^2$$

Required:

(a) Fit a 'zero-intercept' line of the form Y = bX to the data and explain the meaning of the regression coefficient b.

(b) Plot a scatter diagram of the data and draw on your diagram the two regression lines. Comment on the apparent fit of these two lines.

(c) Give two possible reasons why a non-zero intercept (such as the value 1.60) may arise when fitting a general line of the form Y = $a + b$X to a sample of data, even though the true relationship must pass through the origin. What relationship would you advise the management accountant to use in this situation and what precautions, if any, would you suggest.

ACCA; Level 2 – Professional examination; Quantitative analysis, June 1987.

Solutions/hints

This question introduces the idea of 'zero-intercept'. If the facts of the question suggest that the line should intercept at ($x = 0$, $y = 0$) then the equation becomes greatly simplified: The equation of the line is of the form Y = bX, and the coefficient, b, is found using $b = \Sigma XY/\Sigma X^2$, as stated in the question.

Part (a)

$\Sigma XY = (5.2 \times 10.7) + \ldots + (5.8 \times 11.8) = 765.52$

$\Sigma X^2 = (5.5)^2 + \ldots + (5.8)^2 = 382.76$

So, $b = 765.52/382.76 = 2$ and Y = 2X

The coefficient b (2) gives an estimate of the average revenue per customer, in this case £2 per customer.

Part (b)

Drawing the scatter diagram is left as an exercise. It should be noticed from the diagram that the original line (Y = 1.6 + 1.71X) is, not surprisingly, a better fit than Y = 2X. However the difference is not that great and only becomes pronounced as the limits of the data are reached (see section 6.4).

Part (c)

(i) When the least squares method is used to find a line of best fit, the values a and b obtained are only sample estimates and can be expected to contain sampling error.

(ii) The data observed only covers a limited range of values. The equation will not necessarily be valid outside of that range.

Questions 5–7 are further examples of the type of question you are likely to encounter in examinations and are left as exercises.

5 You are working with a colleague in the marketing department trying to establish the relationship, if any, which exists between the sales of your company's main product and the amount of advertising budget (adspend) spent in each quarter. Data is available from 10 television regions from the last quarter. The company assumes that adspend influences sales.

Region	A	B	C	D	E	F	G	H	I	J
Sales ('000 cases)	90	25	45	10	60	20	75	110	130	15
Adspend (£'000)	9	4	6	1	7	3	8	10	11	2

You are required to:

(a) draw a scatter diagram of the data;

(b) estimate by any method you consider appropriate the sales to be expected in response to an adspend of £5000 in a comparable quarter (briefly justify your choice of method);

(c) state an approximate equation which links adspend and sales, bearing in mind that the overall amount of adspend is budgeted to be 25 per cent higher in the next quarter, and comment briefly on your answer.

CIMA; Foundation Stage, section B; Mathematics and statistics; November 1986.

6 A company estimated that the sales of one of its products depended on the price charged. The sales manager considered that the relationship between price and quantity sold was likely to be as shown below:

Price (£)	Quantity sold (hundreds)
17	12
16	13
15	16
14	20
13	28
12	38
11	60

(a) Draw a scatter diagram to represent this data.

(b) Calculate the product moment correlation coefficient between price and sales.

(c) Rank the data and then find Spearman's rank correlation coefficient between price and sales.

(d) Explain briefly why the results of parts (b) and (c) differ.

CIS; Part 1; Quantitative studies; June 1987.

7 The following data is derived from 12 randomly selected workers in a job involving the wiring of miniature electronic components. The data shows, for each worker selected, the number of weeks of experience doing the job and the number of components rejected.

Week 9, 1987

Sampled worker	1	2	3	4	5	6	7	8	9	10	11	12
Weeks of Experience	8	10	7	15	9	13	10	5	3	12	2	9
Number of Rejects	25	20	28	16	24	19	25	27	40	23	33	27

Using the above data:

(a) plot a scatter diagram;

(b) calculate the least squares regression line of number of rejects on weeks of experience;

(c) from the regression line, estimate the number of components rejected by an employee with: (i) 3 weeks' experience; (ii) 15 weeks' experience.

CIPFA; Diploma in Public Sector Audit and Accounting; Quantitative analysis; May 1987.

Recommended reading

Lipsey, R.G., *An Introduction to Positive Economics* (Weidenfeld and Nicholson, 1975), part 8.

Owen, F. and Jones, R., *Statistics* (Pitman, 1988), chapter 22.

Whitehead, P. and Whitehead, G., *Statistics for Business* (Pitman, 1984), chapter 13.

Curwin, J. and Slater, R., *Quantitative Methods for Business Decisions* (Van Nostrand Reinhold, 1986), chapter 16 and 17.

Lucey, T., *Quantitative Techniques* (D.P. Publications, 1980), chapter 4.

7 Probability theory and practical applications

7.1 Introduction

One often hears people make statements such as, 'the likelihood of rain this afternoon is 50 per cent' or 'the chance of the Labour Party winning the next general election is about 1 in 4' or 'there's no chance of catching a bus'. Likelihood, chance, probability, etc, as used in this casual sense, are all words which imply a level of uncertainty. Probability, in the mathematical sense, is also to do with uncertainty but it is used in a much more formal way.

Probability is a measure of likelihood. This measure is often stated as a percentage or as a ratio but it is more usual, in this context, for it to take numerical values between zero and one. A zero probability corresponds to impossibility (or no chance of occurrence) and a probability of one corresponds to certainty (the event will definitely occur). 50 per cent likelihood is equivalent to a probability of 0.5. A chance of '1 in 4' is equivalent to a probability of 0.25. 'No chance' corresponds to a probability of zero; thus 'no chance of catching a bus' would be expressed as

$$P\text{ (catching a bus)} = 0$$

P is read as 'the probability of'.

The ways people think about probability can be grouped into three broad areas, which overlap to a certain extent.

1 Theoretical probability

If an ordinary die, which is believed to be fair, is thrown, the probability of throwing a five is $\frac{1}{6}$. This is because a die is a cube with its six sides numbered one to six all of which, because of symmetry, have an equal chance of turning up. This type of probability is based on a theoretical analysis of the particular circumstances. It can be applied to a number of situations where there are a finite number of possible outcomes which can be defined in advance, e.g. choosing a card from a pack of playing cards, tossing a coin, etc.

2 Statistical probability

Some mathematicians would claim that the only way to attach a probability value to an event is to monitor past outcomes. In considering the probability value to be attached to obtaining a five when throwing a die, we would have to provide experimental evidence showing that the proportion of throws of the die when the five occurred was $\frac{1}{6}$, before it would be acceptable to state

$$P\text{ (the die shows 5)} = \frac{1}{6}$$

There are situations where monitoring past events appears to be the only method to employ in order to obtain a probability. For instance, a shop owner wants to know the probability of more than 80 customers using his shop on a particular Saturday morning. He has kept records of the past 100 Saturday mornings, which show that on 23 of these there were more than 80 customers. From this he is able to state

$$P\text{ (more than 80 customers on Saturday a.m.)} = 0.23$$

This probability could be updated as more Saturdays are monitored.

3 Subjective probability

A subjective probability is a measure of an individual's degree of belief or strength of conviction. For instance, a subjective estimate of the chances of the Labour Party winning the next general election may conclude that

$$P\text{ (Labour victory)} = 0.8$$

For a statement of subject probability to be a reliable estimate, the person making the estimate should be an expert in some sense. Lack of reliability is the main weakness of subjective probability. In practice, it is often necessary for business decisions to be based on subjective (though expert) estimates of probability.

Howsoever an estimate for a probability is reached, it must be remembered that no event can be guaranteed to happen (or not happen). For instance, if a salesman knows from past experience that he succeeds in making a sale in four out of five attempts, he should not expect that the next five sales attempts will necessarily contain exactly four successes and one failure. Similarly, in throwing a die he would not necessarily expect to get two 1s, two 2s, two 3s, two 4s, two 5s and two 6s from 12 throws. The probabilities refer to the expected proportions 'in the long run'.

This chapter will deal with three broad areas of probability:

1 basic probability, including the laws of probability;
2 decision theory, an application of probability; and
3 important probability distributions, including the normal, Poisson and binomial distributions.

The types of questions you are likely to encounter in examinations will also fall into these broad areas, though some questions will contain elements of all three.

7.2 Probability: the basic rules

7.2.1 Simple events

If all the possible outcomes of a situation are equally likely and indivisible then we can define the probability of obtaining a simple event A. It is the ratio of 'the number of outcomes which correspond to obtaining event A' to 'the total number of possible outcomes'. This can be written as follows:

$$P(A) = \frac{\textbf{number of outcomes relating to obtaining the event } A}{\textbf{total number of possible outcomes}}$$

For instance, on throwing a fair die, the probability of obtaining the event 'the number obtained is an even number' is equal to $\frac{3}{6}$ (or 0.5), because there are three possible outcomes which correspond to the event of interest, namely the set {2,4,6}, and a total of six possible outcomes, namely the set {1,2,3,4,5,6}.

Note: The above definition is only valid if the outcomes are equally likely. The outcomes of throwing a die, {1,2,3,4,5,6}, are all equally likely. However, if the outcomes of throwing a die were defined as {'even number', 1,3,6}, then we would not have equally likely outcomes. Once the probabilities of simple events are determined, the probabilities of compound events can be found. These are subject to certain laws of probability which will be explained in the following sections.

7.2.2 Complementary events

It is intuitively obvious that if the Labour party have an 80 per cent chance of winning the next general election, conversely the chance of *not* winning (the **complementary event**) is 20 per cent. In probability terms this is expressed as follows:

If P (the Labour Party wins the next general election) = 0.8, then

P (the Labour Party does *not* win the next general election) = 1 − 0.8 = 0.2.

This illustrates the **first law of probability:**

$$P(\bar{A}) = 1 - P(A)$$
$$\text{or} \quad P(A) + P(\bar{A}) = 1$$

where $\bar{A}$, read as 'not A', represents the complementary event. So, 'the probability of an event happening is equal to one minus the probability of it not happening'; or 'the sum of the probabilities of complementary events is one'.

7.2.3 Simple addition law

Suppose the result of monitoring the number of people in passing cars gave the following results:

P(1 person) = 0.4   (3 persons) = 0.2
P(2 persons) = 0.3 P(4 or more persons) = 0.1

From these results it is possible to calculate the following probabilities:

$$P(\text{1 or 2 persons in a particular car}) = P(\text{1 person}) + P(\text{2 persons}) = 0.4 + 0.3 = 0.7$$

$$P(\text{less than 4 persons in a particular car}) = P(\text{1 person}) + P(\text{2 persons}) + P(\text{3 persons}) = 0.4 + 0.3 + 0.2 = 0.9$$

The individual probabilities are simply added. It is only permissible to do this if the events are **mutually exclusive.**

7.2.3.1 Mutual exclusiveness

Events are mutually exclusive if they can not possibly happen at the same time: it is impossible to have one person *and* two persons in the same passing car. The events 'score a 6' and 'score a 5' can not both occur when a single die is thrown, but the events 'score a 6' and 'an even score' could occur simultaneously. In the latter example therefore the events are not mutually exclusive.

In general, if the events A, B, C, etc, are mutually exclusive

$$P(A \text{ or } B \text{ or } C \text{ or } \ldots) = P(A) + P(B) + P(C) + \ldots$$

7.2.4 Simple multiplication law

The previous section dealt with the simplest case of combined probabilities, the probability of one event *or* another event occurring. Here we look at the probability of one event *and* another happening simultaneously.

Table 7.1 The possible outcomes of throwing a die and tossing a coin simultaneously

Die	1	2	3	4	5	6
Coin						
head	H & 1	H & 2	H & 3	H & 4	H & 5	H & 6
tail	T & 1	T & 2	T & 3	T & 4	T & 5	T & 6

Consider table 7.1. It should be obvious that all of the possible outcomes, 12 in all, are equally likely. It is therefore possible to state the following

$$P(\text{H \& 4}) = \frac{1}{12}$$

$$P(\text{T \& even score}) = \frac{3}{12} = \frac{1}{4}$$

It is not necessary to refer to all the possible outcomes to get the desired result. The above probabilities could be found as follows:

$$P(\text{H \& 4}) = \text{P(Head)} \times P(\text{score a 4}) = \frac{1}{2} \times \frac{1}{6} = \frac{1}{12}$$

$$P(\text{T \& even score}) = P(\text{Tail}) \times P(\text{even score}) = \frac{1}{2} \times \frac{1}{2} = \frac{1}{4}$$

The probabilities are multiplied. In general,

$$P(A \text{ and } B) = P(A) \times P(B)$$

This is only true if the events A and B are **independent.**

7.2.4.1 Independent events

Two events are said to be independent if the occurrence (or non-occurrence) of one event does not affect the occurrence (or non-occurrence) of the other event.

In the above example, the outcome of throwing the die and tossing the coin are independent. An example of two events which are not independent are the events 'score a 5' and 'an even score' when throwing a single die – 'score a 5' is dependent on *not* obtaining 'an even score'.

7.2.4.2 Simple multiplication law: an example

Problem: A machine is subject to failure because of a number of reasons: operator abuse, which accounts for failure on 1.5 per cent of occasions the machine is used; mechanical fault, which occurs on 2 per cent of occasions; and electrical fault, which occurs on 1 per cent of occasions. If it is assumed that these faults are independent, but not mutually exclusive, what is the probability of the machine failing?

Solution: Because the machine can fail for a variety of reasons and these reasons are not mutually exclusive (i.e. the machine could fail for more than one reason simultaneously), it is easier to calculate the probability it will *not* fail. There are a number of stages to the solution. Using the law of complementary events we can calculate:

$$P(\text{no operator abuse}) = 1 - P(\text{operator abuse}) = 1 - 0.015 = 0.985$$

Similarly: $P(\text{no mechanical fault}) = 1 - 0.02 = 0.98$

$$P(\text{no electrical fault}) = 1 - 0.01 = 0.99$$

The question states that the faults are not mutually exclusive, so the simple addition law is not applicable.

Independence is assumed though, so we can use the simple multiplication law:

$$\begin{aligned} P(\text{machine will not fail}) &= P(\text{no operator abuse } \textit{and} \text{ no mechanical failure } \textit{and} \text{ no electrical failure}) \\ &= 0.985 \times 0.98 \times 0.99 \\ &= 0.955647 \end{aligned}$$

By reference once again to the law of complementary events, we can now calculate the probability of the machine failing on a given occasion:

$$\begin{aligned} P(\text{machine fails}) &= 1 - P(\text{machine will } \textit{not} \text{ fail}) \\ &= 1 - 0.955647 \\ &= 0.044353 \\ &\simeq 4.4\% \end{aligned}$$

7.2.5 The general addition law

When events are not mutually exclusive it is not possible to use the simple addition law (introduced in section 7.4).

Table 7.2 Possible outcomes when a card is drawn from a standard 52-card pack (+ indicates a 'red card'; * indicates a 'picture card')

	♣	♦	♥	♠
A		+	+	
2		+	+	
3		+	+	
4		+	+	
5		+	+	
6		+	+	
7		+	+	
8		+	+	
9		+	+	
10		+	+	
J	*	* +	* +	*
Q	*	* +	* +	*
K	*	* +	* +	*

Consider table 7.2; let R be the event 'a red card' (+ on the table) and let P be the event 'a picture card' (* on the table). By counting the possible outcomes on the table, we see that

$$P(\text{R}) = \frac{26}{52} = \frac{1}{2}$$

$$\text{and } P(\text{P}) = \frac{12}{52} = \frac{3}{13}$$

However, it is **not true** that
$P(\mathrm{R}\ or\ \mathrm{P}) = P(\mathrm{R}) + P(\mathrm{P}) = \frac{1}{2} + \frac{3}{13} = \frac{19}{26}$

By counting the possible outcomes (* or +) on the table, we see that

$P(\mathrm{R}\ or\ \mathrm{P}) = \frac{32}{52} = \frac{8}{13}$

That is, there are 32 chances out of 52 of picking a red card *or* a picture card. The reason for this apparent contradiction is that events R and P are not mutually exclusive – it is possible for R and P to occur at the same time, that is, it is possible to choose a card which is *both* red *and* a picture card. The probability of this simultaneous event is easily found by counting the possible (*+) outcomes on the table:

$$P(\mathrm{R}\ and\ \mathrm{P}) = \frac{6}{52}\left(= \frac{3}{26}\right)$$

The probability of the card being either red *or* a picture card (but not both) is:

$$P(\mathrm{R}\ or\ \mathrm{P}) = \frac{26}{52} + \frac{12}{52} - \frac{6}{52}$$
$$= \frac{32}{52} = \frac{8}{13}$$

In general, the law is $\boldsymbol{P(A \text{ or } B) = P(A) + P(B) - P(A \text{ and } B)}$.

If A and B were mutually exclusive, then $P(A \text{ and } B) = 0$ and the general addition law would reduce to the simple addition law.

7.2.6 The general multiplication law

Like the addition law, the multiplication law has a general form. The simple multiplication law can only be applied when the events concerned are independent. Consider the following example: The four kings are extracted from the full pack of cards. When picking a card from this reduced pack the following are true:

$$P(\text{King of Spades}) = \frac{1}{4};\quad P(\text{King of Hearts}) = \frac{1}{4}\quad \text{etc.}$$

Suppose we plan to choose two cards in succession and we wish to know the probability that the first card chosen is the King of Hearts and the second is the King of Spades. We might expect, using the simple multiplication law, that P(1st card chosen is the King of Spades *and* 2nd card chosen is the King of Hearts) is $\frac{1}{16}$ $(\frac{1}{4} \times \frac{1}{4})$, but this is not so: Once a card is chosen there will only be three cards left to choose from. Consequently the probability that the second card chosen is the King of Hearts (given that the first card chosen is the King of Spades) will be $\frac{1}{3}$.

This is more easily seen by reference to the tree diagram in fig. 7.1 (tree diagrams will be considered again later in this chapter). There are a total of 12 possible outcomes, which are all equally likely, so the desired probability is $\frac{1}{12}$, i.e.:

$$P(\text{1st card is King of Spades } and \text{ 2nd card is King of Hearts}) = \tfrac{1}{4} \times \tfrac{1}{3} = \tfrac{1}{12}$$

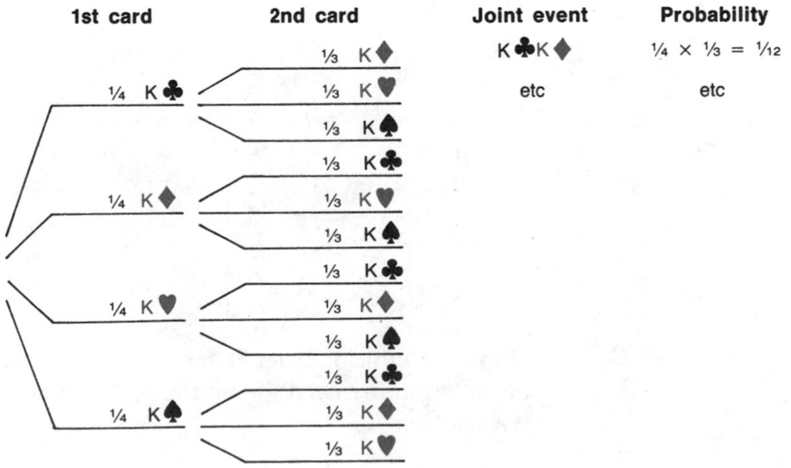

1st card	2nd card	Joint event	Probability
¼ K♣	⅓ K♦	K♣K♦	¼ × ⅓ = 1⁄12
	⅓ K♥	etc	etc
	⅓ K♠		
¼ K♦	⅓ K♣		
	⅓ K♥		
	⅓ K♠		
¼ K♥	⅓ K♣		
	⅓ K♦		
	⅓ K♠		
¼ K♠	⅓ K♣		
	⅓ K♦		
	⅓ K♥		

Fig. 7.1 Tree diagram showing the possible outcomes of choosing two cards (without replacement) from {King of Clubs, King of Diamonds, King of Hearts, King of Spades}

7.2.6.1 Conditional probability

We need to digress for a moment to consider the idea of **conditional probability** which is a necessary factor in understanding the general multiplication law.

The probability of picking the King of Spades from the four kings is $\frac{1}{4}$. This could be written as

$$P(\text{K of S}) = \frac{1}{4}$$

However, if one card has already been taken (the King of Hearts), then the probability of choosing the King of Spades is $\frac{1}{3}$. This could be written as

$$P(\text{K of S} \mid \text{1st card chosen was K of H}) = \frac{1}{3}$$

This would be read as 'the probability that the card chosen is the King of Spades, *given* the 1st card chosen was the King of Hearts, is equal to $\frac{1}{3}$.

Generally, where two events are not independent, a conditional probability is expressed as

$$P(A|B)$$

where | is read as 'given'.

The general multiplication law states that if A and B are two events, then the probability that both A and B will occur is

$$\begin{aligned} P(A \text{ and } B) &= P(A) \times P(B|A) \\ &= P(B) \times P(A|B) \end{aligned}$$

Note: When A and B are independent events, then $P(B|A) = P(B)$, since, by definition, the occurrence of B (and therefore $P(B)$) does not depend on the occurrence of A. Similarly, $P(A|B) = P(A)$.

7.3 Basic decision criteria

Decision-making is a central and essential activity in business life. We are all experienced decision-makers and business decision-making is just an extension of the skills we already use in our everyday lives. This section highlights some of the important factors behind decision-making, it will be useful to use an example, the 'site problem', as a means of focusing on these factors.

7.3.1 The site problem

Sam and Emily's Fish & Chip Shop is a small business. It is run from a mobile van specially designed for the purpose of selling a variety of foods directly to the customer. Over a period of 10 years they have moved their 'shop' between three sites: next to the town cinema; next to the public swimming pool; and next to the council tennis courts. Their main criterion for changing sites is the condition of the weather. They have maintained records of profits over the 10 years and the average profits are shown in table 7.3. You will notice that sales at the three sites appear to be influenced by the weather conditions. To date, their decision-making process has been quite straightforward: If the weather report is 'hot', they site themselves next to the swimming pool, and so on. Fortunately the local weather station is very reliable, so there is no risk of choosing a less favourable site. Their decisions (sometimes called **strategies**) are based on the weather (circumstances outside their control) and table 7.3 represents the possible results of their decisions (sometimes referred to as the **payoff**).

Table 7.3 Sam and Emily's Fish & Chip Shop: average daily profits (£)

Venue	Weather conditions		
	Rain	Fair	Hot
Cinema	150	85	45
Swimming pool	40	70	180
Tennis courts	60	170	50

In 1988 the local council elections resulted in a change of power and one consequence of this was a policy changing the rules for mobile shops. Emily and Sam were required to choose a permanent site. This, as you will see in the following sections, gave Sam and Emily's business different problems to solve.

7.3.2 Maximum payoff: maximax

Emily comments first: 'The decision is easy. The swimming pool site has the greatest potential, as long as the weather is hot, so let's have the shop permanently sited there.' It is true that, weather permitting, the swimming pool gives maximum profits. This type of optimistic approach involves finding the **maximum profit related to each choice**–in this case:

Cinema: £150 per day (weather 'rain')
Swimming pool: £180 per day (weather 'hot')
Tennis courts: £170 per day (weather 'fair')

–then choosing the **maximum of those maximums**. Hence its name: the maximax criterion.

7.3.3 Minimum payoff: maximin

Sam is a little more cautious. He reasons: 'The weather cannot be relied upon. I think the business should be much more cautious. Let's consider the worst outcome in each case and pick the best of those outcomes'. This pessimistic approach involves finding the **worst level of profits associated with each decision**. In this case these are:

Cinema: £45 per day (weather 'Hot')
Swimming pool: £40 per day (weather 'Rain')
Tennis courts: £50 per day (weather 'Hot')

Consequently, Sam chooses the tennis courts site: £50 is the **maximum of the minimums**. This is called the maximin criterion.

7.3.4 Maximum regret: minimax

Sam and Emily are undecided. After some discussion they decide on a compromise solution. They recognize that a wrong decision will result in some regret. For instance, if they decide to site their shop next to the cinema and it rains, they will have no regrets (they will have made the maximum possible profit in the circumstances), but if the weather is 'fair' it would have been better to have sited the shop next to the tennis courts. The regret is measured in terms of the **lost profits**, which in this case would be £170–£85 = £85. Similarly, if the weather is 'hot', the best site to have chosen would be next to the swimming pool, giving profits of £180, £135 better than the profits on the cinema site–hence a 'regret' of £135. A matrix of regrets is shown in table 7.4. Each entry in the matrix is determined by calculating the difference between the profit obtainable and the maximum profit obtainable (given the weather).

Table 7.4

Possible venue	Weather conditions Rain	Fair	Hot	Maximum regret (£)
		Regrets (£)		
Cinema	0	85	135	135
Swimming pool	110	100	0	110
Tennis courts	90	0	130	130

Sam and Emily decide to choose the site which **minimizes the maximum regret**. Consequently, they chose the site next to the swimming pool (maximum regret £110).

In this example the decision criteria were subjective, reflecting lots of instances in real-life decision-making. However, it is sensible to try and make decisions which relate in some way to the available information and are also objective. To do this more information is required; Sam and Emily did the best they could with the information they had.

7.4 Expected values

This section deals with a very important input to decision-making, the idea of expected values (in business situations this is usually expected monetary value).

Suppose, in the above example, more information was available regarding the weather over the past 10 years:

Weather	Percentage of time
Rain	35
Fair	50
Hot	15

If the decision made above related to a short time period, it would be perfectly acceptable to take a chance on the weather being 'hot' and choose the swimming pool site to minimize potential regret. But Emily and Sam are choosing a permanent site, i.e. the decision is for a long period. Consequently, it makes sense to investigate what is likely to occur over a long period. (In the above considerations, only one day's possible results were investigated.) We need to know the expectations over the longer period or the average outcome.

Consider the above example again: The weather condition 'rain' has occurred 35 per cent of the time, over the last 10 years. Therefore we will assume that the probability of rain occurring on any day in the future is 0.35. This requires an assumption of independence: that the event 'rain' does not influence the event 'hot weather', and so on.

We are now in a position to calculate the expected monetary value, E (expected profits in this case), of each choice. The expected profits of choosing the site next to the cinema are calculated as follows:

$$
\begin{aligned}
E(\text{profits: cinema}) &= [\text{profits when there is rain} \times P(\text{rain})] \\
&\quad + [\text{profits when weather is fair} \times P(\text{fair weather})] \\
&\quad + [\text{profits when weather is hot} \times P(\text{hot weather})] \\
&= (£150 \times 0.35) + (£85 \times 0.5) + (£45 \times 0.15) \\
&= £101.75
\end{aligned}
$$

The other expected values can be found similarly. (The calculations are left as an exercise.) The results are:

E(profits: swimming pool) = £76
E(profits: tennis courts) = £113.50

Consequently, the tennis courts site seems to be the one to choose. It will not match the cinema site for highest possible profits on any given day, but it is expected to produce the greatest profits in the long run.

The expected value is summarized in equation form as follows:

$$E(X) = \Sigma XP(X)$$

This is read as 'the expected value of 'X' is equal to the sum of the products of each value of X and the corresponding probability of that value of X occurring'.

7.5 Tree diagrams

There was a brief mention of tree diagrams in section 7.2.6 (fig. 7.1). They are a useful means of investigating probability problems and you should be familiar with their application as they are often included in examination questions.

Once again it will be convenient to use an example to demonstrate their use: A company, which sells double-glazing, keeps a file of sales contacts which is regularly updated. The sales contacts are shared between three salesmen, A, B and C. They are given 50, 30 and 20 per cent of the contacts and fail to secure a sale on 2, 5 and 10 per cent of their contacts, respectively. This information is illustrated in the tree diagram in fig. 7.2.

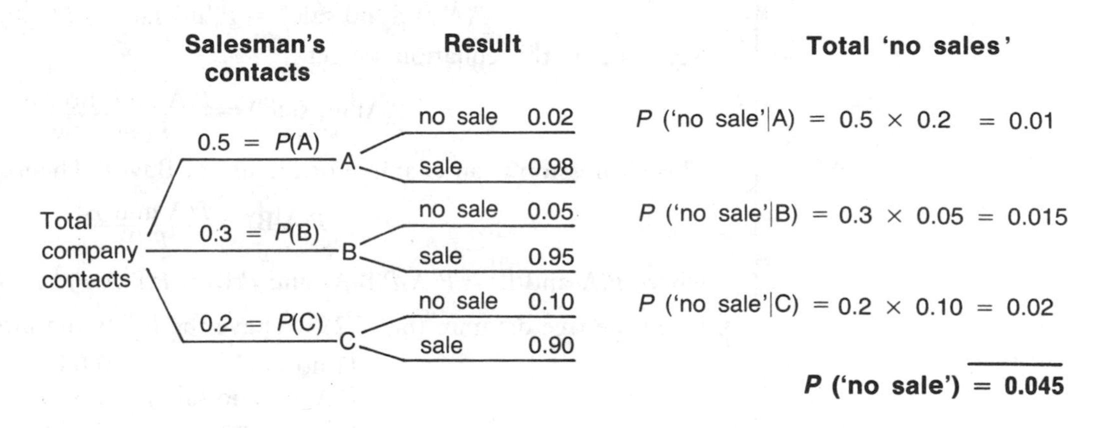

Fig. 7.2 Tree diagram showing sales contacts and related success rates

The company wishes to determine the total percentage of contacts which result in 'no sale'. Be careful! At first sight, one might think the answer is found by simply adding the salesmen's individual failure rates: $2\% + 5\% + 10\% = 17\%$. However this is not the case because these percentages are conditional on which salesman is involved. For instance, salesman A has a failure rate of 2 per cent. This is 2 per cent of *his* sales, not of the total sales. The required information can be written in the form of conditional probabilities as follows:

$$P(\text{'no sale'}|\text{A}) = 0.02$$
$$P(\text{'no sale'}|\text{B}) = 0.05$$
$$P(\text{'no sale'}|\text{C}) = 0.10$$

i.e. 'the probability of 'no sale', *given* that salesman A is involved, is 0.02', etc.

To find what these conditional proportions represent in terms of proportions of total contact is quite straightforward. For instance, 2 per cent of A's contacts (which are 50 per cent of the company's total contacts) result in no sale; 2 per cent of 50 per cent is 1 per cent of the total contacts ($0.02 \times 0.5 = 0.01 = 1\%$). As a probability statement, this can be expressed as:

$$P(\text{'no sale' } \textit{and} \text{ A}) = P(\text{A}) \times P(\text{'no sale'}|\text{A}).$$

You should recognize this as a use of the general multiplication law. It is not necessary to use the formula though – you can follow the same calculations easily on the tree diagram (top row). The total percentage of 'no sales' is found by adding the results from the 'branches' of the tree:

$$0.01 + 0.015 + 0.02 = 0.045 = 4.5\%$$

As a probability statement this can be expressed as:

$$P(\text{'no sale' } \textit{and} \text{ A}) + P(\text{'no sale' } \textit{and} \text{ B}) + P(\text{'no sale' } \textit{and} \text{ C}) = P(\text{'no sale'})$$

7.6 Prior and posterior probabilities

The information we used above concerning A, B and C's sales contacts was based on the company's records, so can be thought of as **prior (or initial) probabilities.** For instance, a sales contact taken from company files at random has a probability of 0.50 of belonging to salesman A.

Suppose further information becomes available. For instance, it might be known that a particular sales contact resulted in 'no sale'. Given this information, we might wish to find the probability that that sales contact belonged to salesman A, i.e. we now want the probability a sales contact is A's *given* that a 'no sale' resulted: $P(\text{A}|\text{'no sale'})$. This probability is a **posterior (or final) probability**. Posterior probabilities are updates or reviews of prior probabilities, given extra information. The idea is similar to that of conditional probability. It is likely that you will know the reverse of the required probability, as is the case in the example being considered. That is, we know $P(\text{'no sale'}|\text{A})$ (see fig. 7.2) but not the required probability $P(\text{A}|\text{'no sale'})$.

7.6.1 Bayes' Theorem

Posterior probabilities can be calculated by manipulating the multiplication law:

$$P(\text{A and 'no sale'}) = P(\text{'no sale'}) \times P(\text{A}|\text{'no sale'})$$

Rearranging this equation we get

$$P(\text{A}|\text{'no sale'}) = \frac{P(\text{A and 'no sale'})}{P(\text{'no sale'})}$$

This expression is an example of the use of **Bayes' Theorem.** The general form is:

$$\mathbf{P(A|B) = \frac{P(A \text{ and } B)}{P(B)}}$$

where $P(\text{A and B}) = P(\text{A})P(\text{B}|\text{A})$ and $P(\text{B}) = P(\text{B}|\text{A})P(\text{A}) + P(\text{B}|\text{C})P(\text{C}) + \ldots$

From the tree diagram (fig. 7.2) we have the following information:

$$P(\text{'no sale'}) = 0.045$$
$$P(\text{A and 'no sale'}) = 0.01$$
$$P(\text{B and 'no sale'}) = 0.015$$
$$P(\text{C and 'no sale'}) = 0.02$$

From this information we can find the posterior probabilities:

$$P(\text{A}|\text{ 'no sale'}) = \frac{0.01}{0.045} = 0.22 \text{ (to 2 d.p.)}$$
$$P(\text{B}|\text{ 'no sale'}) = 0.33 \text{ (to 2 d.p.)}$$
$$P(\text{C}|\text{ 'no sale'}) = 0.44 \text{ (to 2 d.p.)}$$

We can conclude that C is the salesman most likely to have a 'no sale'. He is responsible for 44 per cent of the company's 'no sales'. A has the best sales record, even though he has the most contacts to cope with (50 per cent of the total contacts). He must be a very good salesman or perhaps he just has the best contacts!

7.7 Multi-stage decision analysis

The decision processes just considered each involved only one stage. The situation becomes more complex when there are a number of stages to be considered. Again it is appropriate to use an example to illustrate this.

A company is planning to introduce a new product on to the market. Before doing so, they must decide whether to advertise nationally, at a cost of £12 000, or in one main region only, at a cost of £6500. The initial advertising campaign would last for three months, whichever option is chosen. If the company advertises nationally, the expected sales are as follows:

Sales (no. of units)	Probability
2 000 (poor)	0.15
7 000 (average)	0.75
12 000 (good)	0.10

If the company advertises in one region only the expected sales are as follows:

Sales (no. of units)	Probability
2 000 (poor)	0.35
7 000 (average)	0.60
12 000 (good)	0.05

Later in the year it will be necessary to decide whether to advertise in the second half of the year, at a cost of £9000. If there is no advertising in the second half of the year, sales will be poor. If the company does advertise further, sales will depend on the first six months' performance (see table 7.5).

Table 7.5 Possible sales performance

First 6 months' sales performance (F)	Second 6 months' sales performance (S)	Probability of second 6 months' sales performance given first 6 months' sales performance $P(F|S)$
poor	poor	0.75
	average	0.25
average	poor	0.15
	average	0.50
	good	0.35
good	average	0.25
	good	0.75

The problems facing the company are:

1 Which decisions should be taken at each stage?

2 What will be the expected profit? (Each unit of sale produces £3 profit.)

7.7.1 Decision trees

The best way to approach the problems outlined above is to draw a decision tree. They are reasonably simple to construct, but do need care as they can get complicated. In an examination give yourself plenty of space – do not try to cram a decision tree into the corner of a page.

There are two types of **node** or **joint** on decision trees: **decision nodes**, which in this text will be represented by squares; and **chance nodes**, which will be represented by circles.

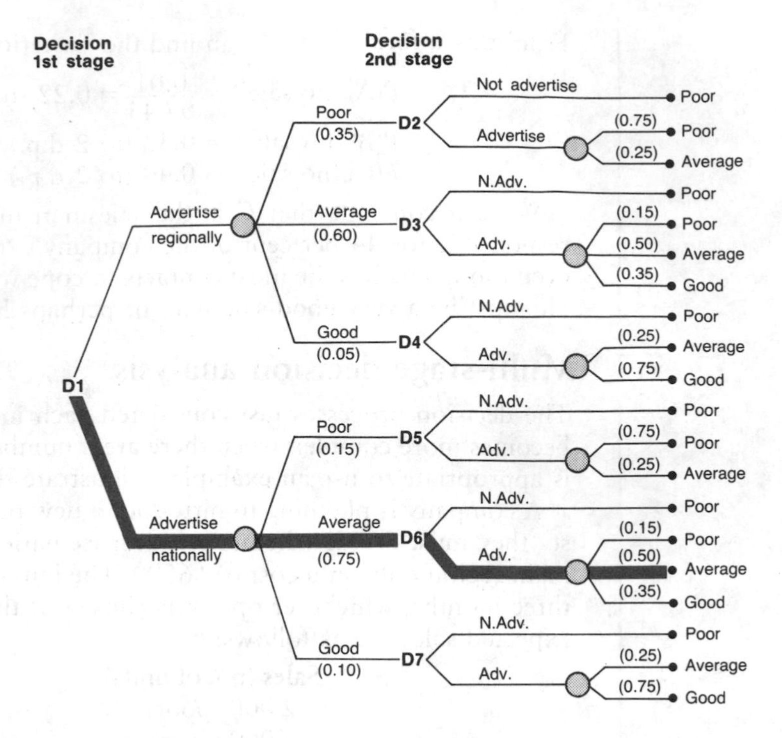

Fig. 7.3

Figure 7.3 illustrates the decision tree for the above example. You should be able to see that all the possible outcomes are shown. Of the 20 possible outcomes, one has been highlighted: The initial decision is to advertise nationally; the outcome that follows this decision is 'average sales'; the second decision, relating to advertising in the second half of the year, is to advertise; the outcome is again 'average sales'. There are 19 other possibilities. This is the first stage of decision tree analysis, the 'left-to-right sweep'. It establishes all the possibilities of the problem.

The next stage is to insert the relevant probability values, sales values and advertising costs, and to calculate expected values for each possibility. This will involve a 'right-to-left sweep' of the decision tree. The completed decision tree is illustrated in fig. 7.4.

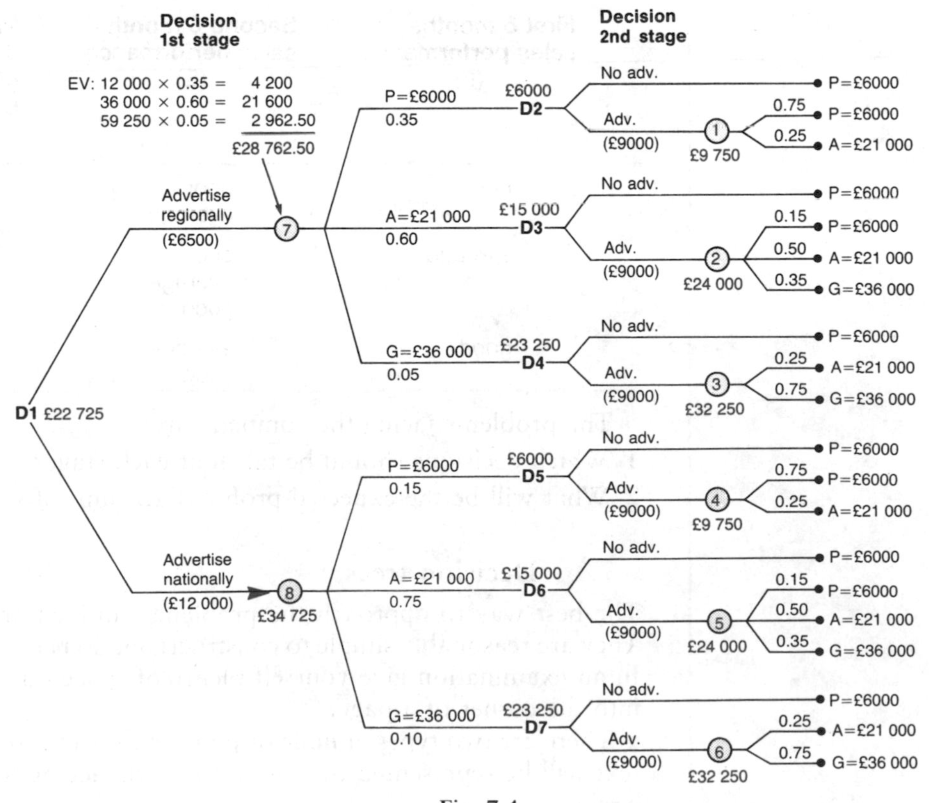

Fig. 7.4

Analysing from right to left, we first calculate the expected value at each chance node 1–6. For example, the expected value at node 1 is calculated as follows:

$$E = (£6000 \times 0.75) + (£21\,000 \times 0.25) = £9750$$

With this information known, we consider decision nodes D2 to D7. For example, at D2 there are two possible decisions:

1 'not to advertise' – resulting in poor sales, profit value £6000
2 'advertise' – at a cost of £9000, resulting in an expected profit value of £9750, a net profit of only £750.

The best option at D2 is to choose not to advertise, giving an expected profit value of £6000 at D2. At decision nodes D3, D4, D6 and D7 it is better to advertise: the corresponding expected values are indicated on the decision tree.

The next stage is to calculate the expected values at chance nodes 7 and 8. Care is needed at this stage: The expected values of each 'path' from node 7 to D2, D3 and D4 include the expected values of the previous analysis: the possible outcomes at node 7 are poor, average and good sales. 'Poor sales' has a profit value of £6000, but we must also take into account the expected value at D2 of another £6000, giving a total expected profit value for that 'path' of £12 000. The calculation for the expected value at chance node 7 is therefore:

$$\begin{aligned} E(7) &= [0.35 \times (£6000 + £6000)] + [0.60 \times (£21\,000 + £15\,000)] + \\ &\quad [0.05 \times (£36\,000 + £23\,250)] \\ &= (0.35 \times £12\,000) + (0.60 \times £36\,000) + (0.05 \times £59\,250) \\ &= £28\,762.50 \end{aligned}$$

Thus the decision to advertise regionally would result in an expected value of £28 762.50, less advertising costs of £6500, giving net expected profit of £22 262.50. Analysing the bottom half of the tree shows that the decision to advertise nationally produces net expected profits of £22 725, £462.50 better than advertising regionally.

Therefore the company should first advertise nationally. If sales in the first half of the year are average or good, they should advertise again in the second half; if sales are poor they should not advertise further.

7.8 Permutations and combinations

The remainder of this chapter will be concerned with a number of important probability distributions. Before looking at them in more detail, it is necessary to be familiar with the arithmetic of permutations and combinations:

7.8.1 Permutations

A company has shortlisted five candidates for three positions of sales manager in the Birmingham, Bristol and Exeter regions. In how many ways can the selection be made? There are five candidates for the first job. Having made the first choice there are now four candidates left for the second position. Finally there are three candidates left to choose from for the third job. The total number of possible selections is 60 ($5 \times 4 \times 3$). If the five candidates are represented by the letters A, B, C, D and E, the list of possible selections is:

ABC	ACB	BAC	BCA	CAB	CBA
ABD	ADB	BAD	BDA	DAB	DBA
ABE	AEB	BAE	BEA	EAB	EBA
ACD	ADC	CAD	CDA	DAC	DCA
ACE	AEC	CAE	CEA	EAC	ECA
ADE	AED	DAE	DEA	EAD	EDA
BCD	BDC	CBD	CDB	DBC	DCB
BCE	BEC	CBE	CEB	EBC	ECB
BDE	BED	DBE	DEB	EBD	EDB
CDE	CED	DCE	DEC	ECD	EDC

The order of the letters is important: The first candidate gets the Birmingham job; the second, the Exeter job; and the third, the Bristol job.

Each possible selection is called a **permutation**, in this case a permutation of 3 chosen from 5.

There is a mathematical 'shorthand' for 'the number of permutations of 3 from 5'. It is written as ${}^5P_3$. We could write ${}^5P_3 = 60$. Similarly:

$$
\begin{aligned}
{}^6P_2 &= 6 \times 5 = 30 \\
{}^7P_2 &= 7 \times 6 = 42 \\
{}^7P_3 &= 7 \times 6 \times 5 = 210 \\
{}^6P_6 &= 6 \times 5 \times 4 \times 3 \times 2 \times 1 = 720
\end{aligned}
$$

The expressions can be made more manageable by the use of **factorial notation,** which is represented by an exclamation mark, !. We can write

$$
\begin{aligned}
1! &= 1 \\
2! &= 2 \times 1 = 2 \\
3! &= 3 \times 2 \times 1 = 6 \\
4! &= 4 \times 3 \times 2 \times 1 = 24
\end{aligned}
$$

and so on. (0! is defined as 1.)

Returning to our original problem, we can write

$$
{}^5P_3 = \frac{5!}{2!} = \frac{5 \times 4 \times 3 \times 2 \times 1}{2 \times 1} = \frac{120}{2} = 60
$$

In general, the number of permutations of r items taken from n items (where the order is important) is

$$
{}^n\boldsymbol{P}_r = \frac{\boldsymbol{n}!}{(\boldsymbol{n}-\boldsymbol{r})!}
$$

7.8.2 Combinations

If the company was not concerned with which candidate had which job, the possible selections would be called **combinations**. In such a situation the choice ABC is the same as CAB, BCD is the same as CBD, and so on – i.e. the order of choice does not matter. The list of permutations in section 7.8.1 was set out in such a way that each row contains the same letters, that is, each permutation in a particular row is equivalent to the same combination. Each combination has been duplicated 6 times. The number of combinations of 3 from 5 is therefore 10 ($60 \div 6$). In mathematical 'shorthand' we write

$$
{}^5C_3 = \frac{60}{6} = 10
$$

In general, the expression for the number of combinations of r items chosen from n (irrespective of order) is written as

$$
{}^n\boldsymbol{C}_r = \frac{\boldsymbol{n}!}{\boldsymbol{r}!(\boldsymbol{n}-\boldsymbol{r})!} \left(= \frac{{}^n\boldsymbol{P}_r}{\boldsymbol{r}!} \right)
$$

7.9 Binomial distribution

The probabilities considered so far have involved very few events. Consequently it has been quite straightforward to carry out the calculations. If there are a large number of events the computation becomes very tedious. The idea of combinations helps to significantly reduce the calculations. Consider the following example:

A company is concerned about the time-keeping of some of its employees. It has calculated that there is a 25 per cent chance of an individual being late for work. They wish to know the probability of two out of five workmen being late.

Table 7.6 Possible combinations and associated probabilities, of two out of five workmen being late (L = 'late'; N = 'not late')

Workmen (1 to 5)					
1	**2**	**3**	**4**	**5**	**Probability (multiplication law)**
L	L	N	N	N	1/4 × 1/4 × 3/4 × 3/4 × 3/4 = 27/1024
L	N	L	N	N	1/4 × 3/4 × 1/4 × 3/4 × 3/4 = 27/1024
L	N	N	L	N	1/4 × 3/4 × 3/4 × 1/4 × 3/4 = 27/1024
L	N	N	N	L	1/4 × 3/4 × 3/4 × 3/4 × 1/4 = 27/1024
N	L	L	N	N	3/4 × 1/4 × 1/4 × 3/4 × 3/4 = 27/1024
N	L	N	L	N	3/4 × 1/4 × 3/4 × 1/4 × 3/4 = 27/1024
N	L	N	N	L	3/4 × 1/4 × 3/4 × 3/4 × 1/4 = 27/1024
N	N	L	L	N	3/4 × 3/4 × 1/4 × 1/4 × 3/4 = 27/1024
N	N	L	N	L	3/4 × 3/4 × 1/4 × 3/4 × 1/4 = 27/1024
N	N	N	L	L	3/4 × 3/4 × 3/4 × 1/4 × 1/4 = 27/1024

Table 7.6 shows all the possibilities and the associated probabilities. It is assumed that the event 'workman 1 is late' is independent of the event 'workman 2 is late', etc, so the probability of each combination can be calculated using the multiplication law. For instance, the probability of the first combination (the first row) is

P('1 is late' *and* '2 is late' *and* '3 is not late' *and* '4 is not late' *and* '5 is not late')
$= P(1 \text{ is late}) \times P(2 \text{ is late}) \times P(3 \text{ is not late}) \times P(4 \text{ is not late}) \times P(5 \text{ is not late})$

The total probability (that *any* two out of five workmen will be late) is found by adding the separate probabilities ($270/1024 = 0.264$).

A simpler approach is to note that all 10 probabilities are the same and equal to $(\frac{1}{4})^2 \times (\frac{3}{4})^3$, and the number of ways two workmen from five can be late is equal to ${}^5C_2$ (choosing 2 from 5). The answer can then be calculated as follows:

$$P(2 \text{ workmen out of 5 are late}) = {}^5C_2 \times \left(\frac{1}{4}\right)^2 \times \left(\frac{3}{4}\right)^3$$
$$= \frac{5!}{2! \times 3!} \times \frac{27}{1024}$$
$$= 0.264$$

Similarly, the probabilities of 0, 1, 3, 4 or 5 late arrivals can be calculated:

$$P(0 \text{ late}) = {}^5C_0 \times \left(\frac{1}{4}\right)^0 \times \left(\frac{3}{4}\right)^5 = 1 \times \frac{243}{1024} = 0.237$$

$$P(1 \text{ late}) = {}^5C_1 \times \left(\frac{1}{4}\right)^1 \times \left(\frac{3}{4}\right)^4 = 5 \times \frac{81}{1024} = 0.396$$

$$P(3 \text{ late}) = {}^5C_3 \times \left(\frac{1}{4}\right)^3 \times \left(\frac{3}{4}\right)^2 = 10 \times \frac{9}{1024} = 0.088$$

$$P(4 \text{ late}) = {}^5C_4 \times \left(\frac{1}{4}\right)^4 \times \left(\frac{3}{4}\right)^1 = 5 \times \frac{3}{1024} = 0.015$$

$$P(5 \text{ late}) = {}^5C_5 \times \left(\frac{1}{4}\right)^5 \times \left(\frac{3}{4}\right)^0 = 1 \times \frac{1}{1024} = 0.001$$

Note:

1 The total probability is one, i.e.

$P(0 \text{ late}) + P(1 \text{ late}) + P(2 \text{ late}) + P(3 \text{ late}) + P(4 \text{ late}) + P(5 \text{ late}) = 1.0$

2 ${}^5C_0 = \frac{5!}{0!5!} = \frac{120}{1 \times 120} = 1$ and ${}^5C_5 = \frac{5!}{5!0!} = \frac{120}{120 \times 1} = 1$

3 By definition, any number to the power zero is 1, e.g. $(\frac{1}{4})^0 = 1$

This problem is an example of the **binomial distribution**, illustrated in fig. 7.5. It is an example of a discrete probability distribution. The variable is the number of workmen who are late. The distribution as shown in fig. 7.5 is similar to frequency distributions met in earlier chapters. In this instance the graph is a probability distribution – the vertical axis represents probability, not frequency.

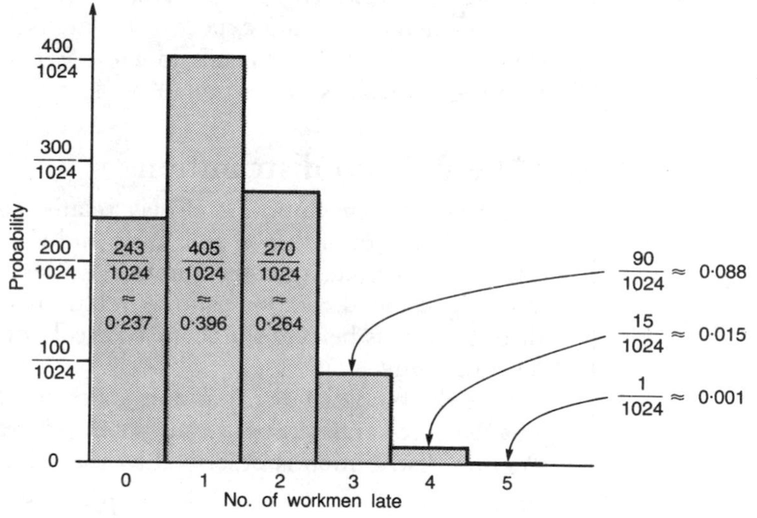

Fig. 7.5 (*Note:* The probabilities add up to 1 and the total area of the rectangles is also 1)

7.9.1 The binomial distribution: definition

The binomial distribution can be derived from a situation which involves the repetition of an event which has only two possible outcomes – e.g. success/failure, win/lose, late/not late, etc. Each repetition (or trial) must be independent. (We assumed, in the above example, that the event 'workman 1 being late' was independent of the event 'workmen 2 being late', etc.)

In n independent trials of an event, each of which has the same probability of success, p, and the same probability of failure, $q = 1 - p$, the probability of obtaining r successes, $P(r)$, is given by the expression

$$P(r) = {}^nC_r p^r q^{n-r}$$

In the lateness example, $n = 5$, $p = \frac{1}{4}$ and $q = \frac{3}{4}$.

7.9.2 Mean and standard deviation

Like frequency distributions, probability distributions have a mean and a standard deviation. The equations for the probability mean and standard deviation look very like those you have already met for frequency distributions:

mean (or expected value), $\mu = \Sigma xP(x)$ (see section 7.4)

and variance, $\sigma^2 = \Sigma(x-\mu)^2 P(x)$

where x represents the values of the variable and $P(x)$ the corresponding probabilities. μ (mu) and σ (sigma) are Greek letters used here instead of $\bar{x}$ and s. This is because the probability mean and standard deviation are theoretical values and define the parameters of the population, whereas $\bar{x}$ and s are sample values.

Table 7.7 shows the mean and standard deviation calculations for the current example, using these formulae.

Table 7.7

No. of workmen late x	Probability $P(x)$	$xP(x)$	$x-\mu$	$(x-\mu)^2$	$(x-\mu)^2P(x)$
0	0.237	0	(0 − 1.25 =) −1.25	1.56	0.37
1	0.395	0.395	(1 − 1.25 =) −0.25	0.06	0.02
2	0.264	0.528	0.75	0.56	0.15
3	0.088	0.264	1.75	3.06	0.27
4	0.015	0.060	2.75	7.56	0.11
5	0.001	0.005	3.75	14.06	0.01
Totals	1.000	1.252			0.93

$\mu = 1.252$

$\sigma = \sqrt{0.93} = 0.96$

(*Note:* The sum of the probabilities $P(x)$ always equals 1.0.)

For the binomial distribution, it is not necessary to do all these calculations. It can be shown that

$$\mu = np \text{ and } \sigma = \sqrt{npq}$$

In the example considered above, $\mu = 5 \times \frac{1}{4} = 1.25$ (i.e. the average number of workmen one would expect to be late is 1.25) and $\sigma = \sqrt{5 \times \frac{1}{4} \times \frac{3}{4}} = \sqrt{0.9375} = 0.97$ (to 2 d.p.). (The result in table 7.7 was $\sigma = 0.96$. The difference is due to rounding of values in the table.)

7.10 The Poisson distribution

The Poisson distribution is closely related to the binomial distribution, to such an extent that if the value of n is large and the value of p is small, so that np remains constant and finite, then the binomial distribution reduces to the Poisson distribution. The algebraic derivation of the relationship between the binomial and Poisson distributions is beyond the scope of this book and is not required knowledge for this level of study.

Like the binomial distribution the Poisson distribution is a probability distribution of a discrete variable, specifically x can only take integer values (i.e. 0, 1, 2, 3,...). The Poisson distribution is described by the following formula:

$$P(x) = \frac{e^{-m} m^x}{x!}$$

where $P(x)$ is the probability of an event occurring x times;
m is the mean number of occurrences, or expected number of occurrences (in the case of the binomial distribution $m = np$);
e is a constant with value 2.718282 (to 6 d.p.).

The standard deviation of the Poisson distribution is equal to $\sqrt{m}$.

The Poisson distribution has two main functions that you should be familiar with:

7.10.1 Probabilities of rare events

The Poisson distribution is particularly used to estimate the probability of isolated or rare events. It is widely used in industrial sampling and in demographic problems.

For example, a certain newspaper has an average of 3 misprints per page. This situation is ideally suited to the use of the Poisson distribution: misprints are (or should be) rare events and the number of misprints on different pages are independent.

1 What is the probability that the sports page is free of misprints?
Using the Poisson distribution formula defined above, we have

$$P(0 \text{ misprints}) = \frac{c^{-3} \times 3^0}{0!} = \frac{0.0498 \times 1}{1} = 0.05 \text{ (to 2 d.p.)}$$

That is, there is a 5 per cent chance of the sports page being free of misprints.

2 What is the probability that the front page has more than two misprints?
In this case it is more convenient to use the complementary law:

$$P(\text{front page has more than 2 misprints}) = P(x \geqslant 3) = 1 - P(x < 3)$$
$$= 1 - [P(0) + P(1) + P(2)]$$

We can calculate the appropriate probabilities:

$$P(0) = e^{-3}.3^0/0! = (0.0498 \times 1)/1 = 0.05 \text{ (2 d.p.)}$$
$$P(1) = e^{-3}.3^1/1! = (0.0498 \times 3)/1 = 0.15 \text{ (2 d.p.)}$$
$$P(2) = e^{-3}.3^2/2! = (0.0498 \times 9)/2 = 0.22 \text{ (2 d.p.)}$$
$$P(0) + P(1) + P(2) = 0.41$$

Therefore $P(x \geqslant 3) = 1 - 0.41 = 0.59$, that is, there is a 59 per cent chance of the front page having more than 2 misprints.

7.10.2 Poisson approximation to the binomial distribution

An equally important function of the Poisson distribution is the capacity for it to be used as an approximation for the binomial distribution. For the approximation to be acceptable certain conditions must exist:

1 n, the number of trials, must be large (greater than 100)

2 p, the probability of the event concerned, must be very small compared to q, the probability of an event not occurring (p should be less than 0.05)

3 np should be less than 5

Conditions 1, 2 and 3 are only a guide. Table 7.8 (overleaf) illustrates quite clearly the similarity of the two distributions and the level of accuracy obtained when using the Poisson distribution to approximate the binomial distribution. It is clear that as n increases and p decreases the probabilities become very close.

The following is an example of the use of the Poisson distribution as an approximation for the binomial distribution:

A life insurance company knows from its records that each of its clients has a probability of dying within a particular year of 0.0005. They wish to determine the probability that there will be:

1 no claims;

2 five claims; and

3 more than three claims;

from the company's 20 000 policy-holders during the next year?

This situation satisfies the criteria for a binomial distribution: the variable (the number of claims in next year) is discrete (only values of 0,1,2,... are possible); there are only two possible outcomes (either a client dies or he/she doesn't); and the events are independent. We have $n = 20\,000$, $p = 0.0002$ and $q = 1 - 0.0002 = 0.9998$. It would therefore be possible to calculate the desired probabilities using the binomial distribution formulae as follows:

1 $P(0) = {}^{20\,000}C_0\ (0.0002)^0 \times (0.9998)^{20\,000}$

2 $P(5) = {}^{20\,000}C_5\ (0.0002)^5 \times (0.9998)^{19\,995}$

3 $P(4 \text{ or more}) = 1 - [P(0) + P(1) + P(2) + P(3)]$

These probabilities would prove cumbersome and difficult to calculate. Since n is large, p is small and $np = 4$ (<5), the Poisson distribution can be used:

The mean, $m = 4$ ($= np$) so:

1 $P(0 \text{ claims}) = e^{-4}.4^0/0! = e^{-4} = 0.018$ (3 d.p.)

2 $P(5 \text{ claims}) = e^{-4}.4^5/5! = e^{-4} \times 1024/120 = 0.156$

3 $P(4 \text{ or more claims}) = 1 - P(3 \text{ or fewer claims})$
$= 1 - (0.018 + 0.073 + 0.147 + 0.195)$
$= 1 - 0.433 = 0.566$

Table 7.8 Extracts from binomial and Poisson probability tables for comparison

Binomial distribution (n, p)			**Poisson distribution ($m = np$)**		
$p =$	0.01	0.02	$m =$	0.02	0.04
$n = 2$					
$P(x = 0)$	0.9801	0.9604	$P(x = 0)$	0.9802	0.9608
1	0.0198	0.0392	1	0.0196	0.0384
2	0.0001	0.0004	2	0.0002	0.0008
$p =$	0.01	0.02	$m =$	0.05	0.10
$n = 5$					
$P(x = 0)$	0.9510	0.9039		0.9512	0.9048
1	0.0480	0.0923		0.0476	0.0905
2	0.0010	0.0037		0.0012	0.0045
3		0.0001			0.0002
$p =$	0.01	0.02	$m =$	0.10	0.20
$n = 10$					
$P(x = 0)$	0.9044	0.8171		0.9048	0.8187
1	0.0913	0.1667		0.0905	0.1637
2	0.0042	0.0153		0.0045	0.0164
3	0.0001	0.0009		0.0002	0.0012
$p =$	0.01	0.02	$m =$	0.20	0.40
$n = 20$					
$P(x = 0)$					
1	0.8179	0.6676		0.8187	0.6703
2	0.1652	0.2725		0.1637	0.2681
3	0.0159	0.0528		0.0164	0.0536
4	0.0010	0.0065		0.0012	0.0071
5		0.0006			0.0007
6					0.0001

7.11 The normal distribution

The final distribution to be considered in this chapter is probably the most important distribution you will encounter. It is different from those considered so far in a number of ways. In particular, it describes a continuous variable, not discrete variables.

An example will be used to introduce this distribution. A production process involves the manufacture of bearings for the car industry and machine A produces a bearing which should have a diameter of 10 cm. In practice, the precision of the machine is not perfect, so there tends to be some error. Experience has shown the diameters to have a mean diameter of 10 cm and a standard deviation of 0.09 cm. The following characteristics of the process will help to describe the resulting distribution:

1 The variable being measured (diameter) is continuous. Because the variable is continuous, the probability of a 'point' event is always zero. When a coin is thrown it is easy to find P('heads'). This is a simple probability situation:

$$\frac{\text{number of possible events resulting in 'heads''}}{\text{total number of possible events}} = \frac{1}{2}$$

However, treating the following probability in the same way is less satisfactory:

$$P(\text{diameter} = 10 \text{ cm}) = \frac{1}{\infty} = 0$$

There is one way of obtaining the desired event and an infinite number of possible events. This would be the result for any 'point' event. Consequently, only probability statements of the type $P(10.5 < \text{diameter} < 12.5)$, can be sensibly calculated.

2 The distribution of the variable is symmetrical about the mean value, in this case 10 cm: It is as probable that the machine would produce bearings with diameter between 2 and 3 mm too small as with diameter between 2 and 3 mm too large.

3 The probability of obtaining a bearing with diameter in a certain interval will decrease as the distance of the interval from the mean increases. For instance, it is more likely that the machine will produce a bearing which has a diameter between 2 and 3 mm too large than it is for the machine to produce a bearing with diameter between 9 and 10 mm too large.

The resulting normal distribution would look like that shown in the top diagram in fig. 7.6.

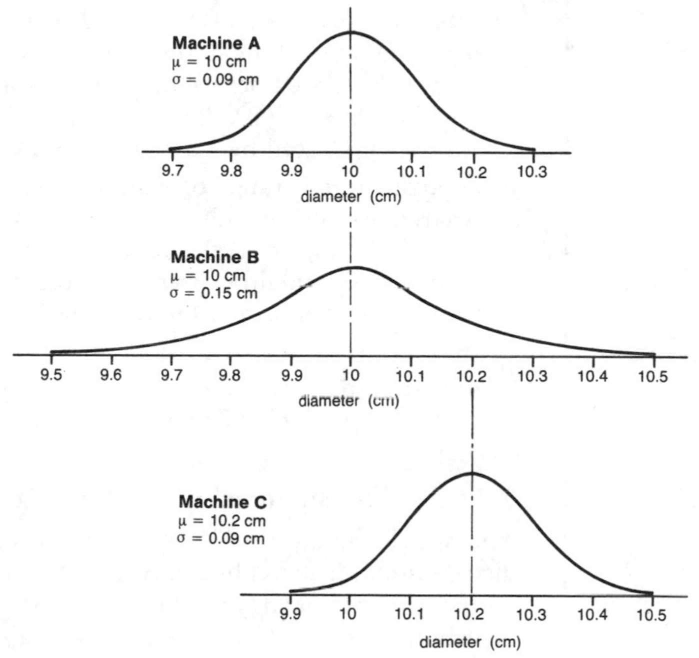

Fig. 7.6 Distributions of diameters (cm) of bearings produced by machines A, B and C

Machine B, in the same factory, produces the same bearings as machine A. However, the level of precision of machine B is not as good as that of machine A. That is, the variable (diameter) is spread wider around the mean: the standard deviation is 0.15 cm. Machine C also produces bearings, but these should have a diameter of 10.2 cm. The precision of machine C is equal to that of machine A ($s = 0.09$ cm), but the setting of the machine is different.

The resulting normal distributions are shown in fig. 7.6. Machine B has a greater dispersion, reflected in the value of its standard deviation. Machine C has a different centre (its mean is 10.2 cm), but the same dispersion as machine A.

7.11.1 Properties of the normal distribution

Examination questions will require you to use the normal distribution to solve particular problems, examples of which are given at the end of this chapter. You may also be required to answer questions which refer to particular properties of the normal distribution. Consequently before looking at typical problems, you need to become familiar with the general characteristics of the distribution: symmetry, continuity, etc.

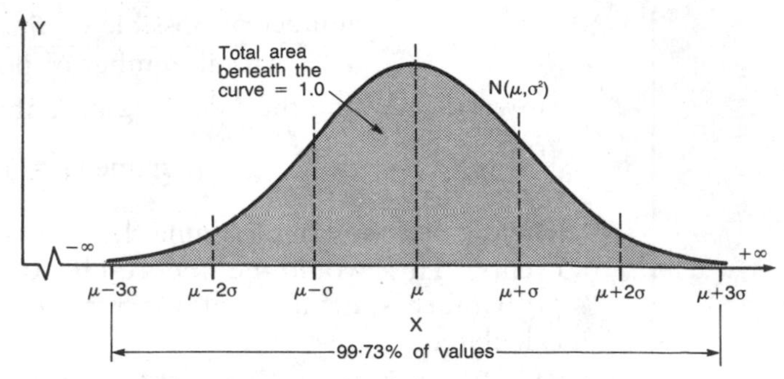

Fig. 7.7 Properties of the normal distribution

The following list summarizes the important properties of the normal distribution, which are also illustrated in fig. 7.7.

1 It is a symmetrical distribution.
2 It is the distribution of a continuous variable.
3 It has a mean, μ, and standard deviation, σ. These two parameters are sufficient to describe a particular example of the normal distribution.
4 The total area under the curve is unity (1.0), corresponding to the total probability.
5 The normal distribution with mean μ and variance σ^2 is referred to as $N(\mu,\sigma^2)$. For example, the mean and standard deviation of the diameter of bearings (machine A) are 10 cm and 0.09 cm, respectively, and the variance is therefore 0.0081 (0.09^2); we can therefore refer to the distribution of the diameters of bearings produced by machine A as N(10, 0.0081). Similarly, the distributions of the diameters of bearings produced by machines B and C are N(10, 0.0225) and N(10.2, 0.0081).
6 Though the full range of values which the normal distribution can take is from negative to positive infinity ($-\infty$ to $+\infty$), most of the possible values (99.73 per cent of values) lie in the range: the mean ± 3 standard deviations ($\mu \pm 3\sigma$). For example, one would expect 99.73 per cent of bearings produced by machine A to have a diameter within the range [9.73, 10.27]:

$$10.0 \pm (3 \times 0.09) = 10.0 \pm 0.27 = [9.73, 10.27]$$

Similarly, machines B and C have 99.73 per cent ranges of [9.55, 10.45] and [9.93, 10.47] respectively. Other important 'intervals' will be considered in the next chapter.

7.11.2 The standard normal distribution

You should be aware that there are an infinite number of examples of the normal distribution, identified by the values of their mean and standard deviation. However, unlike the binomial and Poisson distributions, when referring to the normal distribution it is not easy to use the available equations to calculate probabilities. Consequently statisticians have constructed tables of values for just this purpose. Obviously, it is impossible to tabulate probability values for every one of the infinite examples, but this does not pose a serious problem because it is possible to transform any normal distribution into the standard normal distribution. The values tabulated by statisticians are those of the standard normal distribution.

The standard normal distribution is the normal distribution with mean 0 and standard deviation 1, i.e. N(0,1). We will now consider an example showing its use.

A machine is set to produce components with diameter *d*. A quality control technician checks the components for error. He records the amount that the diameter of a particular component is in error: e.g. 0.0 would indicate 'no error', +1.5 mm would indicate 'diameter 1.5 mm too large' and −2.0 mm would indicate 'diameter 2.0 mm too small'. Suppose these error recordings are known to have a normal distribution with mean value $\mu = 0.0$ mm and standard deviation $\sigma = 1.0$ mm. The components are acceptable as long as the error lies between −2.7 and +2.5. We wish to know the probability that a component chosen at random from this distribution is unacceptable. Figure 7.8 illustrates the situation. The table of values for the area under the curve (which represents appropriate probabilities) is reproduced in full in appendix 4. Table 7.9 shows a part of that table. The following section explains how, in general, the normal distribution tables are used and how they can be used in this particular case.

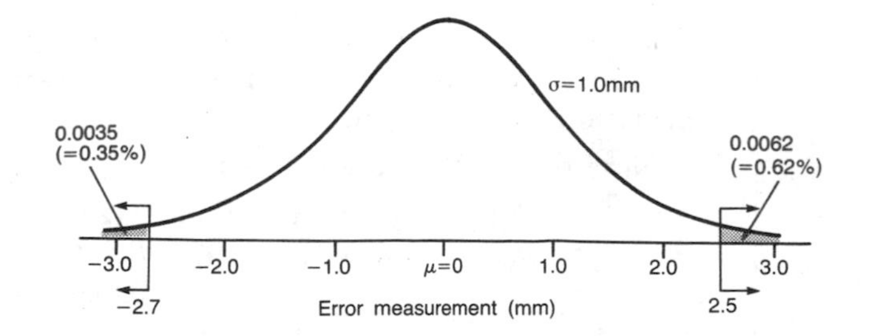

Fig. 7.8 Distribution of component diameter error (mm; mean error 0 mm, standard deviation 1.0 mm)

Table 7.9 A section of the table of values of the area under the standard normal distribution curve (probability values)

$z=\frac{x-\mu}{\sigma}$

σ=1
1−φ(z)
μ=0
z

$z=\frac{(x-\mu)}{\sigma}$	.00	.01	.02	.03	.04	.05	.06	.07	.08	.09
0.0	.5000	.4960	.4920	.4880	.4840	.4801	.4761	.4721	.4681	.4641
⋮										
1.0	.1587	.1562	.1539	.1515	.1492	.1469	.1446	.1423	.1401	.1379
⋮										
2.5	.0062	.0060	.0059	.0057	.0055	.0054	.0052	.0051	.0049	.0048
⋮										
2.7	.0035	.0034	.0033	.0032	.0031	.0030	.0029	.0028	.0027	.0026
⋮										
3.0	.00135									
⋮										

7.11.2.1 Standard normal distribution tables

Referring to table 7.9, 'z' represents the standard normal variable, that is, the values on the horizontal (x) axis. The values in the body of the table represent corresponding probability values. For example, the circled value, 0.1515, is the probability of z being greater than 1.03 (see fig. 7.9).

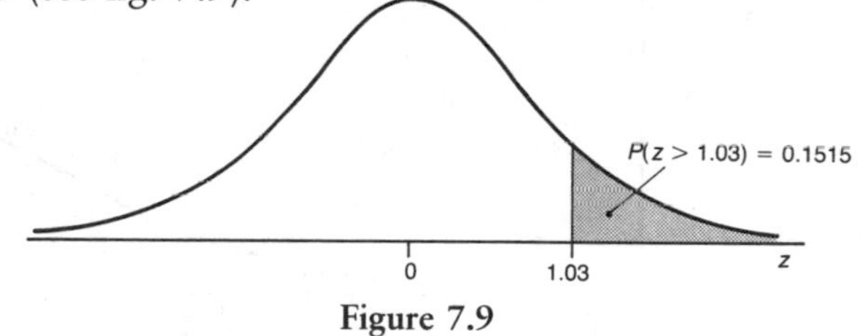

Figure 7.9

Similarly $P(z > 2.5) = 0.0062$, i.e. there is a 0.62 per cent chance that a component will have a diameter which is 'too large' (the acceptable error range was [−2.7, +2.5]; see fig. 7.8).

Because the distribution is symmetrical, $P(z > a)$ is the same as $P(z < -a)$. Using this, we are able to find the probability that a component will have a diameter more than 2.7 mm too small (the lower limit of our acceptable range):

$$P(z < -2.7) = P(z > 2.7) = 0.0035 \ (= 0.35\%)$$

We are now in a position to answer the question 'what is the chance that a component chosen at random from the production line is unacceptable, i.e. has diameter error outside the range [−2.7 mm, 2.5 mm]?' Since the outcomes '< −2.7' and '> 2.5' are mutually exclusive we can use the addition law:

$$\begin{aligned} P(z < -2.7 \text{ or } > 2.5) &= P(z < -2.7) + P(z > 2.5) \\ &= 0.0062 + 0.0035 \\ &= 0.0097 \end{aligned}$$

That is, there is slightly less than 1 per cent chance that a randomly-chosen component will be unacceptable.

We can now confirm one of the properties of the normal distribution given in section 7.11.1 – that 99.73 per cent of possible values will fall within three standard deviations of the mean:

Note that in this case $\mu = 0$ and $\sigma = 1.0$, so $\mu \pm 3\sigma = 0.0 \pm (3 \times 1.0) = 0 \pm 3$.

$$P(z > 3.0) = 0.00135 \text{ and } P(z < -3.0) = 0.00135$$
$$P(z < -3.0 \text{ or } > +3.0) = P(z < -3.0) + P(z > +3.0)$$
$$= 0.00135 + 0.00135 = 0.0027$$

Using the complementary law we can now state that

$$P(-3.0 < z < +3.0) = 1 - 0.0027 = 0.9973$$

This is read as 'the probability that the variable z lies between minus 3 and plus 3 is equal to 0.9973', i.e. 99.73 per cent of the values will lie within three standard deviations of the mean.

A few words of advice on using the standard normal tables:

1 Find out which tables you will have to use in your examination and obtain a copy. Not all tables are set out in the same way.

2 Sketch a picture of the 'area' you require. It will help you to see which probability you are trying to find.

3 Practise using the tables. Be careful you are referring to the correct tables – there are a number of similar ones in most texts.

7.11.3 Standardization of $N(\mu,\sigma^2)$

As we have just seen, tables exist for the standard normal distribution, but what of other normal distributions? It is a relatively straightforward process to transform a non-standard distribution, $N(\mu, \sigma^2)$, to the standard distribution, $N(0, 1)$, and use the tables as before.

Consider the following example: A student sits two examinations, one in English and the other in statistics. He obtains a mark of 40 in the English examination and 36 in the statistics examination. The marks for all students, in both papers, are believed to be distributed as normal distributions. The distribution details are: English $N(38, 2^2)$ and statistics $N(30, 4^2)$. Both papers had the same nominal maximum mark.

The statistics paper appears to have been more difficult – there is a lower mean and a greater dispersion of marks.

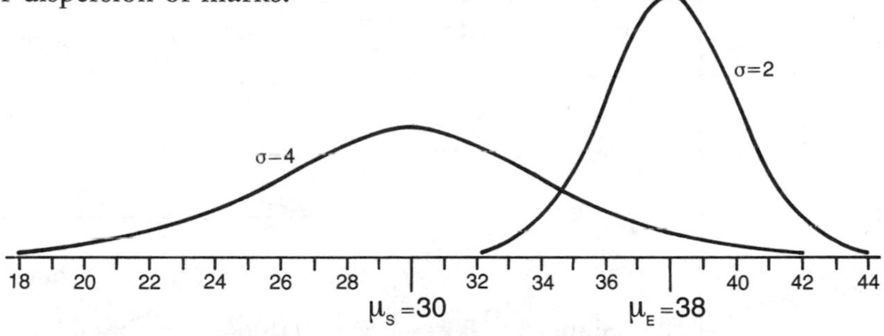

Fig. 7.10 Distributions of the marks gained in English, $N(38, 2^2)$, and statistics, $N(30, 4^2)$

We wish to know the answers to the following questions:

1 What is the probability of a student scoring more than 41 marks in the English examination?

2 What is the probability of a student scoring more than 41 marks in the statistics examination?

3 In which of the two examinations did the above-mentioned student perform better?

The distribution curves for the two examination papers are shown in fig. 7.10. The transformation for the English examination marks distribution [from N(38,4) to N(0,1)] is shown pictorially in fig. 7.11.

Symbolically this transformation is $\quad z = \dfrac{X + \mu}{\sigma}$

where X is the variable under consideration ('marks obtained in English'), μ is the population mean, σ is the population standard deviation and z is the standard normal variable (or the number of standard deviations above or below the mean). It is the

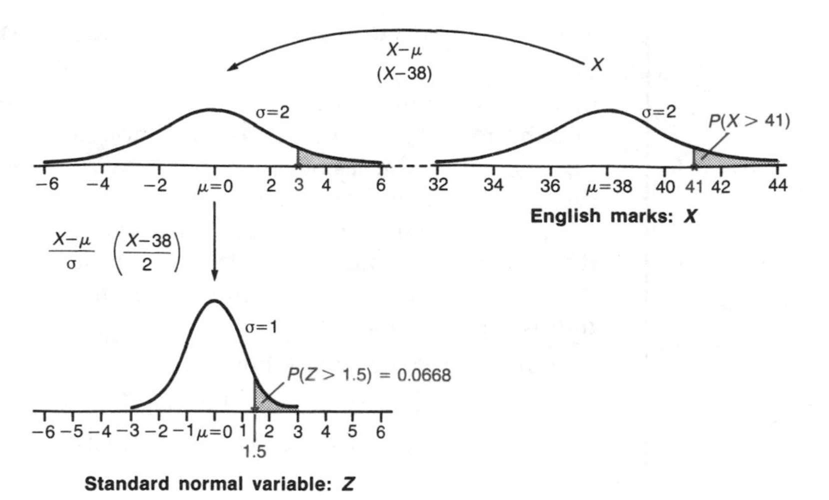

Fig. 7.11 Full transformation of $X \sim N(38,2^2)$ into $z \sim N(0,1)$: $z = \dfrac{X+\mu}{\sigma}$

standard normal variable, z, that you use to consult the tables of probabilities. You should be able to confirm, either from fig. 7.11 or using formula given above, that $P(X > 41)$ is equivalent to $P(z > 1.5)$, and find that the corresponding probability is 0.0668 by reference to the table of normal probabilities (appendix 4).

We have essentially answered question **1**. Formally the solution would be written as follows:

$$\begin{aligned}&P(\text{a student obtained more than 41 marks in English})\\&= P(X > 41)\\&= P\left(z > \frac{41-38}{2}\right) = P(z > 1.5)\\&= 0.0668 \text{ (from tables)}\end{aligned}$$

That is, just under 7 per cent of students would be expected to have scored more than 41 marks in this English examination.

The answer to question **2** is similarly found:

$$P(\text{statistics mark} > 41) = P\left(z > \frac{41-30}{4}\right) = P(z > 2.75) - 0.00298$$

That is, just under 0.3 per cent of candidates would have been expected to obtain a mark of 42 or better in this statistics examination.

To rate the performance (question **3**), we need to compare how much above average his/her two marks are. In the English examination, the student's mark of 40 is 1 standard deviation above the mean $[(40-38)2 = 1$ s.d.]; in the statistics examination, the student's mark of 36 is 1.5 standard deviations above the mean $[(36-0)/4 = 1.5$ s.d.]. Consequently the mark obtained in statistics was a relatively better result, even though it involved a lower number of marks. This comparison is shown diagrammatically in fig. 7.12.

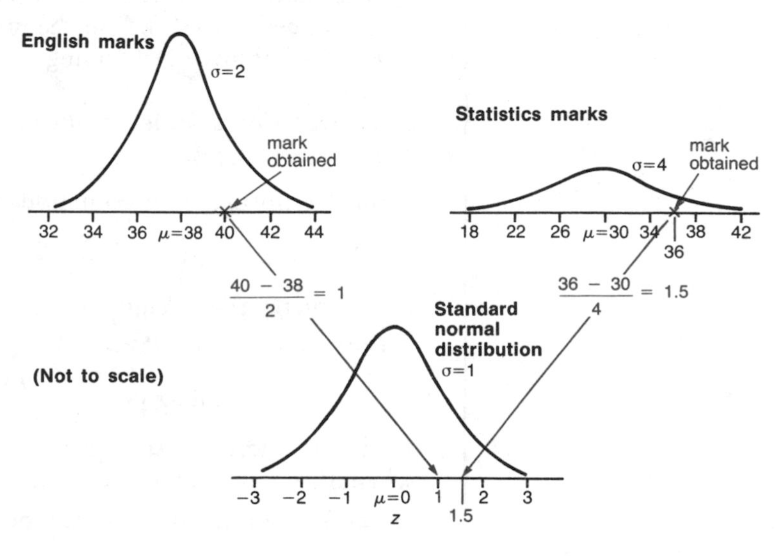

Fig. 7.12

7.11.4 Normal approximation to the binomial and Poisson distributions

Other aspects of the normal distribution will be considered in the next chapter, but before completing this chapter we will look at an important application of the normal distribution – approximation of the binomial and Poisson distributions. This will not figure in all syllabuses but is nevertheless worth being aware of if you plan to pursue the study of quantitative analysis.

Even though the binomial and Poisson distributions represent discrete variables and the normal distribution represents continuous variables, it is possible to use the normal distribution to approximate the binomial and Poisson distributions, under the following conditions:

1 n must be larger than (about) 20

2 p must not be too close to 0 or 1. That is, the nearer p is to 0.5, the better.

3 For values of p close to 0 or 1, n must be larger to remove the skewness of the binomial distributions. (Remember that the normal distribution is always symmetrical.)

Once again, the best way to illustrate this is to consider an example: The probability of obtaining more than 44 heads if a coin is tossed 100 times would prove tedious to calculate using the binomial formula. The Poisson distribution cannot be used as an approximation. Can you see why not? (See section 7.10.2.)

The normal approximation is easy to apply:

$$\mu = np = 100 \times 0.5 = 50$$
$$\sigma^2 = npq = 100 \times 0.5 \times 0.5 = 25$$
$$\text{therefore} \quad \sigma = \sqrt{25} = 5$$

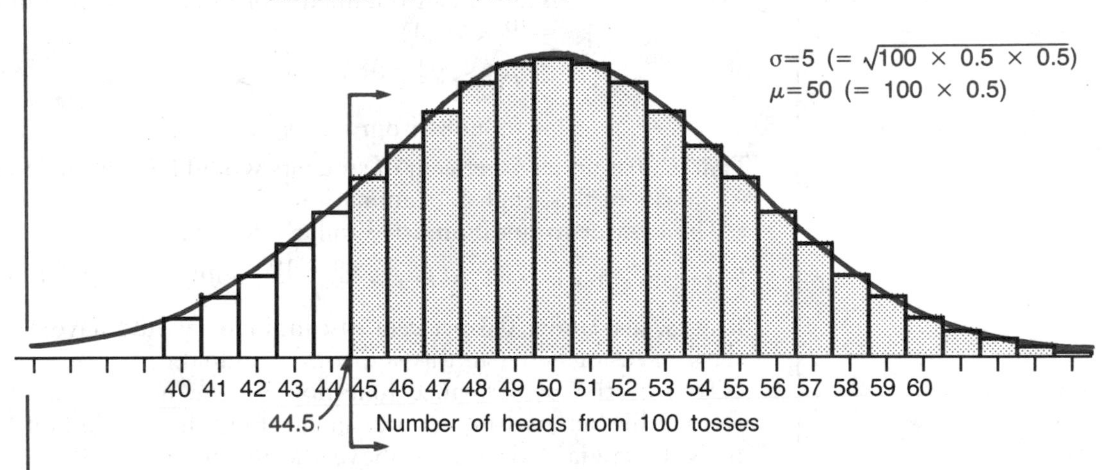

Fig. 7.13 Normal approximation to the binomial distribution

Figure 7.13 illustrates the approximation. Because we are using a continuous variable to approximate a discrete variable, a correction is required to improve the approximation: the discrete variable x in the binomial distribution, becomes the interval $[x-0.5, x+0.5]$ in the normal distribution.

The probability of obtaining more than 44 heads from 100 tosses is calculated as follows:

Since each toss is an independent event

$$P(\text{H} > 44) = P(\text{H} = 45 \text{ or } 46 \text{ or } 47 \text{ or } \ldots \text{ or } 99 \text{ or } 100)$$

Using the normal approximation, then standardizing, this becomes

$$P(\text{X} > 44.5) = P\left(z > \frac{44.5 - 50}{5}\right) = P(z > -1.1)$$

By symmetry this is equal to

$$\begin{aligned} P(z < +1.1) &= 1 - P(z > 1.1) \\ &= 1 - 0.1357 \text{ (from tables)} \\ &= 0.8643 \end{aligned}$$

That is, there is just over an 86 per cent chance of obtaining more than 44 heads when 100 coins are tossed (or 1 coin is tossed 100 times).

The Poisson distribution can be similarly approximated.

Sample questions

1 Two regular six-sided dice are rolled. The probability that a six appears at least once is

A 1/36 B 10/36 C 11/36 D 1/3 E 1/2

CIMA; Foundation stage, Section B; Mathematics and statistics; May 1986.

Solution/hints

Die 1 / Die 2	1	2	3	4	5	6
1						*
2						*
3						*
4						*
5						*
6	*	*	*	*	*	*

* = '6 appears'

Number of possible * outcomes = 11

Total number of possible outcomes = 36

$P(\text{at least one six}) = \frac{11}{36}$ (Answer C)

2 A salesman makes four calls in a day. At each call there is a probability of 1/5 of making a sale. Assuming these probabilities to be independent, the probability the salesman will make at most one sale in the day is

A 1/625 B 16/625 C 256/625 D 4/5 E 512/625

CIMA; Foundation stage, Section B; Mathematics and statistics; May 1986.

Solution/hints

Each of the four calls ($n = 4$) is independent, with only two possible outcomes: a sale $\left(p = \frac{1}{5}\right)$ or no sale $\left(q = \frac{4}{5}\right)$. Using the additive law and the binomial distribution formula:

$$\begin{aligned} P(\text{at most one sale}) &= P(0 \text{ sales or } 1 \text{ sale}) \\ &= P(0 \text{ sales}) + P(1 \text{ sale}) \\ &= {}^4C_0(\tfrac{1}{5})^0\,(\tfrac{4}{5})^4 + {}^4C_1(\tfrac{1}{5})^1(\tfrac{4}{5})^3 \\ &= \left(1 \times 1 \times \frac{256}{625}\right) + \left(4 \times \frac{1}{5} \times \frac{64}{125}\right) \\ &= \frac{256}{625} + \frac{4 \times 64}{625} = \frac{512}{625} \quad \text{(Answer E)} \end{aligned}$$

3 From an analysis of past records, a manufacturer has discovered that 20 per cent of his jackets have imperfections in them. Two inspectors are to be employed to check the jackets before they leave the factory. Each inspector will have to check all jackets and grade them as 'OK' or 'reject' and will make his decision independently. The probability that an inspector will mis-classify a jacket is 0.1. The company has decided to grade as 'fail' any jacket which either inspector rejects and grade as 'pass' the remainder.

You are required to:

(**a**) draw a tree diagram to represent this situation;

(**b**) find the percentage of jackets which will be correctly classified by this inspection system;

(c) find the probability that a jacket is actually imperfect, given that it has been passed by the inspection system.

CIMA; Foundation stage, section B; Mathematics and statistics; May 1986.

Solution/hints

Part (a)

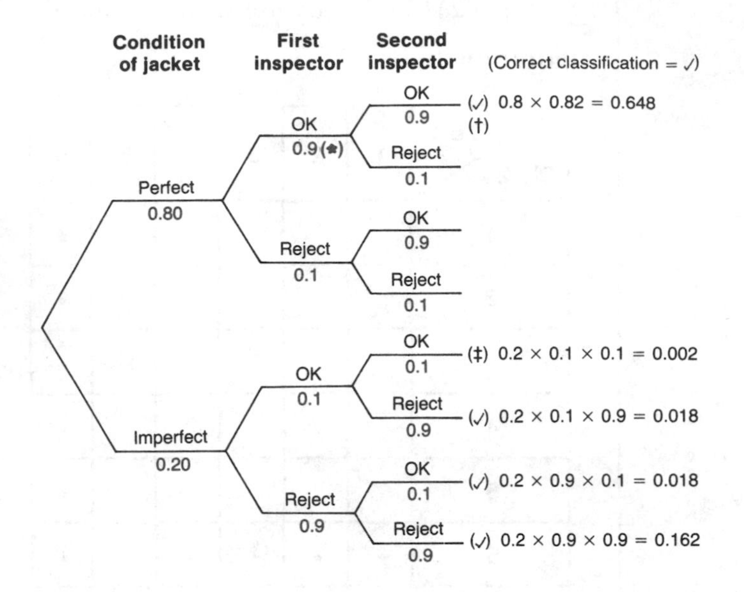

* indicates the probability that the first inspector grades a jacket 'OK' *given* it is OK.
i.e. $P(\text{OK}|\text{perfect}) = 0.9$

† indicates the probability that a perfect jacket is classified OK by both inspectors and is therefore correctly graded as 'pass' by the company.

$$\begin{aligned} \text{i.e. } \dagger = P(\text{per and OK}) &= P(\text{per}) \times P(\text{OK}|\text{per}) \\ &= 0.8 \times (0.9 \times 0.9) \\ &= 0.648 \end{aligned}$$

Part (b) Correct classification (indicated by a tick (✓) on the tree diagram) occurs in either of the following situations:

1 Jacket in perfect condition ('per') and *both* inspectors grade it 'OK' († on the tree diagram).

2 Jacket imperfect ('imp') and *either* inspector grades it as 'reject' ('rej')

$$\begin{aligned} \text{i.e. } P(\text{correct classification}) &= P[(\text{per } \textit{and} \text{ OK}) \textit{ or } (\text{imp } \textit{and} \text{ rej})] \\ &= P(\text{per and OK}) + P(\text{imp and rej}) \\ &= P(\text{per}) \times P(\text{OK}|\text{per}) + P(\text{imp}) \times P(\text{rej}|\text{imp}) \\ &= 0.648 + (0.018 + 0.018 + 0.162) \\ &= 0.846 \end{aligned}$$

Part (c) We require the probability of the event 'imperfect *given* passed by the inspection system' i.e. $P(\text{imp}|\text{OK})$.

Using Bayes' theorem, $P(\text{imp}|\text{OK}) = P(\text{imp } \textit{and} \text{ OK})/P(\text{OK})$

$P(\text{imp } \textit{and} \text{ OK}) = 0.002$ (‡ on tree diagram)

$$\begin{aligned} P(\text{OK}) &= 0.002 + 0.648 \quad (\ddagger + \dagger) \\ &= 0.65 \end{aligned}$$

Therefore $P(\text{imp}|\text{OK}) = \dfrac{0.002}{0.65} = 0.003$ (3 d.p.).

4 (a) Applicants for a certain job are given an aptitude test. Past experience shows that the scores from the test are normally distributed with a mean of 60 points and a standard deviation of 12 points. What percentage of candidates would be expected to pass the test, if a minimum score of 75 is required?

(b) The undercarriage of a Ruritanian aeroplane includes four wheels, two under the left wing and two under the right wing. This system functions perfectly provided at least one tyre on each side remains operational, otherwise the pilot has to use a back-up procedure. The independent probability of any tyre bursting on take-off is 0.01.

You are required to:

(i) draw a tree diagram for this problem;

(ii) find the probability that the pilot has to use the back-up procedure on take-off.

CIMA; Foundation stage, Section B; Mathematics and statistics; November 1986.

Solution/hints

Part (a)

You are required to find $P(X > 75)$. This is a straightforward exercise in standardizing a variable, which we have been told is normally distributed, and using the standard normal tables to calculate the appropriate probability. Note that you are required to present the answer in the form of a percentage.

Answer: 10.56 per cent

Part (b)

(i) There is often no unique solution methodology to a problem. The following tree diagram is the one you are likely to get on your first attempt.

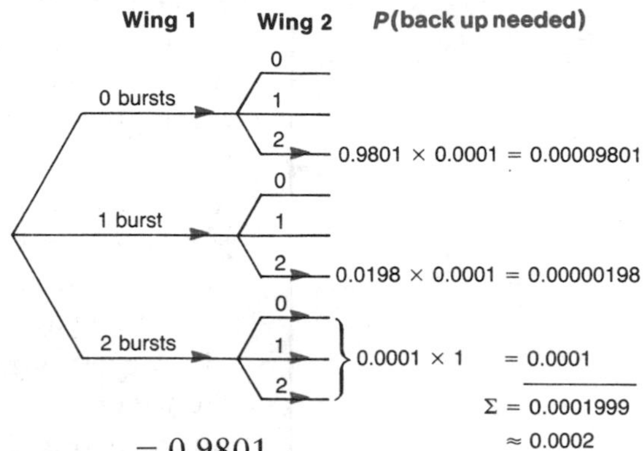

Using the binomial theorem:

$P(0 \text{ burst on 1 wing}) = 0.99^2 = 0.9801$
$P(1 \text{ (out of 2) burst on 1 wing}) = 2 \times 0.01 \times 0.99 = 0.0198$
$P(2 \text{ (i.e. both) burst on 1 wing}) = 0.01^2 = 0.0001$
$\Sigma = 1.0$

As an exercise you should complete the tree diagram and check that you can get the following answers:

(ii) $P(\text{back-up procedure is necessary}) = P(\text{both tyres on at least one wing burst})$

That is:

	wing 1		wing 2		
	wheel 1	wheel 2	wheel 3	wheel 4	Probability
$P(1)$:	burst	burst	both operational		$(0.01)^2 \times (0.99)^2$
$+P(2)$:	burst	burst	burst	operational	$(0.01)^2 \times (0.1 \times 0.99)$
$+P(3)$:	burst	burst	burst	burst	etc.
$+P(4)$:	burst	operational	burst	burst	
$+P(5)$:	both operational		burst	burst	

Answer: 0.0002

It is worth spending some time thinking about a problem before trying to produce your answer. Consider the following alternative tree diagram:

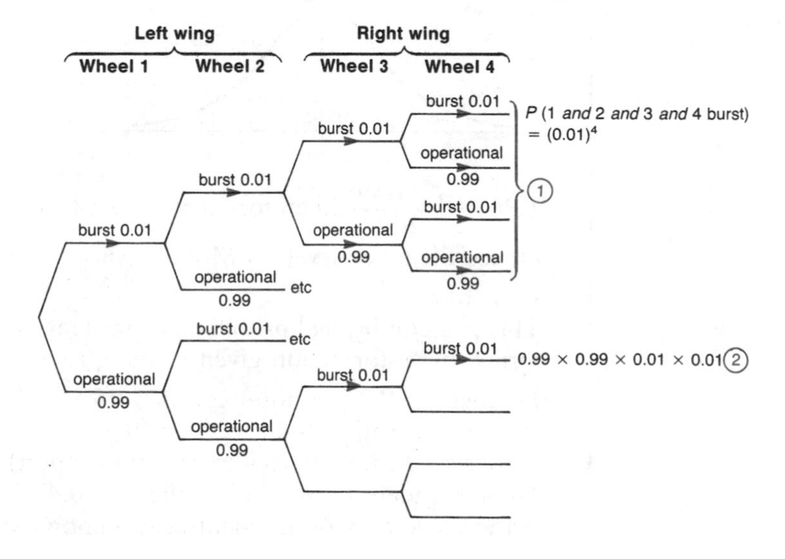

Back-up will be needed in the events marked by arrowheads:

$P(\text{Back-up needed}) = ① + ②$
$= \{(0.01)^2 \times [0.01^2 + (2 \times 0.01 \times 0.99) + 0.99^2]\} + (0.99^2 \times 0.01^2)$
$= (0.01^2 \times 1) + 0.000098$
$= 0.0001999 \approx 0.0002$

5 (a) The weekly sales of a product are found to be normally distributed with mean 1000 and standard deviation 200.

(i) Find the probability that weekly sales are:
  1 less than 700;
  2 between 800 and 1250;
  3 more than 1400.

(ii) Each Monday morning deliveries of the product are received from the factory. Assuming that there is one delivery per week, find the stock level required on Monday if the probability of running out of stock at the end of the week is to be less than 0.025.

(b) A wine exporter holds a wine-tasting evening to promote sales of wine. Of the guests, 25 per cent are known to be wine-tasting experts. If an expert is given a glass of wine, there is an 0.8 chance that he will identify the wine correctly; the corresponding chance for a non-expert is 0.4.

(i) A guest is given a glass of wine: what is the probability he will correctly identify the wine?

(ii) A guest is given three wines to try and he correctly identifies the three wines; what is the probability that he is an expert?

CIS; Quantitative studies; December 1986.

Solution/hints

Part (a)

Another straightforward standardization problem:

(i) (1) *Answer:* 0.0668

(2) $P(800 < X < 1250) = P\left(\frac{800-1000}{200} < z < \frac{1250-1000}{200}\right)$
$= P(-1 < z < 1.25)$
$= P(z < 1.25) - P(z < -1)$
$= 0.8944 - 0.1587 = 0.7357$

(3) *Answer:* 0.02275

(ii) We need to find the level of sales (a) such that the probability of exceeding that level is 0.025. From standard normal distribution tables, $P(z) = 0.025$ when $z = 1.96$.

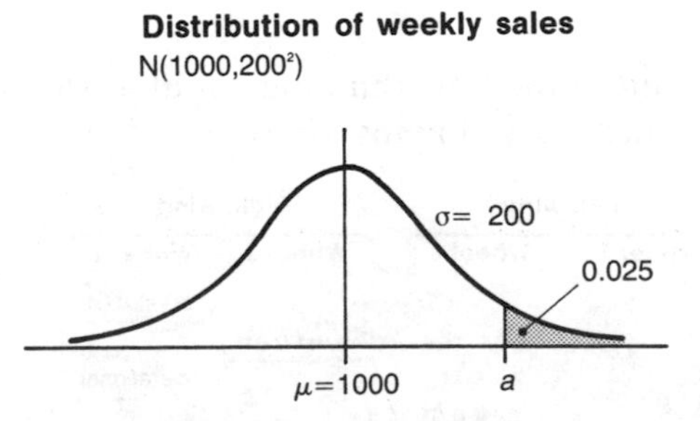

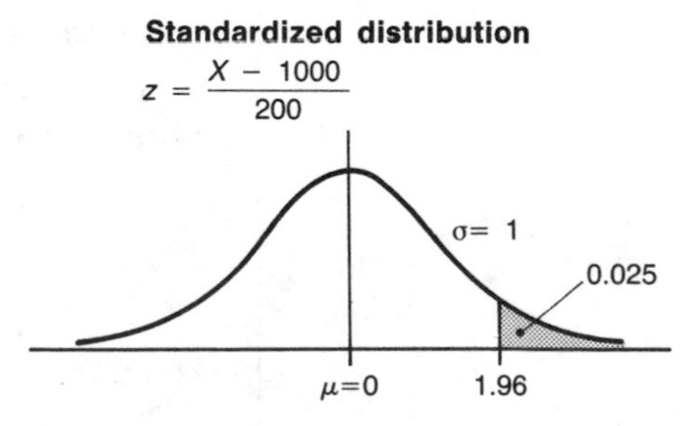

$1.96 = \frac{a - 1000}{200}$, therefore $a = (200 \times 1.96) + 1000 = 1392$.

Hence the stock level on Monday should be not less than 1392.

Part (b)

This is a conditional probability problem. It is useful to summarize, in mathematical terms, the information given in the question:

$P(\text{expert}) = 0.25$, $P(\text{non-expert}) = 0.75$
$P(\text{correct identification}|\text{expert}) = 0.8$
  therefore $P(\text{incorrect identification}|\text{expert}) = 0.2$
$P(\text{correct identification}|\text{non-expert}) = 0.4$
  therefore $P(\text{incorrect identification}|\text{non-expert}) = 0.6$

(i) Require P(a guest correctly identifies a given wine)
$= P(\text{correct identification}|\text{expert}) \times P(\text{expert})]$
$+ [P(\text{correct identification}|\text{non-expert}) \times P(\text{non-expert})]$
$= (0.8 \times 0.25) + (0.4 \times 0.75) = 0.5$

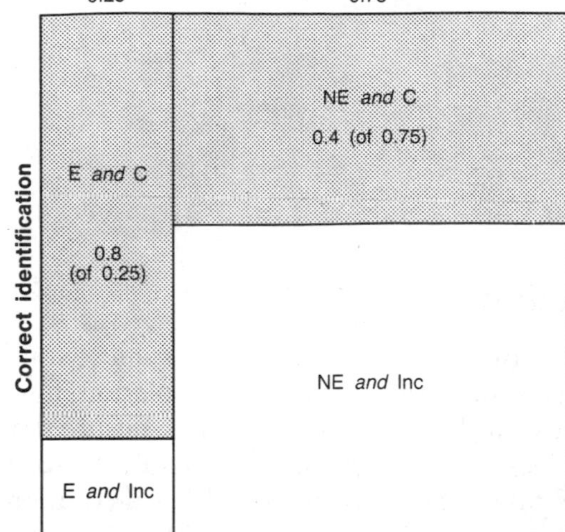

(ii) We require P(expert|3 wines out of 3 correctly identified). This question appears quite tricky but is really a fairly straightforward application of Bayes' theorem. We can quite easily find the reverse of the required probability. This is quite common in these questions.
P(3 wines correctly identified|expert) $= (0.8)^3 = 0.512$

We assume each wine-tasting is independent. Using the above notation and '3C' for '3 wines correctly identified', then the following is true:

$$P(\text{E}|3\text{C}) = \frac{P(3\text{C}|\text{E}) \times P(\text{E})}{P(3\text{C})} = \frac{P(3\text{C}|\text{E}) \times P(\text{E})}{[P(3\text{C}|\text{E}) \times P(\text{E})] + [P(3\text{C}|\text{NE}) \times P(\text{NE})]}$$

$$= \frac{0.512 \times 0.25}{(0.512 \times 0.25) + (0.064 \times 0.75)}$$

$$= \frac{0.128}{0.176} = 0.727 \text{ (to 3 d.p.)}$$

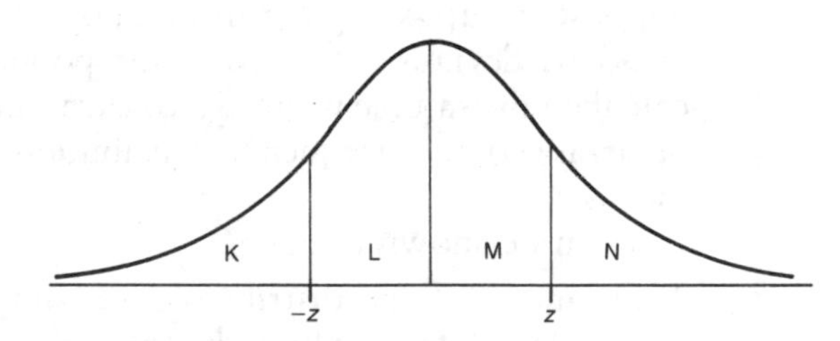

6 Areas K,L,M and N under the normal distribution above are equal. What is the value (to 2 decimal places) of $|Z|$? The diagram is not to scale.

A 0.10 **B** 0.25 **C** 0.67 **D** 0.90 **E** none of these

CIMA; Foundation stage, Section B; Mathematics and statistics; November 1986.

Solution: C

Recommended reading

Owen, F. and Jones, R., *Statistics* (Pitman, 1988), chapters 11–13.

Whitehead, P and Whitehead, G., *Statistics for Business* (Pitman, 1984), chapters 14 and 15.

Thomas, R., *Notes and Problems in Statistics* (Stanley Thornes, 1984), chapter 2.

Curwin, J. and Slater, R., *Quantitative Methods for Business Decisions* (Van Nostrand Reinhold, 1986), chapters 9–11.

Lucey, T., *Quantitative Techniques* (D.P. Publications, 1980), chapter 3.

8 Sampling Theory

8.1 Introduction

You will be aware, from chapter 2, that most statistical information is obtained by examining a sample taken from the population in which one is interested. This can be done because, as long as one is careful to ensure that the sample is representative, the characteristics of the sample will be broadly the same as those of the population from which it is taken. Further, one is often called upon to make decisions based on the sample information. In order to be able to make decisions based upon the sample information, inferences must be made about the population. Such inferences, whether true or not, are called statistical estimates.

In this chapter consideration will be given to:

1 the type of information one tends to seek in taking a sample;

2 ways of measuring the confidence one can have in sample information; and

3 the importance of the normal distribution in sampling theory.

These are the areas you are likely to encounter in examination questions at this level. We will also look at some basic hypothesis theory which, though not present in all syllabuses at this level, will provide you with the basis for more advanced work. We will look at the influence of the normal distribution first, as it is one of the factors central to sampling theory.

8.2 The Central Limit Theorem

Suppose we are sampling from a population which has known mean, μ, and known standard deviation, σ. From this population we can take samples of size n and calculate the sample mean, $\bar{x}$, for each sample. The values of the sample means, $\bar{x}$, can be arranged as a frequency distribution, called the **sampling distribution** of sample means.

It can be shown that:

1 the mean of the distribution of sample means is the same as the mean of the population from which the samples were taken, μ; and

2 the standard deviation of the distribution of sample means is equal to $\sigma/\sqrt{n}$. To avoid confusion with the standard deviation of the population, $\sigma/\sqrt{n}$ is called the **standard error.**

The type of distribution which results is of particular importance. Not only does the distribution of sample means have properties **1** and **2** above, but, if the sample size is reasonably large (30 or greater), the resulting sampling distribution is approximately normal. This is a property of the Central Limit Theorem.

If the population distribution is normal, then the distribution of sample means will be normal, regardless of sample size. Table 8.1 and fig. 8.1 illustrate an example. In this example the population distribution is certainly not normal; it is a rectangular or uniform distribution. The sampling distribution is also shown. It is evident that, with even such a small sample ($n = 3$), the distribution of the sample means is acquiring the familiar characteristics of the normal distribution. It can be confirmed that the mean and standard deviation do have the properties outlined in **1** and **2** above, that is $\mu = E(X) = \Sigma X \times P(X)$ and $\sigma^2 = \Sigma(X-\mu)^2 P(X)$. Similarly, for the sample mean distribution, $\mu = E(\bar{X}) = \Sigma \bar{X} \times P(\bar{X})$ and $\sigma^2/n = \Sigma(\bar{X}-\mu)^2 P(\bar{X})$.

Table 8.1 The probability distribution of (a) X, a uniform distribution with $\mu = 3.0$ and $\sigma^2 = 2.0$ ($\sigma = \sqrt{2}$) and (b) of $\bar{X}$, the sample mean distribution resulting from taking samples of size 3 from the population X.

(a) Population distribution

X	No. of ways of choosing X	Probability
1	1	0.2
2	1	0.2
3	1	0.2
4	1	0.2
5	1	0.2

$\mu = E(X) = 3.0;\ \sigma^2 = 2.0;\ \sigma = \sqrt{2} \approx 1.414$

(b) Sample mean distribution

Sample total	$\bar{X}$	No. of ways of choosing $\bar{X}$	Probability
3	1	1	0.008
4	1.33	3	0.024
5	1.67	6	0.048
6	2	10	0.08
7	2.33	15	0.12
8	2.67	18	0.144
9	3	19	0.152
10	3.33	18	0.144
11	3.67	15	0.12
12	4	10	0.08
13	4.33	6	0.048
14	4.67	3	0.024
15	5	1	0.008
Totals		125	1.0

$\mu = 3.0;\ \sigma^2 = \frac{2}{3};\ \sigma = \sqrt{\frac{2}{3}} \approx 0.816$

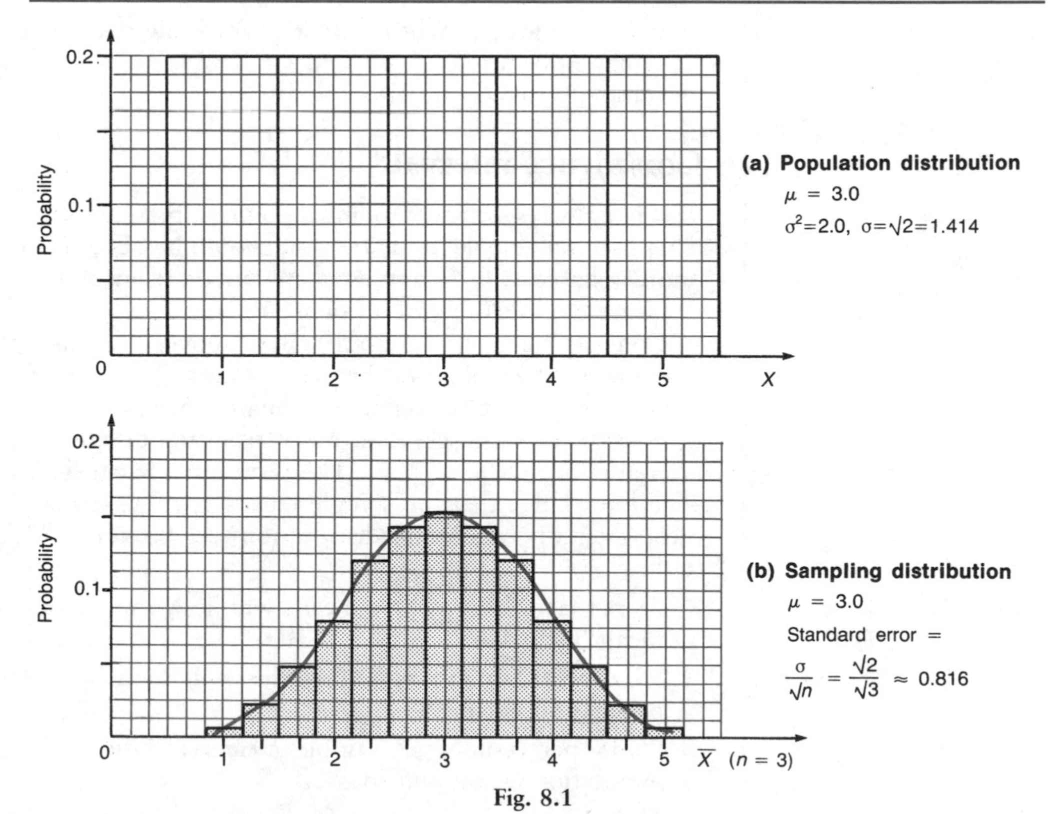

Fig. 8.1

8.2.1 Population and sample characteristics: summary

It can be confusing using the same vocabulary for both sample statistics and the population statistics which the sample is representing. The following symbols are used to distinguish between the sample and the population from which the sample is taken:

1 $\bar{x}$ is the sample mean, the best estimate of the population mean;

2 μ (mu) is the true population mean;

3 s is the sample standard deviation;

4 σ (sigma) is the population standard deviation;

5 s^2 is the sample variance;

6 σ^2 is the population variance.

7 The distribution of the sample means is very close to being a normal distribution (see fig. 8.1). This is true even if the population distribution is skewed. The larger the sample size (n), the better will be the fit of the sample mean distribution to a normal distribution.

8 μ is also the true mean of the sample mean distribution and can therefore be estimated by $\bar{x}$.

9 $\sigma/\sqrt{n}$ is the true 'standard deviation' of the sample mean distribution, called the standard error. If n is large, the standard error can be estimated using $s/\sqrt{n}$.

10 If the sample size is small ($n < 30$), $s \times \sqrt{n/(n-1)}$is a better estimate of the standard error (see section 8.6).

We will now consider how sampling theory can be applied to estimate the mean, variance and standard deviation of a population, when all that is available is a sample from that population.

8.3 Point estimates

Point estimates are single number estimates without confidence intervals. For example, $\bar{x}$, the sample mean, is a point estimate for μ, the population mean.

Consider the following example: Sweet-Tooth Co manufacture boiled sweets. The sweets are packed in 1 kilogram bags. A sample of 100 bags is taken from the production line and weighed. The resulting sample mean is $\bar{x} = 1.005$ kg and the sample standard deviation is $s = 0.02$ kg. $\bar{x} = 1.005$ is a point estimate of the population mean, μ, which is unknown. Similarly, $s^2 = 0.0004$ is a point estimate of the population variance, σ^2, and $s = 0.02$ is an estimate of the population standard deviation, σ, which are also unknown.

8.4 Confidence intervals

It is not always sufficient to find a point estimate; it is usually important to have some idea how reliable the estimate is. It should be obvious that the larger the sample, the more reliable will be any estimates made using that sample. Confidence intervals measure the reliability of a point estimate.

Consider again the example given above: The distribution of the population is unknown but, because of the Central Limit Theorem, the distribution of the sample means is known to be well-approximated by a normal distribution. The sample mean, $\bar{x} = 1.005$, is an observation taken from the distribution of all such sample means, therefore $\bar{X} \sim N(\mu, \sigma^2/100)$. The resulting normal distribution is shown in fig. 8.2. The knowledge that the sample means have a normal distribution and the general properties of the normal distribution (see section 7.11.2), allow us to predict the following:

1 68 per cent of all sample means will lie within 1 standard error of the population mean ($\mu \pm \sigma/10$);

2 95 per cent of all sample means will lie within 1.96 standard errors of the population mean;

3 95.45 per cent of all sample means will lie within 2 standard errors of the population mean; and so on.

These can be checked by referring to the standard normal probability tables.

In fig. 8.2 it is supposed that the sample mean we are considering ($\bar{x}$) is one of the 68 per cent we could expect to observe between the limits $\mu \pm \sigma/10$. Of more practical importance is the associated property that the true mean, μ, must, with 68 per cent certainty, lie between the limits $\bar{x} \pm \sigma/10$. We do not in this case know σ, so we have

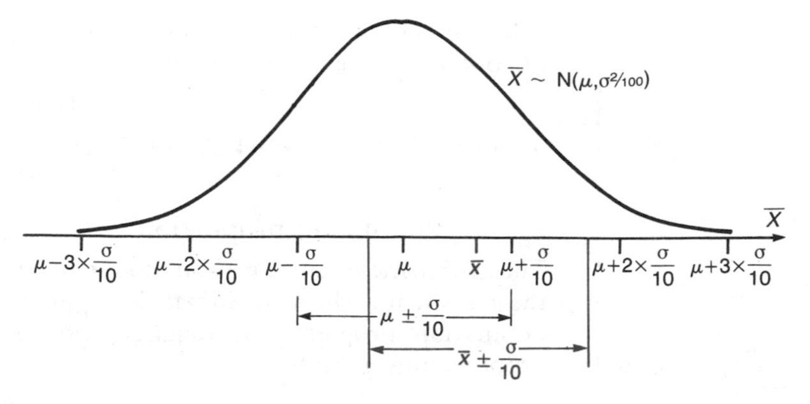

Fig. 8.2 The sampling distribution, $\bar{X} \sim N(\mu, \sigma^2/100)$

to use $s = 0.02$ to estimate it. Consequently, the standard error is estimated to be $0.02/\sqrt{100} = 0.002$. We are now in a position to make the following statement:

'We are 68 per cent certain that the population mean lies within the range $1.005 \pm 0.002 = [1.003, 1.007]$'.

The interval (or range) [1.003, 1.007] is called the **68 per cent confidence interval** for the population mean, μ. It is sometimes referred to as an **interval estimate**, that is, 1.005 ± 0.002 is said to be an interval estimate for the population mean μ.

We can find the 95 per cent confidence interval, similarly:

$$\begin{aligned} &1.005 \pm (1.96 \times 0.02/\sqrt{100}) \\ &= 1.005 \pm (1.96 \times 0.002) \\ &= 1.005 \pm 0.00392 = [1.00108,\ 1.00892] \end{aligned}$$

That is, we can say, with 95 per cent confidence, that the (true) population mean is in the range [1.00108, 1.00892]. It should be noted that there is consequently a 5 per cent chance that the true mean does not lie in the given interval: 2.5 per cent chance that it is less than 1.00108 and 2.5 per cent chance that it is more than 1.00892.

In general, we can be 95 per cent confident that the population mean, μ, lies within the interval $\bar{x} \pm 1.96$ s.e. (s.e. = standard error) and we can be 99 per cent confident that the population mean, μ, lies within the interval $\bar{x} \pm 2.5758$ s.e.

Two factors influence the degree of accuracy of the sample mean as an estimate of the population mean:

1 the standard deviation of the population – the smaller it is, the more accurate the estimate should be; and
2 the size of the sample – as n increases, the standard error ($\sigma/\sqrt{n}$) must decrease, resulting in narrower confidence intervals.

8.5 Sample size

The size of the standard deviation is a property of the population, which cannot be altered, but the sample size is often within the control of the investigator. But, how does he/she decide which sample size is appropriate?

Consider the following example: An engineering company knows from previous experience that the average number of components produced per employee per hour is 50 and the standard deviation is 10. They wish to find an estimate for the true mean (μ) to within 3 components and they want to be 95 per cent confident about this estimate. The sample mean, $\bar{x}$, will provide an estimate; their problem is to decide on the size of the sample which will ensure the required level of accuracy is achieved.

When a sample mean $\bar{x}$ is acquired, we know, with 95 per cent confidence, that the population mean will lie within the interval $\bar{x} \pm 1.96 \times$ s.e. In this case, $\bar{x}$ is required to be no more than 3 units from the true mean, μ, that is, $1.96 \times$ s.e. must not be greater than 3. It is therefore sufficient to find the value of n such that $1.96 \times 10/\sqrt{n} = 3$:

$$\sqrt{n} = \frac{1.96 \times 10}{3} \quad \text{therefore } n = \left(\frac{1.96 \times 10}{3}\right)^2 = (6.53)^2 = 42.68 \text{ (to 2 d.p.)}$$

A sample involving at least 43 employees will give the desired degree of accuracy.

In general, for $\bar{x}$ to be within r units of the population mean, at the 95 per cent level of confidence, we require sample size

$$n = [(1.96 \times \sigma)/r]^2$$

At the 99 per cent level of confidence, we would require

$$n = [(2.5758 \times \sigma)/r]^2$$

8.5.1 Population proportion

The arithmetic mean is, without doubt, a very important parameter and is commonly the statistic for which an estimate is required. However, there are situations where the population proportion is required, not the mean. The standard error for the proportion is given by

$$\text{s.e.} = \sqrt{pq/n}$$

where p is the proportion of times an event occurs;
q is the proportion of times an event does not occur $(1 - p)$; and
n is the sample size.

If r is the level of accuracy required then, at the 95 per cent level of confidence, the necessary sample size is

$$n = (1.96)^2 pq/r^2$$

For example, a random sample of 1000 potential voters were asked who they would vote for in the next general election. 400 replied that they would vote for the Labour Party. You are required to find an estimate for the proportion of voters in the entire population with the intention of voting for the Labour Party and the corresponding 95 per cent confidence interval.

The required estimate is the sample proportion:

$$\text{Sample proportion } (p) = \frac{400}{1000} = 0.4$$

$$\text{s.e.} = \sqrt{\frac{0.4 \times (1 - 0.4)}{1000}}$$

$$= \sqrt{0.0024} = 0.015 \text{ (3 d.p.)}$$

The corresponding 95 per cent confidence interval is

$$0.4 \pm (1.96 \times 0.015) = 0.4 \pm 0.0294 = [0.3706, 0.4294]$$

That is the percentage of voters whose intention it is to vote for the Labour Party is (approximately) between 37 and 43 per cent, at the 95 per cent level of confidence.

8.6 Estimating population variance σ^2: Bessel's correction factor

When estimating the population variance we have tended to use the sample variance, $s^2 = \frac{\Sigma(X - \bar{X})^2}{n}$. However, it can be shown that the best unbiased estimate for the population variance is given by $s^2 \times n/(n - 1)$, i.e. to acquire the best estimate for the population variance, the sample variance is adjusted by a factor of $n/(n - 1)$, known as **Bessel's factor.**

If the sample under consideration is large, there will be little difference between the estimates. For example, suppose a sample variance for a sample of size 100 is $s^2 = 25$. With the Bessel correction this becomes

$$s^2 \times \frac{100}{99} = 25 \times \frac{100}{99} = 25.25 \text{ (2 d.p.)}$$

The estimates for the standard deviations, s, are 5 and 5.025, respectively.

In general, if the sample is small ($n < 30$), Bessel's correction should be used. If in doubt, use Bessel's correction factor – that way you can never be wrong.

The remainder of this chapter will focus briefly on various aspects of hypothesis testing. The extent to which students are expected to be familiar with this topic area depends on the examining body under which they are studying. At foundation level, some examiners expect nothing more than knowledge of confidence limits, while others expect familiarity with t-tests and the Chi-squared (χ^2) distribution and their use in testing hypotheses. Check your appropriate syllabus.

The topic coverage in this text is not intended to be complete. It is merely an introduction to some of the techniques with which you might be expected to be acquainted.

8.7 Hypothesis testing: testing a sample mean

Consider the following problem, which is typical of the type of problem a production manager might face:

A machine packages food items. The nominal mean weight of the packages is 1 kg, with standard deviation 0.005 kg. A package drawn at random from the production line weighs 1.016 kg. Is it reasonable to assume the machine is malfunctioning?

Certain factors need to be considered before a conclusion can be reached:

1 The characteristics of the population: The weights of the packages produced by the machine constitute the population and in this case it would be reasonable to assume that the population has a normal distribution with mean $\mu = 1.0$ kg and standard deviation $\sigma = 0.005$.

2 Having established the characteristics of the population, one has to ask the question, 'Is the sample value a reasonable one?'. In this case the sample (of size 1) is more than three standard deviations from the mean. This can be confirmed by standardizing the sample value:

$$z = \frac{x - \mu}{\sigma} = \frac{1.016 - 1.0}{0.005} = 3.2 \text{ (the critical ratio)}$$

By referring to the standard normal distribution tables, it can be seen that the probability of getting a value as high as 1.016 kg is 0.00069 or 0.069 per cent.

3 Finally, a decision has to be made. If the machine is indeed working correctly, then this occurrence is very rare. The production manager might therefore decide that the machine is in fact faulty.

A similar decision would have been reached if the sample value had been 0.984 kg, which is also 3.2 standard deviations from the mean (less than the mean). Therefore the total probability of obtaining a value so far from the mean, if the population values are correct, is 0.138 per cent (0.069×2), a very rare occurrence.

Steps **1**, **2** and **3** above give a rough outline of hypothesis (or significance) testing.

Consider the production manager's problem again. Questions which you are likely to encounter will involve sample sizes of more than 1; we will consider a sample size of 25. The sample mean weight ($\bar{x}$) is equal to 1.0025 kg. We will carry out a formal hypothesis test focusing on each step.

Step 1: Hypothesis

The problem must be clearly defined in the form of a **null hypothesis** and an **alternative hypothesis.** Without these, it is not possible to carry out a test.

In this case, the problem concerns the functioning of a machine – is it malfunctioning or is it not?

The null hypothesis must assume characteristics of the population which can be tested. In this case, we assume that the machine is not faulty and therefore μ, the population mean weight, is 1.0 kg.

The alternative hypothesis is that the machine is faulty and therefore μ, the population mean weight, is not 1.0 kg. This is set out as follows:

Null hypothesis H_0: $\mu = 1.0$ kg
Alternative hypothesis H_1: $\mu \neq 1.0$ kg

This is called a **two-tailed test.**

If the suspicion is that the machine is not just faulty but, more specifically, faulty and overfilling the packages, then we would have:

Alternative hypothesis H_1: $\mu > 1.0$ kg

This is called a **one-tailed test.**

Step 2: Population characteristics

Characteristics of the population distribution are required. In this case, the weights of the packages produced by the machine are assumed to have a normal distribution with mean 1.0 kg, and standard deviation 0.005 kg.

Step 3: Sample distribution

It is important to emphasize that the sample must be a representative sample, chosen randomly. It is not possible to arrive at an acceptable conclusion if the sample is biased in any way.

In this case, the sample (or test) statistic to be considered is a sample mean, so we need to have information about the distribution of sample means. We have assumed that the population distribution of weights produced by the machine is $X \sim N(1.0, 0.005^2)$. Consequently, the distribution of sample means will be $\bar{X} \sim N(1.0, 0.001^2)$; 0.001 is the standard error $(\sigma/\sqrt{n})$.

We are stating that, as long as the null hypothesis is true, the distribution of sample means is a normal distribution with mean 1.0 kg and standard error 0.001 kg.

Step 4: Level of significance

In the example considered at the start of this section, we found that the probability of observing a value which is more than 3.2 standard deviations from the mean is 0.00138 or 0.138 per cent. The conclusion was that the occurrence of such a value was so rare as to bring into question the initial assumption, which was that the mean value of weights was 1 kg.

In general, the decision to accept or reject the null hypothesis depends on the probability that the sample statistics could have occurred purely by chance if the null hypothesis were indeed true. It is an important part of hypothesis testing to decide at what point a null hypothesis should be rejected. That is, how low should the probability be? The criterion for rejection is called the **level of significance** or **significance level** of the test. 5 per cent is the most commonly used significance level, but 1 per cent or lower might be demanded if the consequences of the test are costly or particularly crucial (tests in medical research or weaponry, for example).

In this example a 5 per cent significance level will be used, that is, if the probability of obtaining the sample mean ($\bar{x} = 1.0025$) (given the assumed population mean of 1 kg) is 5 per cent or less, the null hypothesis will be rejected in favour of the alternative hypothesis (see Step 5).

Step 5: Deciding on the type of test

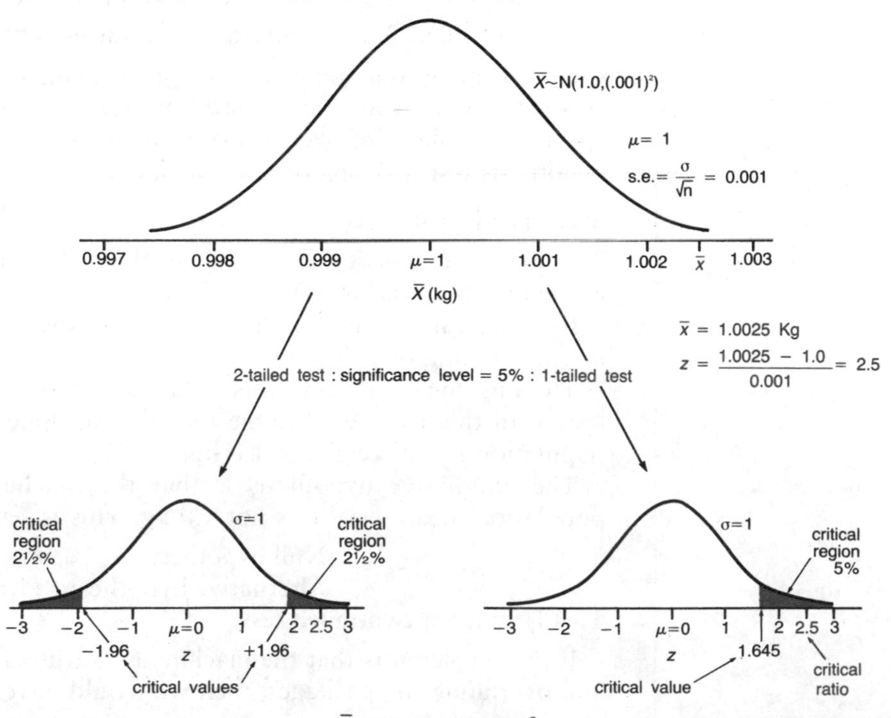

Fig. 8.3 Distribution of sample means $\bar{X} \sim (N(1.0, 0.001^2)$ and standardized distributions showing the critical regions for one- and two-tailed tests at the 5 per cent significance level

The suspicion is that the machine is malfunctioning, that is, it is overfilling or underfilling the bags. This calls for a two-tailed test, as defined above. At the 5 per cent significance level, the critical region (or rejection region) will be in two parts: $z < -1.96$ and $z > +1.96$.

If, before a sample was taken, the production manager suspected the machine to be overfilling then the one-tail test would be appropriate and the critical region would be $z > 1.645$.

Such detail is not necessary every time a test is carried out, as will now be seen.

8.7.1 Hypothesis testing: summary

1 State the problem: Is the machine malfunctioning?

2 State the problem as a hypothesis: (Null hypothesis) H_0: $\mu = 1.0$ kg
(Alternative hypothesis) H_1: $\mu \neq 1.0$ kg

3 Describe the population distribution: $N(1.0, 0.005^2)$

4 Describe the sample distribution: $N(1.0, 0.001^2)$

5 Decide on significance level: 5 per cent

6 Find critical values using normal distribution tables (appendix 4): for 5 per cent significance level, two-tailed test, ± 1.96

7 Find the critical ratio by standardizing the sample statistic:

$$z = \frac{\bar{x} - \mu}{\sigma/\sqrt{n}} = \frac{1.0025 - 1.0}{0.005/\sqrt{25}} = \frac{0.0025}{0.001} = 2.5$$

8 The critical ratio is seen to be in the critical (rejection) region (see fig. 8.3), that is, the sample has a significant value; it is said to be significantly different from that which would be acceptable under the null hypothesis. Consequently the null hypothesis is rejected at the 5 per cent level of significance. However, it would have been accepted at the 1 per cent level. Can you see why?

For a two-tailed test:

5 per cent of the area lies outside the range -1.96 to $+1.96$ (as used in the example; see fig. 8.3);
1 per cent of the area lies outside the range -2.58 to $+2.58$;
0.1 per cent lies outside the range -3.29 to $+3.29$.

For a one-tailed test:

5 per cent of the area lies outside the range $-\infty$ to 1.645 or -1.645 to $+\infty$ (see fig. 8.3);
1 per cent of the area lies outside the range $-\infty$ to 2.326 or -2.326 to $+\infty$;
0.1 per cent of the area lies outside the range $-\infty$ to 3.09 or -3.09 to $+\infty$.

8.7.2 Decision methodology

As noted, in the above example, though the null hypothesis was rejected at the 5 per cent significance level, it would have been accepted at the 1 per cent significance level.

The following is a general guide to decision-making:

Result of test	Decision
Significant at 5% level	Reasonable to reject the null hypothesis (95% confidence)
Significant at 1% level	Confidently reject the null hypothesis (99% confidence)
Significant at 0.1% level	Reject the null hypothesis with a very high degree of confidence (99.9% confidence)
Sample statistic not significant	Accept the null hypothesis
Sample statistic not significant but close to 5 per cent level	Difficult to make a decision: further sampling required

8.8 Sampling error

One can never be certain that the decision taken is the correct one. In rejecting the null hypothesis one could be making a mistake – the sample could have arisen by chance, even though it is highly unlikely. This is known as **Type I error**.

Similarly, one could be making a mistake in accepting the null hypothesis – the sample distribution mean could in fact be larger (or smaller) than the assumed value and the sample still occur within the acceptance region. This is known as **Type II error.**

The following is a summary of the possible types of error:

Null hypothesis	Accept hypothesis	Reject hypothesis
true	correct decision	incorrect decision = type I error
false	incorrect decision = type II error	correct decision

The probability of making a Type I error is the level of significance of the test, therefore the lower the significance level, the lower the probability of making a Type I error. Assessing the probability of making a Type II error (accepting H_0 when it is in fact false) is beyond the scope of this book.

Statistical testing can **never** be certain, it can only infer something, with various degrees of confidence.

The remainder of this chapter will look at a distribution used for hypothesis testing with small samples.

8.9 Small samples: *t*-distributions

The examples considered so far have either had a population distribution which we could confidently assume was normal and/or the sample size has been large, which allowed us to assume the sample distribution was normal. We were able, for example, to find a 95 per cent confidence interval for μ as follows:

$$\bar{x} \pm 1.96 \times s/\sqrt{n}$$

$$\text{or } \bar{x} \pm 1.96 \times s/\sqrt{(n-1)} \text{ (using Bessel's correction)}$$

1.96 is the standard normal variate such that $P(z < -1.96 \text{ or } > 1.96) = 0.05$.

If the sample is small ($n < 30$), one cannot assume that the sample distribution is normal. Estimating the standard error results in increased uncertainty. Consequently the variability or spread of the sample distribution is increased and to maintain a 95 per cent confidence level, 1.96 needs to be replaced by a larger constant; how much larger depends on the sample size. The appropriate value can be found in the table of *t*-distribution values. There is a different *t*-distribution for each sample size with the *t*-distribution is tabulated against **degrees of freedom.** It is convenient to accept that the degrees of freedom are equal to the denominator in Bessel's correction factor, i.e. $n-1$; e.g. for a sample size of 10, the degrees of freedom would be $10-1=9$.

An example will now be considerd:

A sample of 10 items has a mean of 42 and a standard deviation of 12, i.e. $\bar{x} = 42$ and $s = 12$. You are required to calculate an interval estimate for the mean of the population from which the sample was drawn.

The point estimate for μ is $\bar{x} = 42$.

The s.e. of the sample is $s/\sqrt{(n-1)} = 12/\sqrt{(10-1)} = 4$.

Therefore the interval estimate for the mean is $\mu = 42 \pm (t \times 4)$.

The value of t depends on the degrees of freedom in the sample and the confidence level we require. In this example, the degrees of freedom are $10-1=9$. For a 95 per cent confidence interval we consider the appropriate percentage point in the $t-$distribution tables (see appendix 6). An extract from the tables is reproduced below:

The table gives the 100α percentage points of the *t*-distribution with v degrees of freedom. *Note:* The tabulation is for one tail only, i.e. for positive values of t; for two-tailed tests the column headings (α) must be doubled.

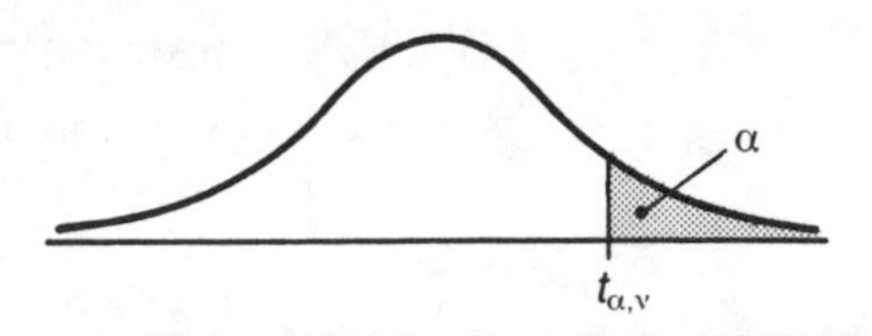

$\alpha =$	0.10	0.05	0.025	0.01	...
$v = 1$	3.078	6.314	12.706	31.821	...
$= 2$	1.886	2.920	4.303	6.965	...
$= 3$	1.638	2.353	3.182	4.541	...
⋮					
$= 9$	1.383	1.833	(2.262)	2.821	...
⋮					
$= 30$	1.310	1.697	2.042	2.457	...
⋮					
$= 120$	1.289	1.658	1.980	2.358	...

The resulting interval estimate for the mean in our example is

$$\mu = 42 \pm (2.262 \times 4) = [32.952, 51.048]$$

Referring back to the table, note how the values get closer to the corresponding normal distribution values as v, the degrees of freedom, increases.

Sample questions

1 AFC Company Limited produce components for the car industry. Random samples of the components undergo a company 'destruction' test and need to withstand loads of up to at least 50 Newtons to satisfy the car manufacturer's safety specifications.

From experience, it is known that the components have an average 'destruction resistance' level of 60 Newtons and a standard deviation of 3.5 Newtons.

Required:

1 Find the probability that a component chosen at random will fail the company test.

At weekly intervals the average 'destruction resistance' of the components is checked. The latest check, involving a sample of 40, gives an average resistance level of 55 Newtons.

2 **(a)** Using the results of this sample, calculate the 95 per cent confidence interval for the mean.

(b) Carry out an appropriate test to see if there is any reason to suppose the quality of the components is falling.

The Polytechnic of North London; Accounting Foundation Course; Quantitative Analysis; 1986.

Solution/hints

Part (1)

Assume 'destruction resistance', D, is distributed as a normal distribution. Then we have:

$$\begin{aligned} &P(\text{randomly chosen component will fail the company test}) \\ &= P(D < 50) = P\left(z < \frac{50-60}{3.5}\right) \\ &\qquad = P(z < -2.86) \\ &\qquad = P(z > +2.86) \text{ (by the symmetry of the normal distribution)} \\ &\qquad = 0.00212 \end{aligned}$$

Part (2)

(a) You need to find the standard error of the sample distribution. The sample is large, so the sample distribution can be assumed to be normal.

Answer: [53.92, 56.08] or 55 ± 1.08 Newtons

(b) (Refer to section 8.7.1) The appropriate test is based on the assumption (under the null hypothesis, $H_0 : \mu = 60$) that the destruction resistance level is normally distributed, *viz.* $N(60, 3.5^2)$.

Test statistic: $\bar{x} = 55$; $n = 40$.

$H_0 : \mu = 60$: 'The average destruction resistance level is unchanged.'

$H_1 : \mu < 60$: 'The average destruction resistance is falling.'

Under H_0, $\dfrac{\bar{X} - 3.5}{3.5/\sqrt{40}} \sim N(0,1)$

At the 5 per cent significance level, the significant value (for a one-tail test) is $-1.645 = z_{5\%}$ (see fig. 8.3).

$$z = \frac{\bar{x} - \mu}{\sigma/\sqrt{n}} = \frac{55 - 60}{3.5/\sqrt{40}} = -9.04 \text{ (to 2 d.p.)} < -1.645$$

The critical ratio (−9.04) is a very significant result. Consequently, we reject H_0 ($z < z_{5\%}$) and conclude that 'there is strong evidence to suggest that the average destruction resistance level is falling'.

2 The Crispie Cookie Company produces a range of pre-packed biscuits which are sold in 500 g packets. A biscuit is produced by baking a measured quantity of the appropriate mixture, and then a specified number of biscuits is placed into each packet. Because the baking process cannot be completely controlled, there is an unavoidable variation in the weights of the biscuits produced. Each packet is therefore automatically weighed, and any packets weighing less than the required 500 g are returned for reprocessing. Whenever a packet is rejected in this way, it is estimated to cost £0.1.

The company is about to start selling a new type of biscuit which has an average weight of 40g and a standard deviation of 6g. At full production, the weekly output of this biscuit will be 500 000 packets and the cost per packet (£) is give by $0.05 + 0.01n$ where n is the number of biscuits in each packet.

Required:

(a) If 13 biscuits are put into each packet, what will be the mean and standard deviation of the weight of the packet?

(b) Explain briefly why the weight of a packet of biscuits will be approximately normally distributed. What is the probability that a packet of 13 biscuits will be rejected as underweight?

(c) Determine the minimum cost number of biscuits per packet by calculating the average weekly production cost for packet sizes of 13, 14 and 15 biscuits.

ACCA; Professional Examination, Level 2; Quantitative analysis; June 1987.

Solution/hints

Part (a)

Each packet contains 13 biscuits. Let the weight of a packet be W. Then

$$W = x_1 + x_2 + x_3 + \ldots + x_{13}$$

where x_i = the weight of the ith biscuit in the packet.

The mean (or expected) weight of a packet of biscuits is

$$\begin{aligned} E(W) &= E(x_1) + E(x_2) + \ldots + E(x_{13}) \\ &= 40 + 40 + \ldots + 40 = 13 \times 40 = 520 \text{ g} \end{aligned}$$

The variance of a packet is

$$\begin{aligned} V(W) &= V(x_1) + V(x_2) + \ldots + V(x_{13}) \\ &= 6^2 + 6^2 + \ldots + 6^2 = 13 \times 36 = 468 \end{aligned}$$

Therefore the standard deviation, s.d.(W) will be 21.63g ($\sqrt{468}$).

Part (b)

The Central Limit Theorem tells us that the distribution of a large number of independent variables is approximately normal. Each packet is made up of 13 independent biscuits; if we consider 13 to be a large number, then the distribution of packet weight can be taken to be normal with mean 520 and standard deviation 21.63.

$$\begin{aligned} P(\text{packet of 13 biscuits will be rejected as underweight}) &= P(W < 500) \\ &= P\left(z < \frac{500 - 520}{21.63}\right) \\ &= P(z < -0.925) \\ &= P(z > +0.925) \\ &= 0.178 = 17.8\% \end{aligned}$$

Part (c)

Average (or expected) weekly production costs:

In each case there are two costs to consider: production cost, p, and rejection cost, r, for the weekly output of 500 000 packets.

(i) Packet size = 13 biscuits:

$p = 500\,000 \times (0.05 + 0.01 \times 13) = £90\,000$
E(cost due to rejected packets)
  = [rejection cost × P(rejection)] + [non-rejection cost × P(non-rejection)]
  = $(0.1 \times 0.178) + (0.0 \times 0.822) = £0.0178$

Therefore, for a production level of 500 000, the average cost due to packets being rejected is $r = 500\,000 \times 0.0178 = £8900$. The total (average) cost is £98 900.

(ii) Packet size = 14 biscuits

$E(W) = 14 \times 40 = 560$ g and s.d.$(W) = \sqrt{14 \times 6^2} = 22.45$ g

Employ the methodology of the second part of (**b**) to find the probability of a packet being rejected (0.00397), then proceed as in (i) above.
Total (average) cost ≃ £95 200.

For 15-biscuit packets, the production cost alone is £100 000. Consequently $n = 14$ is the optimum (or 'minimum cost') number of biscuits per packet.

3 (**a**) A simple random sample of 400 of a large mail-order company's customers showed that the mean value of orders in the first quarter of 1987 was £31 with a standard deviation of £10. Find 95 per cent confidence limits for the population mean and interpret your answer.
(**b**) (i) The sample of 400 comprises 80 pensioners. Find a 99 per cent confidence interval for the population percentage of pensioners and interpret your answer.
(ii) What size of sample would have to be taken to be at least 95 per cent confident that the population percentage of pensioner customers would be estimated to within ±2 per cent?

CIMA; Stage 1; Quantitative methods; May 1987.

Solution/hints

Part (a)
Details of the calculations are left as an exercise.
Answer: $31 \pm 0.98 = [£30.02, £31.98]$; Bessel's correction not necessary as the sample is large.
Interpretation: One would expect, with 95 per cent confidence, that the (true) population mean is contained within the given confidence interval.

Part (b)
(i) Mean proportion, $p = 80/400 = 0.2$ = point estimate.

Proportion s.e. $= \sqrt{0.2 \times 0.80/400} = 0.02$

The normal approximation is applicable, giving a 99 per cent confidence interval for the proportion $0.20 \pm (2.576 \times 0.02)$. The question asks for the population *percentage*– the corresponding 99 per cent confidence interval for the percentage of pensioner customers is

$$20\% \pm 2.576 \times 2\% = [14.85\%, 25.15\%]$$

(ii) Required sample size at 95 per cent confidence level, to within 2 per cent (±0.02):

$$n = (1.96^2 \times 0.2 \times 0.8)/(0.02)^2 = 1536.6$$

Therefore, a sample size of 1537 is needed to be at least 95 per cent confident of being within 2 per cent of the true mean percentage of pensioner customers.

4 A company which sells clothing conducted an advertising campaign at the beginning of 1987 with the aim of selling more of its products to young people and males. In 1986, the company estimated that the average age of its customers was 24 years and that 65 per cent were female. In 1987, a survey of a random sample of 400 customers of the company was conducted. The survey found that the mean and the standard deviation of age were 23.5 and 5 years, respectively, and that 61 per cent were female.
Test whether there has been:

(**a**) a significant change in the age of customers; and

(**b**) a significant increase in the percentage of male customers.

(c) In 1986, three customers entered one of the company's shops. Assuming that these customers were not related or acquaintances, find the probability that they were:

(i) all female;
(ii) all male; and
(iii) at least one male.

CIS; Part 1; Quantitative studies; June 1987.

Solution/hints

Part (a)

Following the steps outlined in section 8.7.1:

1 Has there been a significant change in the age of customers?

2 H_0: $\mu = 24$ years (i.e. no change)
H_1: $\mu \neq 24$ years (i.e. change: older or younger)

3 Population distribution unknown, but . . .

4 Sample mean distribution can be assumed to be normal, as sample is large ($n = 400$).

5 In the absence of other instructions, let the significance level be 5 per cent.

6 The critical values are ± 1.96 (two-tailed test).

7 The critical ratio is

$$\frac{\bar{x}-\mu}{\sigma/\sqrt{n}} = \frac{23.5-24}{5/\sqrt{400}}$$

$$= \frac{-0.5}{0.25} = -0.2$$

8 The critical ratio is seen to be within the critical region ($-2.0 < -1.96$). This is, therefore, a significant result. Hence, the null hypothesis is rejected and we conclude that there has been a significant change in the average age of customers.

Part (b)

To investigate whether the percentage of male customers has increased significantly, we need to compare the sample proportion (1987 sample) with the assumed population proportion (of 1986).

Population proportion (1986) = 0.35
Sample proportion (1987) = 0.39; sample size, $n = 400$.

Test at 5 per cent significance level: H_0: $p = 0.35$
H_1: $p > 0.35$ (one-tail test)

Test statistic (approximate standard normal distribution)

$$= \frac{p_s - p}{\sqrt{\dfrac{p(1-p)}{n}}} \quad \text{where } p_s = \text{sample proportion random variable}, \; p = \text{population proportion}$$

$$= \frac{0.39-0.35}{\sqrt{\dfrac{0.35 \times 0.65}{400}}}$$

$$= +1.677 > +1.645 \; (z_{5\%})$$

The critical ratio (+1.677) is in the critical region (at the 5% significance level). Therefore we reject H_0 and conclude that 'there has been a significant increase in the proportion (and therefore, the percentage) of male customers'.

Part (c)

Applications of the binomial distribution:

(i) P(all 3 customers female) $= 0.65 \times 0.65 \times 0.65$
$= 0.27$ (2 d.p.)

Details of (ii) and (iii) left as an exercise.

(ii) *Answer:* 0.043
(iii) *Answer:* $1 - 0.27 = 0.73$

5 The purchases of a standard packet of Brand X cornflakes by a sample of households over a 24-week period were as follows:

Number of households	Number of purchases
195	0
74	1
49	2
28	3
20	4
20	5
14	6

(*Source: The Open University, Course M245, 1985, adjusted*)

(**a**) Calculate the arithmetic mean, median and standard deviation of this distribution.

(**b**) It is hypothesised that the household purchases of Brand X cornflakes follow a Poisson distribution. Using the value of the mean and CIMA tables [standard tables provided], find the number of these households which would be expected to purchase 0, 1, 2, 3, 4, 5, 6 packets, assuming the Poisson distribution to apply. Comment briefly on how well the Poisson theory appears to correspond to the data for Brand X.

CIMA; Foundation stage; Section B; Mathematics and statistics; November 1986.

Solution/hints

(**a**) Details left as an exercise.
Answer: mean = 1.3; s.d. = 1.71 (2 d.p.); median = 1.0.

(**b**) Use $m = 1.3$ for Poisson distribution. Calculate expected frequencies for $x = 0, 1, 2$, etc.

6

Household size 1983–84

Number in household	Percentage of households
1	23.0
2	32.4
3	16.5
4	18.3
5	6.9
6	2.1
7	0.4
8 or more	0.4
sample size 14 054	

(*Source: Family Expenditure Survey*)

(**a**) Calculate the mean and standard deviation of the size of households.

(**b**) Assuming that the data is based on a simple random sample, find a 95 per cent confidence interval for the mean household size.

(**c**) Explain the rationale of the confidence interval.

CIS; Part 1; Quantitative Studies; December 1986.

Solution

Left as an exercise.

Answers:

(**a**) Mean, $\bar{x} = 2.636$; standard deviation, $s = 1.387$.
(Assumptions: '8 or more' taken as 8.)

(**b**) [2.613, 2.659]

Recommended reading

Owen, F. and Jones, R., *Statistics* (Pitman, 1988), chapters 17–21.

Whitehead, P. and Whitehead, G., *Statistics for Business* (Pitman, 1984), chapters 16 and 17.

Thomas, R., *Notes and Problems in Statistics* (Stanley Thornes, 1984), chapters 3 and 4.

Curwin, J. and Slater, R., *Quantitative Methods for Business Decisions* (Van Nostrand Reinhold, 1986), chapters 12–15.

9 Introducing business mathematics

9.1 Introduction

Mathematics and statistics, in a business context, constitute an applied subject area, for instance, the interpretation of a sample mean is as important as the ability to calculate the sample mean. However, the study of a mathematics-based subject area requires a sound base of mathematical skills. This chapter concentrates on a number of those skill areas. The topic areas which use these skills are many, though some syllabuses require you to show proficiency in the skills themselves – skills such as matrix multiplication and differentiation.

9.2 Number patterns: sequences, series and progressions

Series, or progressions, form an important part of financial analysis; for example, those encountered when compounding or discounting financial data. A number of general examples will be looked at.

9.2.1 Sequences

A sequence is a set of numbers which has some general pattern of increase or decrease.

e.g. 17, 19, 21, 23, 25, ...
9, 4, −1, −6, ...

9.2.2 Series

A series is a sequence of terms joined together, in an ordered fashion, by addition or subtraction.

e.g. $9+13+21+23+25$.
$-9+14-22+30-36$.

9.2.3 Progressions

A progression is a sequence or series which has a constant difference or ratio between successive terms.

e.g. 2,4,6,8,10,...: There is a common difference of 2 between each term in the sequence.

$2+4+8+16+32+\ldots$: There is a common ratio 2 (×2) between each term in the series.

$24+8+\frac{8}{3}+\frac{8}{9}+\frac{8}{27}+\ldots$: There is a common ratio $\frac{1}{3}$ between each terms in the series.

Of particular importance are the arithmetic and the geometric progressions, which will now be considered in some detail.

9.2.4 The arithmetic progression (AP)

An arithmetic progression is a series in which there is a common difference between each successive term.

e.g. $1+5+9+13+17$: This is an AP with first term 1, common difference 4 and 5 terms in total.

$35+26+17+8+(-1)+(-10)$: This is an AP with first term 35, common difference −9 and 6 terms in total.

The algebraic form of an AP is

$$a+(a+d)+(a+2d)+\ldots+[a+(n-1)d]$$

where: a is the first term;
d is the common difference;
n is the number of terms in the series; and
$[a+(n-1)d]$ is the nth term.

In the first example above, $a = 1$, $d = 4$ and $n = 5$ and in the second example, $a = 35$, $d = -9$ and $n = 6$.

The nth term, $[a+(n-1)d]$, is often used to find information about the AP. For instance, given that the 10th term of an AP is 37 and the first term is 1, it is possible to calculate the common difference:

$$1+(10-1)d = 37$$
$$1+9d = 37$$
$$9d = 37-1 = 36$$
$$d = 4$$

If you had difficulty following this calculation refer to section 9.3.1.

9.2.4.1 The sum of an arithmetic progression

If the series is short, such as $1+5+9+13+17$, then it is easy to simply add the terms (= 45). However, when the series is longer, this method would be time-consuming and tedious, and would probably lead to mistakes. Consider the above example again. We have the sum of 5 terms which we will denote S_5:

$$S_5 = 1+5+9+13+17$$

S_5 can also be written as $S_5 = 17+13+9+5+1$

Adding the two equations together gives

$$2 \times S_5 = 18+18+18+18+18 = 90$$

Therefore $S_5 = 90/2 = 45.$

This may seem a long-winded way of finding out the sum of a mere five terms. It is! However, the method leads to formulation of an expression which will enable us to calculate the sum of any AP.

S_5 can be written as $S_5 = 5 \times (\text{1st term} + \text{5th term})/2$

With reference to the algebraic expression for an AP,

$$S_n = a+(a+d)+\ldots+[a+(n-d)d]$$

we can give a general expression for the sum S_n:

$$\boldsymbol{S_n = n \times (\text{1st term} + n\text{th term})/2}$$
$$\boldsymbol{= n(a+L)/2, \text{ where } L = a+(n-1)d}$$
$$\boldsymbol{= n[2a+(n-1)d]/2}$$

Consider the following example: Nearthing Pottery Ltd has a contract to produce a range of dinner plates for a retailer. The contract stipulates that the total agreed supply of 200 000 plates must be delivered within 15 weeks. The production is estimated to be 10 000 in the initial week, rising by 250 per week as the production reaches efficient running capacity. Will the order be delivered on time?

This is a typical AP problem, with $a = 10\,000$, $d = 250$ and n unknown. The total required is 200 000 plates. We need to find out how many weeks it will take to produce this amount. Using the general expression for S_n, we have

$$S_n = 200\,000 = n[(2 \times 10\,000)+(n-1)250]/2$$
$$200\,000 = n(20\,000+250n-250)/2$$
$$400\,000 = n(19\,750+250n)$$
$$400\,000 = 19\,750n+250n^2$$

Dividing the expression by 250, this simplifies to $n^2+79n-1600 = 0$, a quadratic equation. Solving the quadratic equation,* we find

$$n = \frac{-79 \pm \sqrt{79^2 - 4 \times 1 \times (-1600)}}{2 \times 1}$$

$$= \frac{-79 \pm \sqrt{6241+6400}}{2}$$

$$= \frac{-79 \pm 112.43}{2}$$

Therefore $n = (-79-112.43)/2 = -95.72$ or $n = (-79+112.43)/2 = +16.72.$

The first solution is unacceptable (we can not produce anything in −95.72 weeks!). The second solution informs us that the company will need almost 17 weeks to produce the required amount of 200 000 plates, that is, they will fail to meet the contract unless they increase their production rate.

(*Quadratic equations and their solution are considered in section 9.4.)

9.2.5 The geometric progression (GP)

A geometric progression is a series in which there is a common ratio between successive terms. For example,

1 $2+6+18+54+162$ is a GP with a first term 2, common ratio 3 and a total of 5 terms.

2 $27+9+3+1+\frac{1}{3}++\frac{1}{9}+\frac{1}{27}$ is a GP with first term 27, common ratio $\frac{1}{3}$ and 7 terms.

3 $1+\frac{1}{2}+\frac{1}{4}+\frac{1}{8}+\frac{1}{16}+\frac{1}{32}+\frac{1}{64}+\frac{1}{128}+\ldots$ is a GP with first term 1, common ratio $\frac{1}{2}$ and an infinite number of terms. Infinite GPs are considered in section 9.2.5.2.

The algebraic form of a GP is

$$a+ar+ar^2+\ldots+ar^{n-1}$$

where: a is the first term;
r is the common ratio;
n is the number of terms in the series; and
ar^{n-1} is the nth term.

In example **1** above, $a=2$, $r=3$ and $n=5$; in example **2**, $a=1$, $r=\frac{1}{3}$ and $n=7$; in example **3**, $a=1$, $r=\frac{1}{2}$ and $n=\infty$ (infinity).

9.2.5.1 The sum of a geometric progression

As with an arithmetic progression, it is relatively simple to find the sum of a geometric progression if the number of terms is small. Again it is desirable to have a general formula to enable us to calculate the sum of any GP, quickly and accurately. The formula is derived as follows:

$$S_n = a+ar+ar^2+\ldots+ar^{n-2}+ar^{n-1} \qquad (1)$$

Multiplying both sides of this equation by r gives

$$rS_n = ar+ar^2+ar^3+\ldots+ar^{n-1}+ar^n \qquad (2)$$

Subtracting (2) from (1) gives $S_n - rS_n = a - ar^n$, that is

$$(1-r)S_n = a(1-r^n)$$

This gives the general formula for the sum of n terms of a GP:

$$S_n = \frac{a(1-r^n)}{1-r}$$

Consider Nearthing Pottery again: Nearthing Pottery Ltd has revised its production process and claims that, after an initial week's production level of 10 000, production will increase by approximately 4.0 per cent per week. They believe that 15 weeks will not only provide them with the target production level of 200 000, but will leave them with a surplus. Are they correct?

The situation is described by a GP with $a = 10\,000$, $r = 1.04$ (an increase of 4 per cent per week, i.e. each week's production level is 104 per cent times the level of the previous week) and $n = 15$. Applying the general formula for the sum of n terms gives:

$$S_{15} = \frac{10\,000(1-1.04^{15})}{1-1.04}$$

$$= \frac{10\,000(1-1.80)}{-0.04}$$

$$= \frac{10\,000\times(-0.80)}{-0.04} = \frac{-8000}{-0.04}$$

$$= 200\,000$$

If the company's estimate of a weekly improvement of 4 per cent is correct, they will just satisfy the order with no surplus.

9.2.5.2 S_n for a GP with an infinite number of terms

If $|r|$, the common ratio, is greater than or equal to 1 ($|r| \geq 1$), the sum of the GP with an infinite number of terms is infinity. If $|r|$ is less than 1 ($|r| < 1$) it is possible to find a finite value for S_n.

Suppose, for example, the sum of the following GP is required:

$$16 + 4 + 1 + 0.25 + 0.0625 + 0.15625 + \ldots$$

The common ratio is 0.25 ($r < 1$), that is, each successive term is one quarter of the previous term. As n increases, the terms will become less significant.

In the previous section we saw that the formula for the sum of n terms of a geometric progression is

$$S_n = \frac{a(1-r^n)}{1-r}$$

When n is very large and $|r| < 1$, the r^n term is negligible. Thus, when $|r| < 1$, the sum of an infinite number of terms is

$$S_\infty = \frac{a}{1-r}$$

The sum of the above GP can now be found:

$$S_\infty = \frac{16}{1-0.25} = \frac{16}{0.75} = 21.33 \text{ (to 2 d.p.)}$$

Geometric progressions are very important in financial arithmetic and are used extensively in compounding and discounting, as will be seen in chapter 10.

9.3 Linear equations

Linear equations are equations in which the highest power of the unknown variable(s) is one, that is x, y, but not x^2, y^4, etc. A linear equation with only one unknown variable, e.g. $3x + 2 = 3.5$, can be solved. That is, a value of x can be found which satisfies the equation. For the example just mentioned:

$$\begin{aligned} 3x + 2 &= 3.5 \\ 3x &= 3.5 - 2 \quad \text{(2 is subtracted from both sides of the equation)} \\ &= 1.5 \\ x &= 1.5/3 \quad \text{(both sides of the equation are divided by 3)} \\ &= 0.5 \end{aligned}$$

Hence, $x = 0.5$ is the solution to the equation $3x + 2 = 3.5$.

9.4 Simultaneous equations

The solution to the above equation was quite easily found and was, furthermore, unique. This uniqueness of solution was because there was one piece of information, the equation, and only one unknown factor, x. It is not possible to find a unique solution of $y = 7 - 4x$: $x = 1$, $y = 3$ will satisfy the equation, but so will $x = 2$, $y = -1$, and so on. In this instance we have only one piece of information, the equation, but two unknown factors, x and y. In short, to be able to find a unique solution, we need as many independent pieces of information as there are unknown factors. One or more of the pieces of information might be in the form of a given value. For instance, if we are given the information that $x = 3$, then the unique solution to the equation $y = 7 - 4x$ is $y = -5$.

A particular case is when one is given the information in the form of a number of equations, not necessarily linear. For example,

$$\begin{aligned} y &= -5 + 2x \\ \text{and } y &= -4x + 7 \qquad (**) \end{aligned}$$

Such sets of equations, e.g. (**) above, are called simultaneous equations – they can be solved simultaneously.

In general, if we have n independent simultaneous equations containing n variables, it is possible to find single, unique solutions for the n variables.

A single, unique solution could not be found for the two equations

$$\begin{aligned} x + 2y &= 5 \\ \text{and } 4x + 8y &= 20 \end{aligned}$$

They are not independent; $4x + 8y = 20$ is equivalent to $x + 2y = 5$.

The solution to the simultaneous equations (**) can be found graphically, as illustrated in fig. 9.1. The lines represented by the separate equations are plotted and the point at which they intersect gives the values of x and y which satisfy both equations simultaneously. In this case, $x = +2$ and $y = -1$.

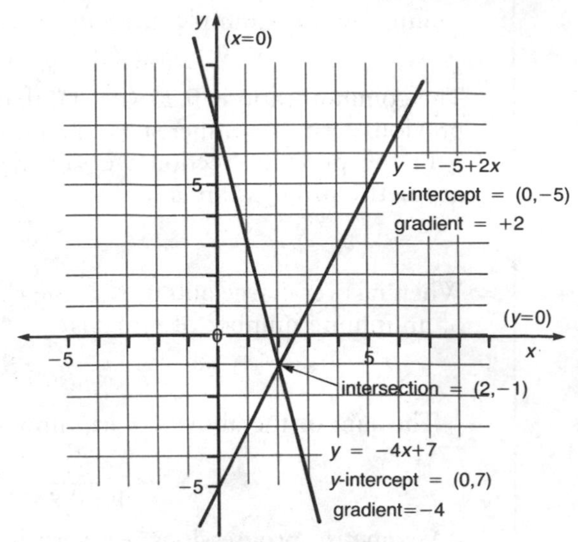

Fig. 9.1 Graphical solution of the simultaneous equations $y = -5 + 2x$ and $y = -4x + 7$

However, it is only practical to use a graphical method when there are two variables. Solutions can also be found algebraically. Consider the same pair of simultaneous equations. It is usual to write the equations in the form

$$2x - y = +5 \qquad (1)$$
$$4x + y = +7 \qquad (2)$$

i.e. with the unknown variables in the same order on the left-hand side of the equations and the constant terms on the right. The aim is to reduce the simultaneous equations to one equation containing only one unknown variable. The value of this variable can be found (its solution) and can then be substituted back into the original equation(s) in order to find the values of the other unknowns.

This example is quite straightforward. We add equations (1) and (2) together to eliminate y:

$$(2x + 4x) + (-y + (+y)) = +5 + (+7) \qquad (1) + (2)$$
$$6x \quad + \quad 0 \quad = +12$$

Therefore $x = 12/6 = +2$. To find y, substitute this value of x back into either of the original equations (1) or (2). Let's use (2):

$$(4 \times 2) + y = +7$$
$$\text{Therefore} \qquad y = +7 - 8 = -1$$

Use the other original equation to check your solution.

9.5 Quadratic equations

Quadratic equations are second order equations, that is, one variable varies with the square of another. The following are some examples of quadratic equations:

$$y = x^2$$
$$y = 2x^2 - 3x$$
$$y = 3x^2 + 4x + 6$$
$$3y - 4x^2 + 2x + 2 = 0$$

The general algebraic expression for quadratic equations is

$$\boldsymbol{y = ax^2 + bx + c}$$

Most quadratic equations have two values of x (two solutions) corresponding to a particular value of y.

Consider the equation $y = 3x^2 - 2x - 8$. For given values of x, it is quite straightforward to calculate the corresponding value of y. For example, if $x = -2$,

$$y = 3(-2)^2 - 2(-2) - 8$$
$$= (3 \times 4) + 4 - 8 = +8$$

It is more likely that you will be required to solve this equation for given values of y, that is, you will have to find the values of x (usually two and never more than two) which satisfy the equation. This can be done graphically, as shown in fig. 9.2, but it is more usual to use **factors**, or the **general formula.**

9.5.1 Quadratic equations: solution by factors

$(3x+2)$ and $(x-2)$ are algebraic expressions which, when multiplied together, are seen to be factors of $3x^2-4x-4$:

$$(x-2)(3x+2) = 3x^2+2x-6x-4 = 3x^2-4x-4$$

To solve a quadratic equation, using the factor method, the reverse process needs to be carried out.

For example, we wish to solve the following equation: $y = 3x^2-2x-8$, when $y = 0$, that is, $3x^2-2x-8 = 0$.

First, we need to 'factorize' the left-hand side of the equation, that is, we need to find two factors, such that we can write

$$(\qquad)(\qquad) = 3x^2-2x-8 = 0$$

In general, it is useful to concentrate on the x^2 term ($3x^2$, in this example) and the constant term (-8, in this example). The only simple factors of $3x^2$ are $3x$ and x, so we can begin to formulate the required factors: $(3x \qquad)$ and $(x \qquad)$. There are several possible pairs of factors of -8: $+1$ and -8; -1 and $+8$; -2 and $+4$; $+2$ and -4. We have to guess which are appropriate. Suppose we try -1 and $+8$; we still have to decide which brackets to put the -1 in and the $+8$ in. Suppose we tried $(3x-1)$ and $(x+8)$. These factors give $3x^2$ and -8 but the fully expanded expression is $3x^2+23x-8$, which is not the one we started with. The correct factorization is

$$(3x+4)(x-2) = 3x^2-2x-8 = 0$$

The equation can now be solved: Since the product of the two factors is zero, one or the other must be equal to zero. In this case

$$(3x+4)(x-2) = 0, \text{ therefore } (3x+4) = 0 \text{ or } (x-2) = 0$$

Solving these two simple equations, we find $x = -\frac{4}{3}$ or $+2$. (Test these values in the original equation.)

It must be apparent that this method is

1 long-winded – the constant term may have many factors and many false attempts may be made before the equation is correctly factorized; and
2 limited in application – it will not always be possible to find factors.

9.5.2 Quadratic equations: the formula method

If a quadratic equation is expressed in the standard algebraic form, $ax^2+bx+c = 0$, then the solutions can be found by using the following formula:

$$x = \frac{-b \pm \sqrt{b^2-4ac}}{2a}$$

This formula will enable you to solve all quadratic equations you are likely to encounter. There are some quadratic equations which do not have a real number solution, but these need not concern us.

The example considered in section 9.5.1, $3x^2-2x-8 = 0$, is solved as follows:

$$3x^2-2x-8 = 0\text{: } a = 3,\ b = -2 \text{ and } c = -8$$

Substituting these values into the given formula, we have

$$x = \frac{-(-2) \pm \sqrt{(-2)^2-4\times3\times(-8)}}{2.3}$$

$$= \frac{+2 \pm \sqrt{4+96}}{6}$$

$$= \frac{+2 \pm 10}{6} = \frac{+12}{6} \text{ or } \frac{-8}{6} = +2 \text{ or } \frac{-4}{3}$$

The solution is shown on a graph (fig. 9.2, overleaf) for illustrative purposes only; though the graph could have been used to solve the problem, the algebraic methods are quicker and more accurate.

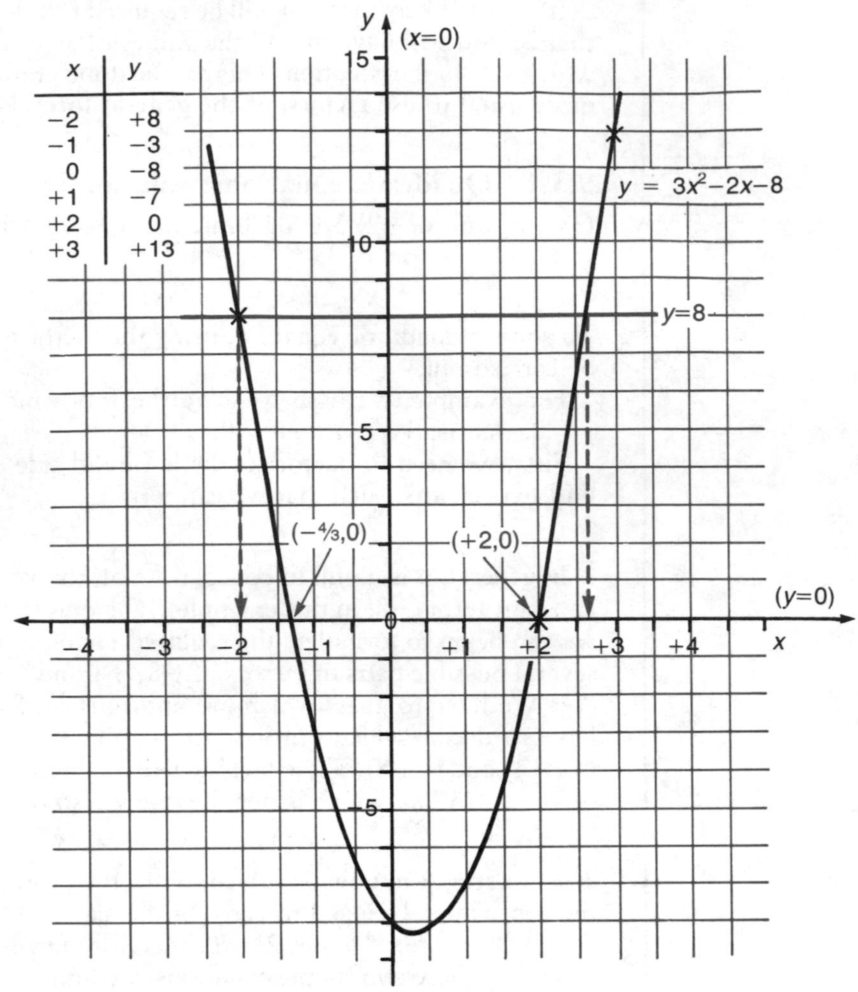

Fig. 9.2 Graphical solution of $3x^2 - 2x - 8 = 0$ and the simultaneous equations $y = 3x^2 - 2x - 8$ and $y = +8$

One more example should suffice: For the quadratic equation, $y = 3x^2 - 2x - 8$, find the x-values when $y = +8$.

That is, solve $3x^2 - 2x - 8 = +8$

$$3x^2 + 2x - 16 = 0$$

The graphical solution is shown in fig. 9.2, but the formula gives us a more accurate solution: This time $a = 3$, $b = -2$ and $c = -16$:

$$x = \frac{-(-2) \pm \sqrt{(-2)^2 - 4 \times 3 \times (-16)}}{2 \times 3}$$

$$= \frac{+2 \pm 14}{6}$$

$$= \frac{+16}{6} \text{ or } \frac{-12}{6} = +\frac{8}{3} \text{ or } -2$$

9.6 Matrix algebra and arithmetic

Table 9.1 East Town Dairy: Record of sales for 1 week (sample of 3 houses) and prices of three commodities

Commodity		1 Dean Street	**Address** 2 Dean Street	3 Dean Street
milk		6	10	3
butter		2	4	1
cheese		2	3	1
Commodity prices:	milk	£0.25		
	butter	£0.50		
	cheese	£1.25		

A matrix is an array or table of numbers, as shown in table 9.1 above. It is usual for a matrix to be displayed in large square brackets, e.g.

$$\begin{bmatrix} 6 & 10 & 3 \\ 2 & 4 & 1 \\ 2 & 3 & 1 \end{bmatrix} = \mathbf{A} \quad \begin{bmatrix} 0.25 \\ 0.50 \\ 1.25 \end{bmatrix} = \mathbf{B} \quad \begin{bmatrix} 2 & 0 \\ 3 & 4 \\ 5 & -2 \end{bmatrix} = \mathbf{C} \quad \begin{bmatrix} 6 & 5 & -2 \\ 0 & -3 & 2 \\ 2 & 5 & 1 \end{bmatrix} = \mathbf{D}$$

Matrices **A** and **B** represent the information given in table 9.1. Matrix **B** is an example of a **column matrix**, usually referred to as a **column vector.**

The numbers in a matrix are called the **elements** of the matrix and are identified by reference to the row and the column positions which they occupy. For instance, the number 10 in matrix **A** is in row 1 and column 2, so it could be identified as element $\mathbf{A}_{12}$. Similarly, $\mathbf{A}_{23} = +1$, $\mathbf{B}_{21} = 0.50$, and so on.

Matrix **A** has 3 rows and 3 columns and is therefore said to be a '3-by-3' matrix, or to have **size** 3×3. **B** is a 3-by-1 column vector and **C** is a 3-by-2 matrix.

9.6.1 Matrix addition

Matrices can be added (or subtracted) as long as they are the same size. Of our examples above, matrices **A** and **D** can be added – they are both 3-by-3. In adding matrices the corresponding elements are added together. Subtraction of matrices can be carried out in a similar way.

$$\mathbf{A}+\mathbf{D} = \begin{bmatrix} 6+6 & 10+5 & 3+(-2) \\ 2+0 & 4+(-3) & 1+2 \\ 2+2 & 3+5 & 1+1 \end{bmatrix} = \begin{bmatrix} 12 & 15 & 1 \\ 2 & 1 & 3 \\ 4 & 8 & 2 \end{bmatrix}$$

$$\mathbf{A}-\mathbf{D} = \begin{bmatrix} 6-6 & 10-5 & 3-(-2) \\ 2-0 & 4-(-3) & 1-2 \\ 2-2 & 3-5 & 1-1 \end{bmatrix} = \begin{bmatrix} 0 & 5 & 5 \\ 2 & 7 & -1 \\ 0 & -2 & -0 \end{bmatrix}$$

9.6.2 Scalar multiplication

To multiply a matrix by a scalar (a scalar is simply a single number), multiply each element of the matrix by that number. For example, 4**A** is matrix **A** multiplied by 4:

$$4\mathbf{A} = \begin{bmatrix} 4\times 6 & 4\times 10 & 4\times 3 \\ 4\times 2 & 4\times 4 & 4\times 1 \\ 4\times 2 & 4\times 3 & 4\times 1 \end{bmatrix} = \begin{bmatrix} 24 & 40 & 12 \\ 8 & 16 & 4 \\ 8 & 12 & 4 \end{bmatrix}$$

9.6.3 Matrix multiplication

9.6.3.1 Matrix multiplication: general rules

The multiplication of matrices by other matrices is a little more complicated, but with practice it should prove to be quite straightforward. Before attempting an example, it is worth being aware of general rules which apply to all matrix multiplication:

1 If **X** and **Y** are matrices with sizes $m \times n$ and $n \times p$ respectively, then it is possible to calculate the matrix product **XY**, as **X** has the same number of columns as **Y** has rows. Referring back to our example matrices, it is possible to work out the product **AB** (3×3 and 3×1), but not **BA** (3×1 and 3×3).

2 If rule **1** is satisfied, the resulting matrix will have size $m \times p$ (the number of rows of **X**) × (the number of columns of **Y**), e.g. **AB** will have size 3×1.

3 If $\mathbf{XY} = \mathbf{Z}$, then element $\mathbf{Z}_{ij}$ is the result of multiplying row i of matrix **X** with column j of matrix **Y**:

$$\mathbf{Z}_{ij} = \mathbf{X}_{i1}.\mathbf{Y}_{1j} + \mathbf{X}_{i2}.\mathbf{Y}_{2j} + \mathbf{X}_{i3}.\mathbf{Y}_{3j} + \ldots + \mathbf{X}_{in}.\mathbf{Y}_{nj}$$

9.6.3.2 Matrix multiplication: examples

Consider the 'dairy data' shown in table 9.3. The matrix and column vector of the commodity quantities and the commodity prices are **A** and **B**, respectively. As has been mentioned, it is possible to calculate **AB**, but not **BA**. However, vector **B** can be **transposed** into the row vector **B**′, [0.25 0.50 1.25]. It is possible to calculate the product **B′A** (**B**′ has size 1×3 and **A** has size 3×3). The result of this matrix multiplication is a 1×3 row vector, **Z**.

$$\mathbf{B}' \times \mathbf{A} = \underset{(1\times 3)}{[0.25\ \ 0.50\ \ 1.25]} \times \underset{(3\times 3)}{\begin{bmatrix} 6 & 10 & 3 \\ 2 & 4 & 1 \\ 2 & 3 & 1 \end{bmatrix}} = \underset{(1\times 3)}{[Z_{11}\ \ Z_{12}\ \ Z_{13}]}$$

$$Z_{11} = B'_{11}.A_{11} + B'_{12}.A_{21} + B'_{13}.A_{31}$$
$$= (0.25\times 6) + (0.50\times 2) + (1.25\times 2) = 5.00$$
$$Z_{12} = (0.25\times 10) + (0.50\times 4) + (1.25\times 3) = 8.25$$
$$Z_{13} = (0.25\times 3) + (0.50\times 1) + (1.25\times 1) = 2.50$$

The row vector **Z** represents the total money owed to the dairy by the three houses. For instance,

(milk) 0.25 × 6 + (butter) 0.50 × 2 + (cheese) 1.25 ×2 = £5.00

is the total amount owed to East Town Dairy by no. 1 Dean Street. Similarly, £8.25 and £2.50 are the amounts owed by numbers 2 and 3 Dean Street, respectively.

Note: It would have been possible to calculate **AB**, but the product would have been meaningless in this context.

Further examples of matrix multiplication, including examples of the identity matrix (I_3, see section 9.6.3.3) and inverse matrices (X^{-1}; see section 9.6.3.4), are shown in fig. 9.3.

$$\mathbf{X} = \underset{(3\times 3)}{\begin{bmatrix} 3 & 2 & -1 \\ 4 & 0 & 1 \\ -2 & 4 & 3 \end{bmatrix}} \qquad \mathbf{Y} = \underset{(3\times 2)}{\begin{bmatrix} 1 & 0 \\ 0 & 4 \\ 3 & 2 \end{bmatrix}} \qquad \mathbf{I} = \underset{(3\times 3)}{\begin{bmatrix} 1 & 0 & 0 \\ 0 & 1 & 0 \\ 0 & 0 & 1 \end{bmatrix}}$$

$$\mathbf{X}^{-1} = \frac{1}{56}\underset{(3\times 3)}{\begin{bmatrix} 4 & 10 & -2 \\ 14 & -7 & 7 \\ -16 & 16 & 8 \end{bmatrix}}$$

YX and **YI** are not possible.

$$\mathbf{XY} = \begin{bmatrix} (3\times 1+2\times 0+(-1)\times 3) & (3\times 0+2\times 4+(-1)\times 2) \\ (4\times 1+0\times 0+1\times 3) & (4\times 0+0\times 4+1\times 2) \\ ((-2)\times 1+4\times 0+3\times 3) & ((-2)\times 0+4\times 4+3\times 2) \end{bmatrix} = \underset{(3\times 2)}{\begin{bmatrix} 0 & 6 \\ 7 & 2 \\ 7 & 22 \end{bmatrix}}$$

$$\mathbf{XX}^{-1} = \frac{1}{56}\begin{bmatrix} (3\times 4+2\times 14+(-1)\times(-16)) & (3\times 10+2\times(-7)+(-1)\times 16) & \dots \\ (4\times 4+0\times 14+1\times(-16)) & (4\times 10+0\times(-7)+1\times 16) & \dots \\ ((-2)\times 4+4\times 14+3\times(-16)) & ((-2)\times 10+4\times(-7)+3\times 16) & \dots \end{bmatrix}$$

$$= \frac{1}{56}\begin{bmatrix} 56 & 0 & 0 \\ 0 & 56 & 0 \\ 0 & 0 & 56 \end{bmatrix} = \begin{bmatrix} 1 & 0 & 0 \\ 0 & 1 & 0 \\ 0 & 0 & 1 \end{bmatrix} = \mathbf{I}$$

$$\mathbf{XI} = \begin{bmatrix} (3\times 1+2\times 0+(-1)\times 0) & (3\times 0+2\times 1+(-1)\times 0) & (3\times 0+2\times 0+(-1)\times 1) \\ (4\times 1+0\times 0+1\times 0) & (4\times 0+0\times 1+1\times 0) & (4\times 0+0\times 0+1\times 1) \\ ((-2)\times 1+4\times 0+3\times 0) & ((-2)\times 0+4\times 1+3\times 0) & ((-2)\times 0+4\times 0+3\times 1) \end{bmatrix}$$

$$= \begin{bmatrix} 3 & 2 & -1 \\ 4 & 0 & 1 \\ -2 & 4 & 3 \end{bmatrix} = \mathbf{X} = \mathbf{IX}$$

Figure 9.3

9.6.3.3 The identity matrix

Particular note should be made of the product involving I_3, the 3 × 3 identity matrix. It plays the same role as does 1 in the ordinary number system. That is, **XI** = **IX** = **X**, for all matrices **X**. I_n is the $n \times n$ matrix with all elements on the diagonal, I_{nn}, equal to 1 and all the other elements zero.

9.6.3.4 X^{-1}, the inverse of X

Most (square) matrices, **X**, have an inverse, denoted by $\mathbf{X}^{-1}$. $\mathbf{X}^{-1}$ is the matrix such that the product of it and the original matrix **X** is the identity matrix, that is,

$$\mathbf{XX}^{-1} = \mathbf{X}^{-1}\mathbf{X} = \mathbf{I}$$

The inverse of a matrix is an important concept. It can be used in a number of situations, particularly to solve simultaneous equations. Before considering how it is used, you need to be familiar with a method of finding the inverse of a matrix, and also with the type of matrix which does not have an inverse.

Consider the matrix **X**:

$$\mathbf{X} = \begin{bmatrix} 3 & 2 & -1 \\ 4 & 0 & 1 \\ -2 & 4 & 3 \end{bmatrix}$$

There are a number of methods you can employ to find the inverse of a matrix, but the method of **'row reduction'** is relatively simple and involves skills which can be employed in a wider context, particularly in the area of linear programming. The matrix must be written in the form of a 'partitioned' matrix adjacent to the identity matrix, which forms a combined matrix, as shown in table 9.2. Row operations are carried out on the combined matrix, as will now be explained. The aim is to end up with the identity matrix in place of **X**, and the inverse of **X**, $\mathbf{X}^{-1}$, in place of the initial identity matrix.

Table 9.2 Calculation of $\mathbf{X}^{-1}$, the inverse of matrix **X** by the row-reduction method

Calculation instruction							**Row no.**
X	3	2	−1	1	0	0	R_1
	4	0	1	0	1	0	R_2
	−2	4	3	0	0	1	R_3
$^*R_1 \times \frac{1}{3} = MR_1$	1	$\frac{2}{3}$	$-\frac{1}{3}$	$\frac{1}{3}$	0	0	R_4
$R_2 - 4 \times MR_1$	0	$-\frac{8}{3}$	$\frac{7}{3}$	$-\frac{4}{3}$	1	0	R_5
$R_3 + 2 \times MR_1$	0	$\frac{16}{3}$	$\frac{7}{3}$	$\frac{2}{3}$	0	1	R_6
$R_4 - \frac{2}{3} \times MR_2$	1	0	$\frac{1}{4}$	0	$\frac{1}{4}$	0	R_7
$^*R_5 \times \left(-\frac{3}{8}\right) \times MR_2$	0	1	$-\frac{7}{8}$	$\frac{1}{2}$	$-\frac{3}{8}$	0	R_8
$R_6 - \frac{16}{3} \times MR_2$	0	0	7	−2	2	1	R_9
$R_7 - \frac{1}{4} \times MR_3$	1	0	0	$\frac{1}{14}$	$\frac{5}{28}$	$-\frac{1}{28}$	
$R_8 + \frac{7}{8} \times MR_3$	0	1	0	$\frac{1}{4}$	$-\frac{1}{8}$	$\frac{1}{8}$	$\mathbf{X}^{-1}$
$^*R_9 \times \frac{1}{7} = MR_3$	0	0	1	$-\frac{2}{7}$	$\frac{2}{7}$	$\frac{1}{7}$	

This example involves three stages:

Stage 1

Aim: To have the number 1 in the top left-hand corner, with zeros elsewhere in the first column. This will produce the first column of the desired identity matrix.

Method: First, identify the master row, *: R_1 (row 1) is divided by 3 (or multiplied by $\frac{1}{3}$, an equivalent operation), to achieve the first part of our aim, a 1 in the top left-hand corner. The resulting new row (R_4) becomes the master row for this stage of the operation, denoted by MR_1. All other operations in this stage are relative to the master row, MR_1.

$$^*R_1 \times \frac{1}{3} = MR_1$$

Next, we must carry out the necessary operations to get zeros elsewhere in the first column. R_1 has been dealt with, so we move on to R_2. To get rid of the 4, presently occupying the first column position, we need to subtract $(4 \times MR_1)$ from R_2 as shown. That is:

$$\begin{aligned} R_5 &= R_2 - (4 \times MR_1) \\ &= (4\ \ 0\ \ 1\ \ 0\ \ 1\ \ 0) - 4(1\ \ \tfrac{2}{3} - \tfrac{1}{3}\ \ \tfrac{1}{3}\ \ 0\ \ 0) \\ &= (4-4\ \ 0-\tfrac{8}{3}\ \ 1-(-\tfrac{1}{3})\ \ 0-\tfrac{4}{3}\ \ 1-0\ \ 0-0) \\ &= (0 - \tfrac{8}{3}\ \ \tfrac{7}{3} - \tfrac{4}{3}\ \ 1\ \ 0) \end{aligned}$$

Similarly, to get rid of −2 in the the third row of the first column, we need to add $(2 \times MR_1)$ to R_3, that is $R_6 = R_3 + (2 \times MR_1)$.

The aim of stage 1 has now been achieved.

Stage 2

Aim: To have the number 1 in position X_{22} (row 2, column 2) of the original matrix **X** and zeros elsewhere in the second column.

Method: First, identify the master row for the stage, *. R_5 is divided by $-\frac{8}{3}$ (multiplied by $-\frac{3}{8}$ to give the master row for the second stage, MR_2:

$$^{*}R_5 \times (-\tfrac{3}{8}) = -\tfrac{3}{8}(0 \;\; -\tfrac{8}{3} \;\; \tfrac{7}{3} \;\; -\tfrac{4}{3} \;\; 1 \;\; 0)$$

$$= (0 \;\; 1 \;\; -\tfrac{7}{8} \;\; -\tfrac{4}{8} \;\; -\tfrac{3}{8} \;\; 0) = MR_2$$

As in stage 1, we need to get zeros elsewhere in the column. The necessary operations are similar to those carried out in stage 1 and are summarized in table 9.2.

Stage 3

Stage 3 deals with the third column. We divide R_9 by 7 (multiply by $\frac{1}{7}$ to get a 1 in the bottom left-hand corner, then we use the resulting master row, MR_3, to obtain zeros elsewhere in the column. The necessary operations are summarized in table 9.2.

Once the identity matrix has been arrived at, in the left-hand portion of the combined matrix, the procedure is complete, the matrix in the right-hand portion being the required inverse matrix $\mathbf{X}^{-1}$. To confirm that it is the inverse of **X** that we have found, refer to fig. 9.3 which shows the result of multiplying **X** and $\mathbf{X}^{-1}$. Note that $\mathbf{X}^{-1}$, as shown in fig. 9.3, has been simplified. Scalar multiplication will confirm it is the same matrix as that found in table 9.3.

9.6.3.5 Singular matrices

One thing to notice about **X**, the matrix just considered, is that its rows are independent, that is, it is impossible to get any row by scalar multiplication of another. If you consider the following matrix, **Y**, which does have dependent rows, you will find it is impossible to find its inverse.

$$\mathbf{Y} = \begin{bmatrix} 1 & -1 \\ 6 & -6 \end{bmatrix} \quad (6 \times \text{row } 1 = \text{row } 2)$$

Y is said to be **singular.**

9.6.4 Matrix solutions of simultaneous equations

Consider the following pair of simultaneous equations:

$$4p + 2q = 8$$
$$p - 2q = 7$$

They can be written in matrix form:

$$\underset{2\times 2}{\begin{bmatrix} 4 & 2 \\ 1 & -2 \end{bmatrix}} \underset{2\times 1}{\begin{bmatrix} p \\ q \end{bmatrix}} = \underset{2\times 1}{\begin{bmatrix} 8 \\ 7 \end{bmatrix}}$$

This forms a matrix equation:

$\mathbf{A}x = y$, where $\mathbf{A} = \begin{bmatrix} 4 & 2 \\ 1 & -2 \end{bmatrix}$, $x = \begin{bmatrix} p \\ q \end{bmatrix}$ and $y = \begin{bmatrix} 8 \\ 7 \end{bmatrix}$

There exists an inverse of **A**, $\mathbf{A}^{-1}$, so it is possible to multiply both sides of the matrix equation to get

$$\mathbf{A}^{-1}\mathbf{A}x = \mathbf{A}^{-1}y$$

$\mathbf{A}^{-1}\mathbf{A} = \mathbf{I}$, the identity matrix, so the equation reduces to

$$\mathbf{I}x = \mathbf{A}^{-1}y$$

which, as multiplying by **I** will leave x unchanged, gives

$$x = \mathbf{A}^{-1}y$$

the solution to the simultaneous equations.

The full working for the example is as follows:

To solve the following simultaneous equations: $4p + 2q = 8$
$p - 2q = 7$

1 Write the simultaneous equations in matrix format:

$$\begin{bmatrix} 4 & 2 \\ 1 & -2 \end{bmatrix} \begin{bmatrix} p \\ q \end{bmatrix} = \begin{bmatrix} 8 \\ 7 \end{bmatrix} \quad \text{or } \mathbf{A}x = y$$

2 Find the inverse of **A**:

					Row no.
	4	2	1	0	R_1
	1	−2	0	1	R_2
$MR_1 = R_1 \div 4$	1	0.5	0.25	0	R_3
$R_2 - MR_1$	0	−2.5	−0.25	1	R_4
$R_3 - (0.5 \times MR_2)$	1	0	0.2	0.2	R_5
$MR_2 = R_4 \div (-2.5)$	0	1	0.1	−0.4	R_6

$$\mathbf{A}^{-1} = \begin{bmatrix} 0.2 & 0.2 \\ 0.1 & -0.4 \end{bmatrix}$$

3 Use $\mathbf{A}^{-1}$ to solve the simultaneous equations – as $\mathbf{A}x = y$, then $x = \mathbf{A}^{-1}y$:

$$\begin{bmatrix} p \\ q \end{bmatrix} = \begin{bmatrix} 0.2 & 0.2 \\ 0.1 & -0.4 \end{bmatrix} \begin{bmatrix} 8 \\ 7 \end{bmatrix} = \begin{bmatrix} (0.2 \times 8) + (0.2 \times 7) \\ (0.1 \times 8) + (-0.4 \times 7) \end{bmatrix} = \begin{bmatrix} 3.0 \\ -2.0 \end{bmatrix}$$

i.e. $p = 3.0$ and $q = -2.0$

9.7 Calculus in a business context: measure of change

Calculus is one of the most important branches of mathematics and has applications in many different areas. Calculus is the study of the way one variable changes with respect to another – a central issue in business. For instance,

1 the change in sales revenue, with respect to changes in the volume of goods sold;

2 the change in production costs, with respect to changes in output;

3 the change in demand, with respect to changes in commodity pricc;

4 the change in production levels, with respect to changes in level of worker experience; and so on.

9.7.1 Rate of change

An important aspect of change is the rate at which it occurs. Calculus helps us to identify these changes. Of particular interest are:

1 the identification of the rate at which revenue is changing at a given volume of sales, which would enable us to get an idea of the marginal revenue possible from extra sales;

2 the identification of the rate at which costs are changing at a given volume of sales, which would enable us to get an idea of the marginal costs likely for increased output; and

3 the identification of points when change is at a maximum (or a minimum). For instance, profits might increase in relation to volume of sales up to a certain point, after which the marginal revenue is such that profits decline. Identifying this point of maximum profit is a common problem area; similarly, identifying points of minimum costs.

To carry out the necessary analyses involved in finding marginal revenues, maximum profits, etc., it is necessary to be familiar with certain techniques of calculus, particularly differentiation. Consider the following example which focuses on some of the areas where calculus will prove particularly useful.

9.7.2 Gradients and stationary points: an example

Figure 9.4, (overleaf) illustrates two examples of the possible relationship between production output, x, and the associated costs, C. The first is a linear relationship:

$$C(x) = 100 + 0.75x$$

The left-hand side of the equation is written in the form $C(x)$, informing us that C, production costs, is a function of x, production output. Similarly, $y = 3x + 2$ can be written as $f(x) = 3x + 2$, indicating that y is a function of x (or y is dependent on x, the independent variable).

A number of factors of the relationship between C and x can be seen from the equation and the graph:

1 There are fixed costs of £100, i.e. costs which exist even when the production level is zero. In practice these could be due to capital outlay on machinery or investment in premises, etc.

2 The gradient of any straight line is, by definition, constant and represents the rate of change of one variable with respect to another – in this case, the rate of change of C with respect to x. We note that for each unit increase in production there is an increase in cost of £0.75 – i.e. the 'marginal cost' of the process, the cost entailed in producing one more unit, is £0.75.

The second cost/production relationship shown in fig. 9.4 is represented by a cubic equation:

$$C(x) = \frac{x^3}{3000} - \frac{x^2}{20} + 2.5x + 100$$

Certain observations can be made about this relationship:

1 Fixed costs are, again, £100. (Observe that the intercept point on the vertical axis, the C-axis, is 100 on both graphs.)

2 The rate of change of costs, with respect to production, is variable. As production increases from 0 to 50, the gradient of the function $C(x)$ can be seen to be getting shallower (less steep), indicating that the marginal costs involved in increasing production are falling. At point B the gradient is zero – this is called the **stationary point**. (Stationary points will be looked at in more detail in section 9.7.5.) As production levels increase from 50, the gradient of the graph increases, indicating an increase in marginal costs once production passes 50.

3 The gradient of a curve at a particular point, which value represents the rate of change of the associated function, can be estimated from the graph. (By the end of the next section you will see that there are much easier and more efficient methods available.) A line is drawn, which just touches the point of interest (e.g. point A on fig. 9.4); this line is called a **tangent**. The gradient of the tangent has the same value as the gradient of the curve at the point of contact. We can thus estimate the gradient at a particular point, as shown on fig. 9.4 (at point A, the gradient is 0.9).

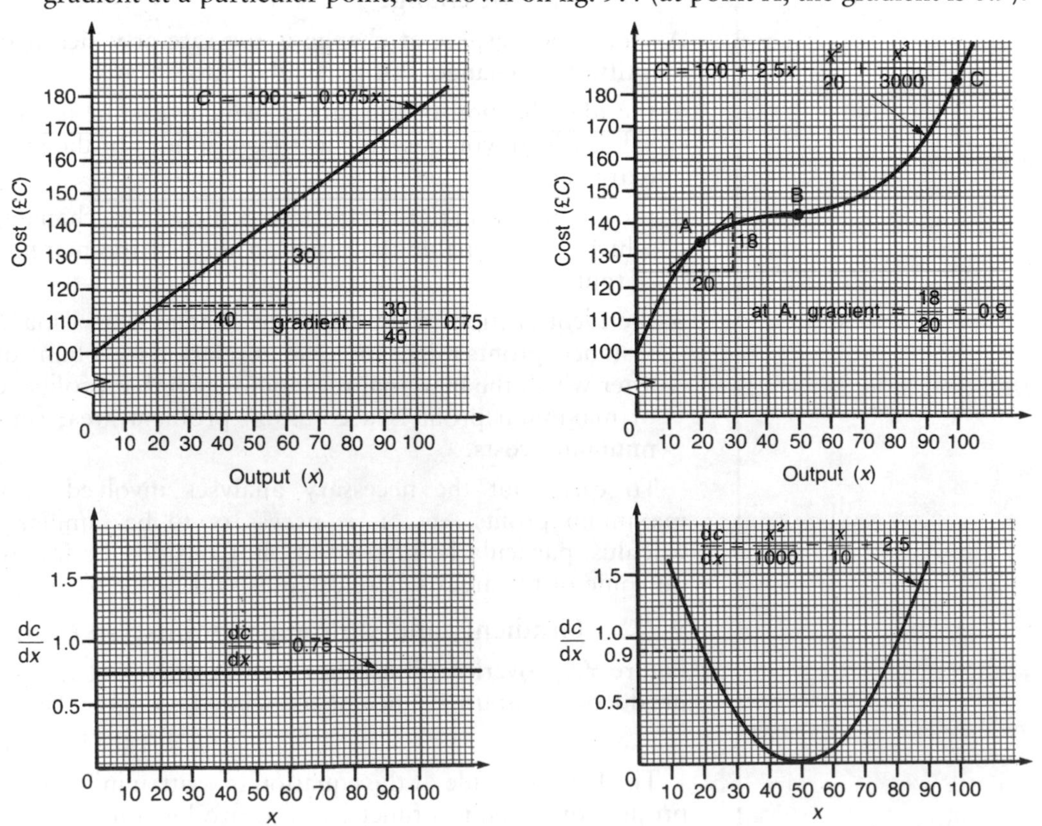

Fig. 9.4 Graphs illustrating two possible relationships between production level (x) and production costs (C)

9.7.3 Differentiation

As just shown, we can estimate the gradient of a curve graphically. We now consider the methodology necessary to calculate gradients simply and efficiently, using an important section of calculus, namely differentiation.

The rate of change of a function, $y = ax^n$, is found by differentiating the function, to acquire a **differential function**, $\frac{dy}{dx} = nax^{n-1}$.

$\frac{dy}{dx}$ (read 'dee-y by dee-x') is called the **derivative of** y.

E.g. if $y = 2x^5$, then $\frac{dy}{dx} = 5.2.x^{5-1} = 10x^4$

and if $C = 5P^3$, then $\frac{dC}{dP} = 3.5.P^{3-1} = 15P^2$

You should be aware of the following aspects of differentiation.

1 Constant terms, when differentiated, are 0.

E.g. if $C = 3P^2 + 2$, then $\frac{dC}{dP} = 2.3.P^{2-1} + 0 = 6P^1 = 6P$.

2 Variables to the power 1 are seldom explicitly written as such and might cause confusion. When differentiating a variable to the power 1, the solution is a constant, in fact just the coefficient of that variable.

E.g. if $y = 4x$, we can write $y = 4x^1$

Hence $\frac{dy}{dx} = 1.4.x^{1-1} = 4x^0 = 4 \times 1 = 4$

3 The general rule for differentiating also applies to variables with negative and fractional powers.

E.g. if $y = 3x^{-2}$, then $\frac{dy}{dx} = -2.3.x^{-2-1} = -6x^{-3}$

and if $y = 3\sqrt{x} = 3x^{\frac{1}{2}}$, then $\frac{dy}{dx} = \frac{1}{2}x^{-\frac{1}{2}} \left(= \frac{1}{2\sqrt{x}}\right)$

4 If a function is made up of several expressions added together, each expression is differentiated separately.

E.g. if $y = -3x^2 + 4x + 200$, then $\frac{dy}{dx} = 2.-3.x^{2-1} + 1.4.x^{1-1} + 0 = -6x + 4$

9.7.4 Costs and marginal costs

Consider again the example illustrated in fig. 9.4:

$$C(x) = 100 + 2.5x - \frac{x^2}{20} + \frac{x^3}{3000}$$

Differentiating, we get:

$$\frac{dC}{dx} = 0 + 1 \times 2.5x^{1-1} - 2 \times \frac{1}{20}x^{2-1} + 3 \times \frac{1}{3000}x^{3-2}$$

$$= 2.5 - \frac{x}{10} + \frac{x^2}{1000}$$

In this context, $\frac{dC}{dx}$ represents the marginal cost function and can be interpreted as the extra cost incurred in producing another item at production level x.

We can now find the gradient of the function (the marginal cost) for a given value of x (i.e. a given production level):

1 In section 9.7.2 we saw from the graph that the gradient at point A ($x = 20$) is 0.9. We could also find the gradient by substituting the value of x into the marginal cost function, $\frac{dC}{dx}$:

$$\frac{dC}{dx} = 2.5 - \frac{20}{10} + \frac{20 \times 20}{1000}$$

$$= 2.5 - 2 + 0.4 = 0.9$$

So, when production is 20, the marginal cost of increasing production by one unit will be £0.9 (90 pence).

2 At point B (fig. 9.4), $x = 50$, therefore

$$\frac{dC}{dx} = 2.5 - \frac{50}{10} + \frac{50 \times 50}{1000}$$
$$= 2.5 - 5 + 2.5 = 0.0$$

That is, the gradient at point B is zero and the marginal cost of increasing production by one unit, when production is 50, is estimated to be £0.

3 At point C (fig. 9.4), $x = 100$, therefore

$$\frac{dC}{dx} = 2.5 - \frac{100}{10} + \frac{100 \times 100}{1000}$$
$$= 2.5 - 10 + 10 = 2.5$$

Strictly speaking, the marginal cost at a general production level, x, is the extra cost involved in producing the $(x+1)$th item. Consequently, the marginal cost at a production level of 50 is

$$C(51) - C(50) = 141.667 + 141.66\dot{6} = 0.00033\dot{3}$$

This suggests that $\frac{dC}{dx}(x = 50) = 0$ is a good estimate. Similarly, $\frac{dC}{dx}(x = 100) = 2.5$ compares favourably with

$$C(101) - C(100) = 185.8836\dot{6} - 183.33\dot{3} = 2.55033\dot{3}$$

9.7.5 Profit maximization

In the above example, we saw how to get an expression for the marginal cost by differentiating the expression for the total cost. Both of these expressions were functions of the production level. We could similarly differentiate the revenue expression to get an expression for the marginal revenue.

It is common in examinations for candidates to be required to deduce the **profit equation,** given equations for total cost and total revenue, and to use this to calculate the maximum profit.

Given expressions for revenue and costs it is easy to compute the related profit equation:

Profit = Revenue − Costs

If we take the cost equation to be the same as that considered above,

$$C(x) = 100 + 2.5x - \frac{x^3}{20} + \frac{x^3}{3000}$$

and we are told that the revenue equation is

$$R(x) = 5x - \frac{x^2}{32}$$

we can get an expression for the profit, $P(x) = R(x) - C(x)$:

$$P(x) = \left(5x - \frac{x^2}{32}\right) - \left(100 + 2.5x - \frac{x^2}{20} + \frac{x^3}{3000}\right)$$
$$= -100 + (5 - 2.5)x + \left(-\frac{1}{32} + \frac{1}{20}\right)x^2 - \frac{x^3}{3000}$$
$$= -100 + 2.5x + \frac{3x^2}{160} - \frac{x^3}{3000}$$

The three expressions, $C(x)$, $R(x)$ and $P(x)$ are graphed together in fig. 9.5 (opposite), from which it would appear that the profit is maximized when the production output is approximately 70 units. Point M is called a **local maximum.** It is important to notice that at a maximum point (or a minimum point) the gradient of the curve is zero. Maximum and minimum points and **points of inflection** (e.g. point B, fig. 9.4), at which the gradient of the curve is zero are called **stationary points.**

9.7.5.1 Finding stationary points

We have an equation for the profits, $P(x)$. $P(x)$ can be differentiated to give the gradient function:

$$\frac{dP}{dx} = 2.5 + \frac{3x}{80} - \frac{x^2}{1000}$$

As stated above, stationary points occur where the gradient is zero. Therefore, to find the stationary points, we need to equate dP/dx to zero and solve the resulting equation. In this case, we get $2.5 + 3x/80 - x^2/1000 = 0$, a quadratic equation which we can solve using the general formula, as follows:

Rearranging the equation into the form $ax^2 + bx + c = 0$:

$$-\frac{x^2}{1000} + \frac{3x}{80} + 2.5 = 0$$

we see that $a = -\frac{1}{1000}$, $b = \frac{3}{80}$ and $c = 2.5$.

Substituting these values into the formula, we get:

$$x = \frac{-3/80 \pm \sqrt{(3/80)^2 - (4 \times -1/1000 \times 2.5)}}{2(-1/1000)}$$

$$= \frac{-0.0375 \pm \sqrt{0.0014062 - (-0.01)}}{-0.002}$$

$$= \frac{-0.0375 \pm 0.1068}{-0.002}$$

$$= +72.15 \text{ or } -34.65$$

The values $x = +72.15$ and $x = -34.65$ correspond to stationary points of the profit function $P(x)$. As has been stated, a stationary point can take a number of forms: a minimum, a maximum or a point of inflection. To determine what type of stationary points we have, a quick and easy method is to differentiate the derivative.

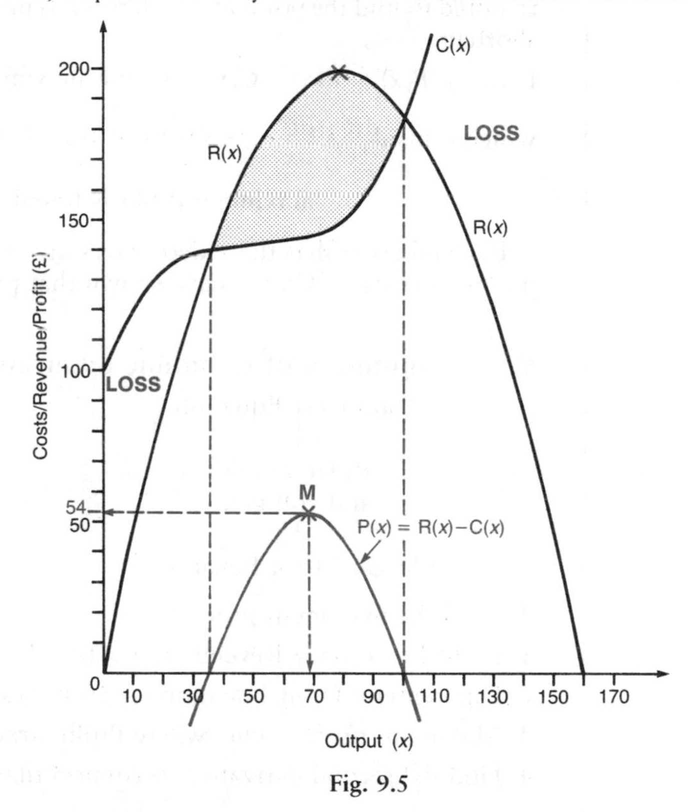

Fig. 9.5

9.7.5.2 The second derivative: maximum, minimum or point of inflection?

In our example, so far we have

$$P(x) = -100 + 2.5x + \frac{3x^2}{160} - \frac{x^3}{3000} \text{ and } \frac{dP}{dx} = 2.5 + \frac{3x}{80} - \frac{x^2}{1000};$$

and $\frac{dP}{dx} = 0$ when $x = +72.15$ or -34.65

Is $x = +72.15$ at a minimum point, a maximum point or a point of inflection? To find out we differentiate $\frac{dP}{dx}$, to get the 'second derivative', $\frac{d^2P}{dx^2}$:

$$\frac{d^2P}{dx^2} = \frac{3}{80} - \frac{2x}{1000} = \frac{3}{80} - \frac{x}{500}$$

The rule is: if $\frac{d^2P(x)}{dx^2} < 0$ **then x is a maximum point;**

if $\frac{d^2P(x)}{dx^2} = 0$ then x is a point of inflection;

if $\frac{d^2P(x)}{dx^2} > 0$ then x is a minimum point.

For $x = +72.15$, $\frac{d^2P}{dx^2}(72.15) = \frac{3}{80} - \frac{72.15}{500}$

$$= -0.1068 < 0$$

Therefore $x = +72.15$ represents a maximum point. (You will find that the negative solution, $x = -34.65$, is a minimum, but it is outside of the range we are considering–we cannot have a production level of -34.65–and is therefore an unacceptable solution.)

We have not yet found the maximum profit. We have found the output (x) at which the maximum profit occurs, $x = 72.15$. The whole number 72 is substituted back into the profit equation, $P(x)$, since we cannot assume that it is possible to manufacture partial units:

$$P(72) = -100 + (2.5 \times 72) + \frac{3 \times 72 \times 72}{160} - \frac{72 \times 72 \times 72}{3000}$$

$$= £52.78$$

Note: If you are not explicitly asked to calculate the profit function, but are still required to find the point at which profit is maximized, the following provides a useful shortcut:

Because $P(x) = R(x) - C(x)$ and the maximum profit is found by solving $\frac{dP}{dx} = 0$, which is just $\frac{dR}{dx} - \frac{dC}{dx} = 0$, then it is true that

maximum profit is found when $\frac{dR}{dx} = \frac{dC}{dx}$.

Remembering that these derivatives are estimates for marginal revenue (MR) and marginal costs (MC), we have shown that profit is maximized when MR = MC.

9.7.6 Summary of economic applications of differentiation

Total Cost Equation	Total Revenue Equation
|	|
differentiate and you get:	differentiate and you get:
↓	↓
Marginal Cost Equation	Marginal Revenue Equation

To find the maximum profit:

1 Profit Equation = Revenue Equation − Cost Equation
2 Differentiate Profit Equation (= profit gradient)
3 Maximum profit occurs where Profit Gradient = 0
4 Find the second derivative to confirm that the maximum has been found.

9.7.7 Calculus: stock-holding problem

Phil-it Storage Ltd has monitored the various costs involved in ordering stock and keeping stock in storage. The total costs of the company can be expressed by the following function:

$$C(Q) = \frac{Qh}{2} + \frac{cD}{Q} + f$$

where: Q = the size of each order;
D = the annual demand for goods;
h = the cost of holding one unit of stock in store per annum;
c = the cost of placing an order with the company's suppliers;
f = the annual fixed costs.

Phillip Robinson, the company accountant, has estimated that for 1989 the following will be the case:

$D = 100\,000$ units;
h = £1.00 per annum (a combination of capital costs and warehousing costs);
c = £20 per order, irrespective of order quantity;
f = £50 000.

He wishes to find the optimum order size to minimize total costs for the year. C, which is a function of Q (all other factors being known), is the expression for the total costs involved. Using the estimated figures:

$$C(Q) = \frac{Q \times 1.00}{2} + \frac{20 \times 100\,000}{Q} + 50\,000$$

$$= 0.5Q + 2\,000\,000Q^{-1} + 50\,000 \qquad \left(Q^{-1} = \frac{1}{Q}\right)$$

To find the value of Q, the order quantity, which will minimize costs, it is necessary to differentiate C, with respect to Q:

$$\frac{dC}{dQ} = 0.5 - 2\,000\,000\ Q^{-2}$$

The minimum cost occurs at the value of Q which makes $\frac{dC}{dQ} = 0$:

That is, $0 = 0.5 + 2\,000\,000\ Q^{-2} = 0.5 = \frac{2\,000\,000}{Q^2}$

Therefore $Q^2 = \frac{2\,000\,000}{0.5} = 4\,000\,000$

$Q = \sqrt{4\,000\,000} = 2000$

Thus, the optimum order size for minimization of total costs is 2000 units. If Mr Robinson's estimates are correct, the total cost for 1989 will be £52 000, $C(2000)$.

9.7.8 Integration

Integration is the reverse of differentiation. If $y = ax^n$ is integrated, then the resulting integral, $\int y\mathrm{d}x$, read as **'the integral of y with respect to x'**, is given by

$$\int y\mathbf{dx} = \int ax^n\mathbf{dx} = \frac{ax^{n+1}}{n+1} + c \qquad (c \text{ is a constant})$$

For example:

1 If $y = 3x^2 + 4x - 5$, we know the differential function is $\frac{dy}{dx} = 6x + 4$.

If we start with the function, $f(x) = 6x + 4$, then the integral is $\int f(x)dx$, ('the integral of function f with respect to the variable x'):

$$\int f(x)dx = \int (6x + 4)dx = \frac{6x^2}{2} + \frac{4x^1}{1}$$

$$= 3x^2 + 4x + c$$

The 'constant of integration', c, can only be found if more information is available. In this instance we know it to be -5, but *any* constant, when differentiated becomes 0. If y had been equal to $3x^2 + 4x + 150$, $\frac{dy}{dx}$ would still have been $6x + 4$. The process of integration must take into account the existence of any possible constant.

2 If $f(x) = 3x^2 + 4x - 5 + 2x^{-2}$, $\int f(x)dx = \int (3x^2 + 4x - 5 + 2x^{-2})dx$

$$= \frac{3x^3}{2+1} + \frac{4x^2}{1+1} - \frac{5x}{0+1} + \frac{2x^{-1}}{-2+1} + c$$

$$= x^3 + 2x^2 - 5x - 2x^{-1} + c$$

Whenever you integrate a function check your result by differentiating it—you should get back to the function you started with.

9.8 Rounding of data

Often data in its raw state is unnecessarily precise, making it difficult to read and remember, which is often the primary objective of presenting data. In such cases it is acceptable to 'round' the data. For example, if a company's profit were £4 231 100.35, for ease of reading, and to give an immediate idea of the scale of profits, this information would in some circumstances be better rounded to the nearest million pounds: 'profits for 1987 were just over £4 million'. Obviously, for accounting purposes, the company would be required to be exact in their profit report.

9.8.1 Unbiased rounding of data

The above example illustrates the desirability of rounding data. As long as this is done in a reasonable manner the resulting data can be confidently used, though one should always find out if rounding has taken place (see the 'painting sale' report example in section 2.7).

Suppose the following data represents the actual attendance figures for a number of football matches:

56 231	998
1 465	15 431
43 501	9 702
2 786	11 100
12 157	7 009

Rounded to the nearest thousand, the data will be:

56 000	1 000
1 000	15 000
44 000	10 000
3 000	11 000
12 000	7 000

The sum of the rounded data is 160 000, compared with the sum of the actual data of 160 470 – an **unbiased error** of 470.

9.8.2 Biased rounding of data

If the data is **rounded upwards (downwards)**, then an **upward (downward) bias** is introduced to the resulting calculations. Rounding the football attendance data upwards to the nearest thousand gives the following:

57 000	1 000
2 000	16 000
44 000	10 000
3 000	12 000
13 000	8 000

The sum is now 166 000; a **biased error** of 5530.

9.8.3 Types of error

9.8.3.1 Absolute error

The errors arrived at in sections 9.8.1 and 9.8.2 are both absolute errors, that is, errors measured in the same units as the original data.

9.8.3.2 Relative error

Relative error is expressed as a proportion (usually a percentage) of the original data. In our example, rounding the data to the nearest thousand (in an unbiased manner) gave an absolute error of 470. In this case the relative error is:

$$\frac{470}{160\,470} \times 100 = 0.293\% \text{ (to 3 d.p., a rounding process in itself)}$$

For the example where the data was rounded up, resulting in a (biased) absolute error of 5530, the resulting relative error can be calculated as follows:

$$\frac{5530}{166\,000} \times 100 = 3.331\% \text{ (to 3 d.p.)}$$

9.8.3.3 Average error

The data shown above was initially rounded to the nearest thousand. This would result in a maximum error (for each piece of data) of ± 500. For example, given the information that the piece of data 56 000 had been rounded to the nearest thousand, all one could say about the true value is that it is 56 000 ± 500, that is, somewhere in the range 55 500 to 56 500. Since the maximum error is ± 500 the average error is 250 (per piece of data). Rounding the data upwards could have resulted in a maximum error of 1000, giving an average error per piece of data of 500.

9.8.3.4 Accumulated error

In adding rounded data (or carrying out any other calculation) the error content of the final statistic will be an accumulation of the separate errors of the individual figures. In the above unbiased-rounding example, the errors would be expected to cancel out to some extent, unlike the case of continuously rounding up which is bound to produce a positive accumulation of error.

For **unbiased approximations**:

$$\textbf{total average error} = \sqrt{n} \times \textbf{average error of individual data}$$

where n = number of pieces of data.

In the example considered above,

$$\text{total average error} = \sqrt{10} \times 250 = 791 \text{ (to the nearest whole number)}$$

The actual error was 470.

For **biased approximations**:

$$\textbf{total average error} = n \times \textbf{average error of individual data}$$

In the example considered above,

$$\text{total average error} = 10 \times 500 = 5000$$

The actual error was 5530.

Sample questions

1 The quantity 10^{-2} means

A -100 B $-\frac{2}{10}$ C $\frac{1}{100}$ D $\frac{2}{10}$ E $\frac{1}{10}$

CIMA; Foundation Stage, section B; Mathematics and statistics; May 1986.

Answer: C

2 The fraction $\frac{x^6}{x^3}$ equals

A 2 B 3 C $\frac{1}{x^2}$ D x^2 E x^3

CIMA; Foundation Stage, section B; Mathematics and statistics; November 1986.

Answer: E

3 Two lines are given by the equations

$$4Y = 9X + 1$$
$$Y = X - 1$$

At which of the following coordinates do these lines intersect?

A $X = -1$, $Y = -2$ B $X = \frac{3}{5}$, $Y = -\frac{2}{5}$ C $X = 1$, $Y = 2\frac{1}{2}$

D $X = 2$, $Y = 1$ E none of these

Solution/hints

The lines intersect at the coordinates which satisfy both equations.

Answer: A

4 The sum of the first six terms of a geometric progression which has a second term of 6 and a common ratio of 2 is

A 54 B 72 C 96 D 189 E 378

Solution/hints

We are given $ar = 6$ and $r = 2$. Use this to find a, the first term of the series and substitute that value in the equation $S_n = \frac{a(1-r^n)}{1-r}$, to find S_6 (see section 9.2.5.1).

Answer: **D**

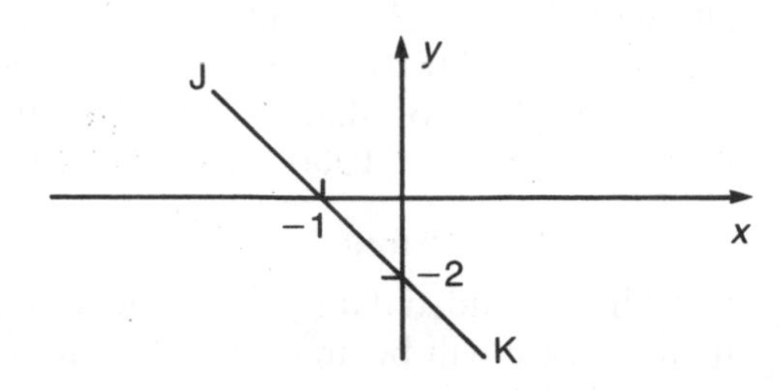

5 The equation of the straight line JK is

A $y = -\frac{1}{2}x$ **B** $y + 2x + 2 = 0$ **C** $y = 2x + 2$ **D** $y = 2x - 2$

E none of these

Solution/hints

1 The equation of any straight line can be written in a 'standard form': $y = a + bx$, where a is the y-axis intercept and b is the gradient of the line.

2 To establish the values of a and b, either:

(a) note the values of the vertical (y-) axis intercept and the gradient of the line from the graph; or

(b) use any two points to form simultaneous equations which can be solved to find a and b.

In this case, (a) is the more convenient method: We can immediately see from the graph that the y-axis intercept gives $a = -2$. Given the two points $(-1,0)$ and $(0,-2)$, the gradient b, can be found as follows:

$$b = \frac{0-(-2)}{-1-0} = \frac{2}{-1} = -2$$

Hence the equation of JK is $y = -2 - 2x$, which can be rearranged to $y + 2x + 2 = 0$.

Answer: **B**

6 The geometric mean of three numbers is 6. A fourth number is included so that the geometric mean for all four numbers is 12. The fourth number is

A 4 **B** $\sqrt{96}$ **C** 16 **D** 30 **E** 96

Solution/hints

The geometric mean for the three numbers can be represented as follows:

$$6 = \sqrt[3]{a \times b \times c}$$

For the geometric mean of the four numbers we can write:

$$12 = \sqrt[4]{a \times b \times c \times d}$$

Rearranging these expressions, we can write:

$$a \times b \times c = 6^3 = 216$$

and $$a \times b \times c \times d = 12^4 = 20\,736$$

Therefore $$d = \frac{a \times b \times c \times d}{a \times b \times c} = \frac{20\,736}{216} = 96$$

Answer: **E**

7 Given that $(x-4)(x-5) = 2$, the roots of the equation are

A $x = -4$ and $x = -5$ **B** $x = -2$ and $x = -3$

C $x = 2$ and $x = 2\frac{1}{2}$ **D** $x = 3$ and $x = 6$

E $x = 4$ and $x = 5$

Solution/hints

Either substitute the given values of x into the equation until you find the pair of values which make the equation a true statement or solve the equation as follows:

1 Multiply out the brackets to give $x^2-9x+20=2$.

2 Rearrange this into the form $ax^2+bx+c=0$: $x^2-9x+18=0$.

3 Factorize: $(x-3)(x-6)=0$.

4 The solutions (roots) of the equation are given by $x-3=0$ and $x-6=0$ i.e. $x=3$ and $x=6$ (*Answer:* **D**).

Alternatively, having rearranged the equation into the form ax^2+bx+c, you could solve it using the general formula (see section 9.5.2).

8 (**a**) A company subsidizes a certain rail journey for some of its employees. When the price of the ticket is increased by £6, the number of tickets which the company can purchase for £2850 is reduced by 36.

You are required to find the percentage increase in the price of the ticket.

(**b**) A mathematically-minded street trader with no overheads has found that the weekly volume of sales of a toy is approximately $100/p^2$, where £p is the fixed price of the toy. The toy costs the trader 15 pence.

You are required, by using a graph, or any other method you consider appropriate, to find:

(i) the level of p which maximizes profit;

(ii) the level of this maximum profit;

(iii) the weekly volume of toys sold at this level.

CIMA; Foundation stage B; Mathematics and statistics; November 1986.

Solution/hints

Part (a)

Let the price and number of tickets sold before the price increase be x (£) and y (tickets), respectively. The amount spent by the company is £2850, i.e.

$$xy=2850 \qquad (1)$$

When the price of tickets is increased by £6 to £$(x+6)$, the number of tickets that can be bought for £2850 decreases by 36 to $(y-36)$. Therefore,

$$(x+6)(y-36)=2850 \qquad (2)$$

To find the percentage increase in ticket prices, we need to know the value of x, the original price which will satisfy both equations (1) and (2).

From (1), we have $y=\dfrac{2850}{x}$ (3)

Substituting this expression for y into (2) we get:

$$2850=(x+6)\left(\frac{2850}{x}-36\right)$$

$$=(x+6)\left(\frac{2850-36x}{x}\right)$$

Multiply both sides of the equation by x to get:

$$(x+6)(2850-36x)=2850x \qquad (4)$$

Equation (4) can be expanded and rearranged to give the quadratic equation:

$$x^2+6x-475=0 \qquad (5)$$

Equation (5) can be solved using the general formula (see section 9.5.2) or by factorizing as follows:

$$(x+25)(x-19)=0$$

Hence, $x=19$ (-25 is unacceptable – the ticket price cannot be minus £25).

We can now answer the question: The percentage increase in ticket price is

$$\frac{\text{price increase}}{\text{original price}}\times 100=\frac{6}{19}\times 100=31.58\%$$

Part (b)

The profit made by the street trader is given by the expression

$$(\text{unit price}-\text{unit cost})\times\text{volume of sales}$$

$$=(p-0.15)\times\frac{100}{p^2}=\frac{100}{p}-\frac{15}{p^2} \qquad (1)$$

The question suggests you could use a graph: Calculate profit values for various prices (p) using the profit equation (**1**). Plot profit versus price as shown:

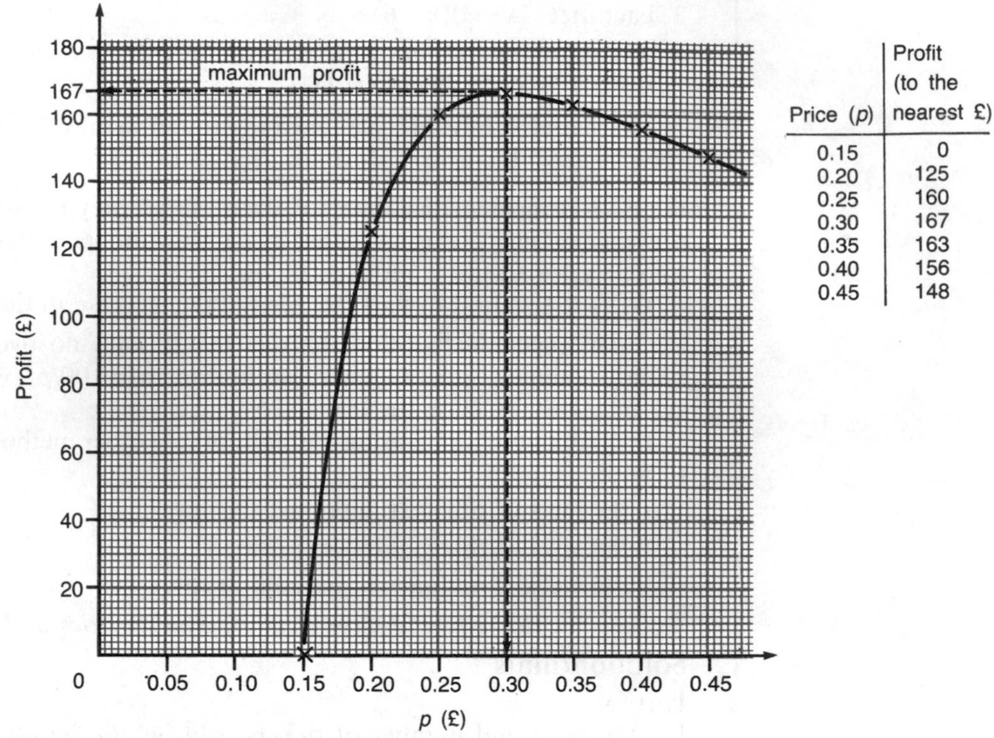

Price (p)	Profit (to the nearest £)
0.15	0
0.20	125
0.25	160
0.30	167
0.35	163
0.40	156
0.45	148

From the graph:

(i) $p = £0.30$

(ii) maximum profit = £167

(iii) weekly volume of toys sold (at $p = 0.30$) $= \dfrac{100}{0.3^2} = 1111$ toys

Alternatively, you could use calculus:

Differentiate the profit equation (**1**): $-\dfrac{100}{p^2} + \dfrac{30}{p^3}$

(i) The maximum profit will occur when the derivative (the gradient) of the profit equation equals zero, that is, when

$$-\frac{100}{p^2} + \frac{30}{p^3} = 0 \qquad \text{or} \qquad \frac{100}{p^2} = \frac{30}{p^3}$$

Rearranging, we get $\dfrac{p^3}{p^2} = \dfrac{30}{100}$, i.e. $p = 0.3$

(ii) Substituting $p = 0.3$ into the profit equation, (**1**), we find

$$\text{maximum profit level } = \frac{100}{0.30} - \frac{15}{(0.3)^2} = £166.67$$

(iii) Find weekly sales volume as above.

9 The following data relate to the retail trade in category Z:

Table 1: Number of different grades of employees

Grade	Type of outlet		
	Branch of multiple store	**Supermarket**	**Independent**
Supervisors	1	5	nil
Cashiers	4	10	2
Storekeepers	3	15	1

Table 2: Number of outlets in different areas

Outlet	Midlands	Other areas
Branch of multiple store	36	104
Supermarket	18	75
Independent	25	30

You are required to:

(**a**) state the purpose of using matrix algebra and give two business situations in which it may be used;

(**b**) present the data in table 1 in a matrix format (subsequently referred to as **A**);

(**c**) present the data in table 2 in matrix format (subsequently referred to as **B**);

(**d**) calculate by means of matrix multiplication, showing clearly all the steps, **AB**, and explain what the products mean.

CIMA; Professional Stage, Part 1; Quantitative techniques; November 1968.

Solution/hints

Part (a)

Matrix algebra enables data to be expressed in a compact form suitable for computer-based storage and computation. They can be used for solving linear-programming problems, to show the relationship between inputs and outputs of related industries, and so on.

Parts (b) and (c)

$$\mathbf{A} = \begin{bmatrix} 1 & 5 & 0 \\ 4 & 10 & 2 \\ 3 & 15 & 1 \end{bmatrix} \qquad \mathbf{B} = \begin{bmatrix} 36 & 104 \\ 18 & 75 \\ 25 & 30 \end{bmatrix}$$

Part (c)

The calculation is left as an exercise.

Answer: $\mathbf{AB} = \begin{bmatrix} 126 & 479 \\ 374 & 546 \\ 403 & 1467 \end{bmatrix}$

The elements of column 1 of AB are the total number in each grade of employee in the Midlands. Similarly, column 2 shows the total number in each grade in 'other areas', e.g. 126 is the total number of supervisors in the Midlands.

(*Questions 10 to 12: CIMA; Quantitative methods: Stage 1; May 1987*)

10 $\int_0^2 x^3\,dx$ equals

A 4 **B** 8 **C** 12 **D** 16 **E** none of these

Solution/hints

This is a definite integral:

$$\int_0^2 x^3\,dx = \left[\frac{x^4}{4}\right]_0^2 = \frac{2^4}{4} - \frac{0^4}{4} = \frac{2^4}{4} = \frac{16}{4} = 4$$

Answer: **A**

11 The relationship between costs of production, y, and units of production, x, is given by the following equation:

$$y = x^2 + 3x + 5$$

The value of $\frac{dy}{dx}$ when $x = 5$ equals

A 9 **B** 13 **C** 45 **D** 141⅔ **E** none of these

Solution/hints

$\frac{dy}{dx} = 2x + 3$

Answer: **B**

12 Two matrices are multiplied together:

$$\begin{bmatrix} 3 & 5 \\ 6 & 7 \end{bmatrix} Y = \begin{bmatrix} 2 & 5 \\ 1 & 7 \end{bmatrix}$$

Which represents **Y**?

A $\begin{bmatrix} 1 & 0 \\ 5 & 0 \end{bmatrix}$ **B** $\begin{bmatrix} 2/3 & 1 \\ 1/6 & 1 \end{bmatrix}$ **C** $\begin{bmatrix} 1 \\ 5 \end{bmatrix}$ **D** $\begin{bmatrix} 16/9 & -5/9 \\ 35/9 & -16/9 \end{bmatrix}$ **E** $\begin{bmatrix} -1 & 0 \\ 1 & 1 \end{bmatrix}$

Solution/hints

Substitute the possible matrices for y and find which gives the true equation. Try the matrices without fractions first.

Answer: **E**

13 A company introduced a new product. Experience showed that if the price was £20, then 100 units of the product could be sold each week. However, if the price were reduced to £15, sales would increase to 150 per week. The company estimates that the fixed costs each week are £190 and the variable costs per product are £10. You may assume that linear relationships exist between price and sales and between costs and quantity produced.

(a) Show that the relationship between sales and price is:

$$p = 30 - 0.1q$$

where p is the price and q is the quantity sold.

(b) Obtain an expression for the company's revenue in terms of q.

(c) Obtain an expression for the total costs in terms of q.

(d) Determine the point(s) where the company breaks even.

(e) Obtain an expression for profits and by drawing a graph, or otherwise, advise the company on its pricing policy if it wishes to maximize profits.

CIS; Part 2, Quantitative studies; December 1986.

Solution/hints

Part (a)

Given 'a linear relationship exists between sales and price', it is possible to write:

$$p = a + bq \text{ (} a \text{ and } b \text{ are constants)}$$

Also given: when $p = 20$, $q = 100$
and when $p = 15$, $q = 150$

We can substitute these p and q values into the general linear equation to get the simultaneous equations

$$20 = a + 100b \quad (1)$$
$$\text{and} \quad 15 = a + 150b \quad (2)$$

(1) − (2) gives $5 = -50b$, therefore $b = \frac{5}{-50} = -0.1$

Substituting $b = -0.1$ into (1) gives

$$20 = a + 100 \times (-0.1)$$

Therefore, $a = 30$, giving the desired price/quantity equation, $p = 30 - 0.1q$.

Part (b)

$$\text{Revenue} = \text{Price} \times \text{Quantity} \quad \text{i.e. } R = p \times q$$
$$= (30 - 0.1q) \times q$$
$$= 30q - 0.1q^2$$

Part (c)

Given: costs and quantity are linearly related. Hence,

$$C = a' + b'q \text{ (} a' \text{ and } b' \text{ are constants)}$$

Also given: fixed costs are £190 and variable costs are £10 per product. Therefore

$$C = 190 + 10q$$

Part (d)
The company breaks even at the point where revenue is equal to costs, that is, where $R = C$:

$$30q - 0.1q^2 = 190 + 10q$$

Rearrange and solve the quadratic equation. Break-even occurs when $q = 10$ or $q = 190$.

Part (e)

$$\begin{aligned} \text{Profits} &= \text{Revenue} - \text{Costs} \\ &= 30q - 0.1q^2 - (190 + 10q) \\ P &= -0.1q^2 + 20q - 190 \end{aligned}$$

The rest of the working is left as an exercise. The question says 'by drawing a graph, *or otherwise*': the calculus for this example is quite simple ($\frac{dP}{dx} = -0.2q + 20$).

Answer: Profits are maximized when $q = 100$, that is when price is £20. To maximize profit the company should price the product at £20 per item, even though this will not achieve the highest possible sales.

14 (a) Sandford Co Ltd works a nominal 37 hour week, but because of overtime and occasional short-time the real figure is 37 ± 2 hours. The production level works out at 30 (± 3) items per hour. The cost of production (to the nearest 10p) is £5 per item and the selling price is £8 (again to the nearest 10p).
Estimate:
(i) the weekly profit;
(ii) the percentage profit, based on cost price, per item sold (to the nearest 0.5 per cent).

(b) Fred states he expects to earn £175 per week, 'give or take' £8. Give this information in the form of:
(i) an estimate with maximum error;
(ii) an estimate with maximum relative error (2 d.p.);
(iii) the appropriate interval.

(c) Fred, still discussing his wages, says 'My weekly pay has been increased by 45 per cent and Jim's has only increased by 20 per cent'. His friend Ann replies, 'That tells me nothing unless I know what the absolute figures are, and what the calculations are based on'.

Explain, giving examples of what it is that Ann is referring to.

The Polytechnic of North London; BTEC HND, Business studies; 1986.

Solution/hints

Part (a)
(i) Given: no. of hours worked = 37 ± 2
no. of items per hour = 30 ± 3

Therefore production level = (hours worked) × (items per hour)
= 1110 ± 117*

Also given: selling price = £8 ± 0.05
cost price = £5 ± 0.05

Therefore profit per item = (selling price) − (cost price)
= (£8 ± 0.05) − (£5 ± 0.05)
= £3 ± 0.1**

We are now in a position to calculate the weekly profit:
Weekly profit = (no. of items produced per week) × (profit per item)
= £3330 ± 659.7 (see * for method)

(ii) The percentage profit (based on cost) is calculated as follows:

$$\frac{\text{profit per item}}{\text{cost per item}} \times 100 = \frac{£3 \pm 0.1}{£5 \pm 0.05} \times 100$$

Method: smallest possible value $= \frac{£3 - 0.1}{£5 + 0.05} \times 100$

$= \frac{2.9}{5.05} \times 100 = 57.4$ (1 d.p.)

largest possible value $= \frac{£3 + 0.1}{£5 + 0.05} \times 100$

$= \frac{3.1}{4.95} \times 100 = 62.6$ (1 d.p.)

Therefore the percentage profit is between the limits [57.4, 62.6] or 60 ± 2.5 per cent (to the nearest 0.5 per cent).

*__Weekly production level__ will be 1110 $(37 \times 30) \pm$ error. Notice that when values are multiplied, the amount of error is not equally distributed about the expected value (unlike in the examples involving adding or subtracting values, **):

minimum hours worked	$= 35\ (37-2)$
maximum hours worked	$= 39\ (37+2)$
minimum hourly production level	$= 27\ (30-3)$
maximum hourly production level	$= 33\ (30+3)$
minimum weekly production	$= 35 \times 27 = 945$
maximum weekly production	$= 39 \times 33 = 1287$

This gives an interval of {945, 1287}. So the production level lies between $1110 - 165$ (= 945) and $1110 + 177$ (= 1287). It is customary to refer to the larger error, so we write 1110 ± 177.

**__Profit per item__:

minimum cost price	= £4.95 (£5 − 0.05)
maximum cost price	= £5.05 (£5 + 0.05)
minimum selling price	= £7.95 (£8 − 0.05)
maximum selling price	= £8.05 (£8 + 0.05)
minimum profit per item	= £7.95 − £5.05 = £2.90
maximum profit per item	= £8.05 − £4.95 = £3.10
Therefore, profit per item	= £3 ± 0.1

Part (b)

(i) £175 ± £8

(ii) £175 ± 4.57% (2 d.p.) $(4.57 = \frac{8}{175} \times 100)$

(iii) [£167, £183]

Part (c)

Fred seems to believe he has enjoyed the greater benefit because his percentage increase is greater than Jim's. When using percentages it is essential to know the full story. Suppose the following was the case:

Jim's wages:	1987	£250
	1988	£312.50 (includes 20 per cent increase based on 1987 figure)
Actual increase		£62.50 (over one year)
Fred's wages:	1984	£121
	1987	£175 (includes approximate 45 per cent increase based on 1984 figure)
Actual increase		£54 (over three years)

What Fred said was correct; his (percentage) increase was higher than Jim's but Jim started with more so his actual increase is larger. Also, Jim's increase was based on the previous year, whereas Fred's increase is based on his wage three years before, giving Fred an average annual increase of only 15 per cent (based on 1984 data).

Recommended reading

Curwin, J. and Slater, R., *Quantitative Methods for Business Decisions* (Van Nostrand Reinhold, 1986), chapters 6 and 7.

Owen, F. and Jones, R., *Modern Analytical Techniques* (Pitman, 1988), chapters 1, 2 and 5.

10 The time value of money and financial appraisal

10.1 Introduction

We value money for its purchasing power now and in the future. As individuals we are continually faced with decisions of whether to purchase an item now, wait until we have saved the purchase cost or, perhaps, borrow money to purchase now and pay back the loan over a period of time. The last is the most common option when it comes to a financial transaction considered by many people to be the most important they will undertake, the purchase of their home. Not many people are in a position to buy their house for cash, consequently they usually save a deposit and borrow the rest.

On a larger scale, monetary value is central to the performance of a business. Whether the business is involved in providing a service or a product, it will need to make money to survive. This chapter is concerned with the analysis involved in financial decision-making, whether domestic in nature or to do with large business transactions. The type of question you are likely to encounter will reflect this divide, ranging from examples to do with personal finance to the sometimes complex analysis of business projects. As many financial calculations involve the concept of paying out or receiving money at regular intervals, you should make sure you are familiar with arithmetic and geometric progressions. Also, it will be useful, in what follows, for you to be familiar with the following terminology and notation.

P = an amount of money at the present time, often referred to as the principal;
S = an amount arising at some future time;
n = the number of periods over which interest is calculated, (usually, but not always, years);
r = the rate of interest, usually per annum;
I = the total amount of interest.

10.2 Simple interest

If an amount of money is invested such that interest is accrued only on the initial amount invested and not on any subsequent reinvested interest, the interest thus earned is called simple interest.

For instance, how much will £12 000 earn in three years at an interest rate of 8 per cent per annum? The answer is quite easy to calculate:

Year (n)	Amount after year n
0	£12 000
1	£12 000 + 0.08 × £12 000 = £12 000 + 960 = £12 960
2	£12 960 + 0.08 × £12 000 = £12 960 + 960 = £13 920
3	£13 920 + 0.08 × £12 000 = £13 920 + 960 = £14 880

At 8 per cent simple interest, £12 000 would **accrue** (increase by) £2880 interest. The total amount available after n years is the $(n+1)$st term of an arithmetic progression, with first term $a = P$ and common difference $d = rP$. The amount present after n years, S, and the interest earned in that period, I, are expressed as

$$S = P(1 + nr) \text{ and } I = nrP$$

With the same conditions, how long would it take for the £12 000 to double?

$$P = 12\,000;\ S = 24\,000;\ r = 0.08;\ n = ?$$

We have $24\,000 = 12\,000(1 + n \times 0.08)$
$= 12\,000 + 960n$

Therefore $n = \dfrac{24\,000 + 12\,000}{960} = 12.5.$

It is assumed in financial appraisal of this type that interest is paid at the end of each year. Consequently, it will be necessary to leave the investment of £12 000 for 13 years before the initial investment is doubled.

10.3 Compound interest

Of more practical relevance is the use of compound interest. Compound interest is accrued on the principal plus the reinvested interest.

For instance, if £12 000 is invested for three years, earning compound interest at a rate of 8 per cent, what will be the total interest accrued?

Year (n)		Amount after year n
0	£12 000	
1	£12 000 + 0.08 × £12 000	= £12 960.00
2	£12 960 + 0.08 × £12 960	= £13 996.80
3	£13 996.80 + 0.08 × £13 996.80	= £15 116.54

Hence, the interest accrued is £3 116.54 (1)

The general formula is derived from a geometric progression with the first term $a = 12\ 000$ and the common ratio $r = 0.08$. The amount accrued after n years is equal to the $(n+1)$st term:

$$S = P(1+r)^n \text{ and } I = P(1+r)^n - P$$

These formulae assume that the interest is paid annually and at the end of each year. If, in the above example, the interest was paid four times a year, then we would have 2% (8%/4) paid 12 (3 × 4) times during the three-year period. The amount accrued would be

$$S = £12\ 000\left(1+\frac{0.08}{4}\right)^{3\times 4}$$

$$= £15\ 218.90$$

$$\text{Hence } I = £3\ 218.90 \qquad (2)$$

In general,

$$S = P(1+r/m)^{nm}$$

where n = the number of years of investment
and m = the number of times interest is calculated each year.

10.3.1 APR: the annual percentage rate

You will notice, by comparing the results (1) and (2), that (not surprisingly) the more often compound interest is calculated, the larger will be the amount earned. This is worth bearing in mind if one is comparing, for instance, competing building society savings accounts. They might all claim an annual interest rate of, say, 8 per cent, but, depending on the frequency that the interest is added to the particular account, the amount earned could vary significantly.

To ensure that you are comparing like with like you need to know the APR, the annual percentage rate (or effective rate of interest). The APR is the rate of interest an investment would earn if the interest were calculated only once a year. Finding the APR is relatively simple:

Consider how the APR for result (2) is found. Interest (annual rate of 8 per cent) is paid four times a year, for a period of three years. We have, with quarterly calculations, the following:

$$£15\ 218.90 = £12\ 000(1+0.08/4)^{3\times 4}$$

We wish to find the APR such that:

$$£15\ 218.90 = £12\ 000(1+\text{APR})^3$$

Hence we can write $12\ 000(1+0.08/4)^{12} = 12\ 000(1+\text{APR})^3$

$$(1+0.08/4)^{12} = (1+\text{APR})^3$$

Take the cube root of both sides to get

$$(1+0.08/4)^4 = (1+\text{APR})$$

Therefore, $\text{APR} = (1+0.02)^4 - 1 = 0.0824 = 8.24\%$

The **general equation for APR** is

$$\text{APR} = (1+r/m)^m - 1$$

where m is the number of times interest is calculated each year.

10.4 Increasing/decreasing the sum invested

So far we have looked at situations which involve investing a sum of money and calculating the amount of interest payable after n years. However, in practice one might want to periodically add or withdraw amounts from the investment. In such cases the formulae change significantly. It is worthwhile being aware of how to adapt a small number of equations, rather than trying to remember lots of similar, and therefore probably confusing, equations. Consider an example:

A sum of £1000 is deposited in an account on 1 January 1988. The interest rate is 10 per cent. It is decided to add an amount of £100 to this account on January of the next n years. The situation will look like this:

End of year 1: $1000(1+0.1)+100$

End of year 2: $1000(1+0.1)^2+100(1+0.1)+100$

$\vdots$

End of year n: $1000(1+0.1)^n+100(1+0.1)^{n-1}+100(1+0.1)^{n-2}+\ldots+100$

This simplifies to $1000(1+0.1)^n+\dfrac{100[(1+0.1)^n-1]}{0.1}$

(refer to the formula for a geometric series).

So, the total amount present in the investment account after, say, five years will be

$$1000(1.1)^5+\frac{100[(1.1)^5-1]}{0.1} = 1000\times 1.61051+\frac{100(0.61051)}{0.1}$$

$$= 1610.51+610.51 = £2221.02$$

In general,

$$S = P(1+r)^n+a_1(1+r)^{n-1}+a_2(1+r)^{n-2}+\ldots+a^n \qquad (3)$$

where a_i is the amount added (or subtracted) each year.

If the amount added (or subtracted) is constant over the n years,

$$S = P(1+r)^n+a\frac{[(1+r)^n-1]}{r} \qquad (4)$$

10.5 Depreciation, including the sinking fund formula

Depreciation is an accounting term which describes the technique of spreading the cost of a capital asset over a number of accounting periods as a charge against profits in those periods. There are a number of methods of calculating the depreciation.

10.5.1 Straight line depreciation

Suppose a machine costing £36 000 is expected to last 5 years and to have a scrap value of £5 000. The annual depreciation is found as follows:

$$£(36\,000-5000)/5 = £6200$$

Thus, the amount depreciated is spread evenly over the life of the machine.

10.5.2 The 'sum of digits' method

By weighting in favour of recent years, this method charges more in the earlier years. Consider the above example again:

Year	Weighting (digits)	Depreciation (£)
1	5	10 333.33 = 31 000/15 × 5
2	4	8 266.67 = 31 000/15 × 5
3	3	6 200.00
4	2	4 133.33
5	1	2 066.67
	Total 15	31 000.00

The depreciation per digit is £31 000/15 = £2066.67. Hence, the depreciation in the first year is £2066.67 × 5 as shown, and so on.

10.5.3 The reducing balance method

Suppose a machine costs £50 000. What will the depreciation charge each year for five years be, if it is to be depreciated by 25 per cent each year? The solution is calculated as follows:

		Depreciation (£)	End of year balance (£)
Year 0	initial cost		50 000.00
Year 1	depreciation (25%)	12 500.00	37 000.00
Year 2	depreciation (25%)	9 375.00	28 125.00
Year 3	depreciation (25%)	7 031.25	21 093.75
Year 4	depreciation (25%)	5 273.44	15 820.31
Year 5	depreciation (25%)	3 955.08	11 865.23

You will notice that a higher depreciation charge is made in earlier years. The net value of the machine at the end of each year can be found by reference to a geometric progression, which is similar to the formula for compound interest:

Year	Net value
0	50 000.00
1	$37\,500.00 = 50\,000(1-0.25)$
2	$28\,125.00 = 50\,000(1-0.25)^2$
3	$21\,093.75 = 50\,000(1-0.25)^3$
4	$15\,820.31 = 50\,000(1-0.25)^4$
5	$11\,865.23 = 50\,000(1-0.25)^5$

In general, the balance after n years is given by $50\,000(1-0.25)^n$.

This formula can be used to solve another form of the same problem:

A machine costs £36 000. In five years it depreciates to £5000. What is the annual rate of depreciation?

$$\text{We have} \quad 5000 = 36\,000(1-r)^5$$

$$\text{Therefore } (1-r)^5 = \frac{5000}{36\,000} = 0.1389$$

Find the 5th root of this to get $1-r = 0.6738$.

Therefore $r = 0.3262$, that is, the annual rate of depreciation is 32.62 per cent.

10.5.4 The sinking fund method

Suppose, at the end of each year, a company wishes to put aside an amount, a, to replace an asset, an important machine for instance. They estimate that in five years' time the machine will cost £50 000 and that the money invested will increase at 10 per cent per annum.

The problem can be interpreted as one of an investment account having a zero initial investment followed by constant additions of £a at the **end** of each year for five years, resulting in a final sum of £50 000. Using formula (**4**):

$$S = P(1+r)^n + \frac{a[(1+r)^n - 1]}{r}$$

$$50\,000 = 0 + \frac{a[(1+0.1)^5 - 1]}{0.1}$$

Rearranging we get

$$a = \frac{0.1 \times 50\,000}{(1+0.1)^5 - 1}$$

$$= £8189.87$$

The general **sinking fund formula** is

$$\boldsymbol{a = \frac{rS}{(1+r)^n - 1}}$$

If, as is likely, the company wishes to start the fund immediately and thereafter at the beginning of each year, the formula is slightly different. We have

$$a(1+r)^n + a(1+r)^{n-1} + \ldots + a(1+r) = S$$

Writing this expression in reverse, the sum can be found using the formula for the sum of a geometric progression, with first term $a(1+r)$ and common ratio $(1+r)$:

$$S = \frac{a[(1+r)^{n+1} - (1+r)]}{r}$$

Rearranging this we get

$$\boldsymbol{a = \frac{rS}{(1+r)^{n+1} - (1+r)}} \qquad (6)$$

The problem can now be solved:

$$a = \frac{0.1 \times 50\,000}{1.1^6 - 1.1} = \frac{5000}{0.67156} = £7445.34$$

10.5.5 Mortgages, loans and trust funds

A common form of mortgage is one which requires one to make regular repayments in return for an amount borrowed. Consider the following problem: What annual repayment is required to repay a mortgage of £50 000 over 25 years if the annual rate of interest if 8 per cent. Using the notation introduced at the beginning of this chapter we have

$S = 0$ (value of mortgage after 25 years)
$P = -£50\,000$ (initial 'investment', a negative amount)
$r = 8\%$
$n = 25$ (number of repayments)
$a = ?$ (annual repayments)

Using formula (**4**):

$$S = P(1+r)^n + \frac{a[(1+r)^n - 1]}{r}$$

$$0 = -50\,000(1+0.08)^{25} + \frac{a[(1+0.08)^{25} - 1]}{0.08}$$

$$\text{Therefore } a = \frac{50\,000 \times 0.08(1+0.08)^{25}}{(1+0.08)^{25} - 1} = 50\,000 \times 0.0936787^*$$

$$= £4683.94$$

*0.0936787 is an example of what is called the **capital recovery factor.**

Another adaptation, similar to that of mortgage repayments, is the idea of **trust funds.** Trust funds involve depositing a certain amount now to allow for fixed amounts to be withdrawn periodically. Suppose you wished to provide for your mother an annual payment of £5000 for the next 10 years. How much will you need to invest now at 8 per cent interest? We have

$S = 0$ (zero balance at the end of 10 years)
$P = ?$ (sum to be invested)
$r = 8\%$
$n = 10$ years
$a = -£5000$ (annual sum mother will withdraw – note that it is a negative amount, as withdrawing is equivalent to investing a negative amount)

Using formula (**4**) again, we get

$$0 = P(1+0.08)^{10} + \frac{(-5000)[(1+0.08)^{10} - 1]}{0.08}$$

Rearranging we get

$$P = \frac{5000[(1+0.08)^{10} - 1]}{0.08(1+0.08)^{10}} = £33\,550.41$$

10.6 Present values and discounting

If you were offered the choice between £1000 now or £1000 in one year from now, there is little doubt that you would choose to take the money now. One reason is probably to do with the influence of inflation. £1000 received in one year from now will not have the same purchasing power as £1000 received now. Another reason is that £1000 received now can be invested and consequently earn interest. Suppose the interest rate is 10 per cent. The growth of an investment of £1000 can be calculated using the compound interest formula:

Investment at year 0 £1000
After year 1 $£1000(1+0.1) = £1100$
After year 2 $£1000(1+0.1)^2 = £1210$, and so on.

For instance, £1100 received one year from now is equivalent to £1000 received now. £1100 is said to have a **present value** of £1000.

It follows that £1000 received one year from now has a present value of less than £1000. The present value of £1000 received one year from now can be calculated using the compound interest formula as follows:

$$1000 = P(1+0.1)$$

$$\text{Hence, } P = \frac{1000}{1+0.1} = 909.09$$

That is, £909.09 is the present value of £1000 received in one year's time or, £909.09 invested now at 10 per cent interest would result in a return of £1000 in one year's time. Similarly, for £1000 received two years from now:

$$1000 = P(1+0.1)^2$$

$$\text{Hence, } P = \frac{1000}{(1+0.1)^2} = 826.45$$

That is, £826.45 is the present value of £1000 received in two years' time or, £826.45 invested now at 10 per cent interest would result in a return of £1000 in two years time.

10.6.1 Present value: definition

Present value is a very useful concept and can be thought of in two ways:

1 the value today of an amount to be received some time in the future; or

2 the amount which would have to be invested today to produce a given amount at some future date.

The general formula for the present value of S, received in n years' time with an interest rate of r, is

$$S = \frac{P}{(1+r)^n} = P(1+r)^{-n} \text{ therefore } P = S(1+r)^n$$

$1/(1+r)^n = (1+r)^{-n}$ is called the **present value factor** and is available in table form (see appendix 7). A section of the tables is reproduced in table 10.1.

Table 10.1 Present value factors: the table gives the present value of a single payment received n years in the future, discounted at x% per year

	x = 1%	...	7%	8%	9%	10%	...
n = 1	0.9901		0.9346	0.9259	0.9174	0.9091	
2	0.9803		0.8734	0.8573	0.8417	0.8264	
3	0.9706		0.8163	.	.	0.7513	
4	0.9610		0.7629	.	.	.	
5	0.9515		0.7130	.	.	.	
6	0.9420		0.6663	.	.	.	
⋮							

10.6.2 Discounting: an example

The process of finding the present value of future sums of money is referred to as discounting.

Suppose an investment produces the following returns over a three year period:

After 1 year: £1000
After 2 years: £1500
After 3 years: £1500

If the interest rate can be taken to be 10 per cent over this period, what is the present value of the individual amounts and what is the present value of the total return? Using the tabulated values we get the result shown in table 10.2, including the total present value of the returns, £3275.65.

Table 10.2 Present value of investment returns over a three year period

Year	Present value factors (10%)	Return (£)	× factor	=	Present value of returns (£)	
1	0.9091	1000	× 0.9091	=	909.10	
2	0.8264	1500	× 0.8264	=	1239.60	Total =
3	0.7513	1500	× 0.7513	=	1126.95	£3275.65

10.6.3 Annuities

In cases where the cash flows are equal amounts for a period, the series of cash flows is called an annuity.

In the above example, if the year 1 return was £1500, then the series would be an annuity. The result could be calculated as before (see table 10.2), but it is more efficient to use the following formula to calculate the sum arising after n years:

$$S = \frac{P[(1+r)^n - 1]}{r(1+r)^n} = \frac{P[1-(1+r)^{-n}]}{r}$$

Because annuities are often encountered, the **annuity factor** $[1-(1+r)^{-n}]/r$ is tabulated as present value annuity factors or as cumulative present values, a section of which is reproduced in table 10.3. If the series of annuity payments goes on forever (in perpetuity) the formula reduces to

$$S = \frac{P}{r}$$

Table 10.3 Cumulative present value factors: The table gives the present value of n annual payments of £1 received for the next n years with a constant discount of x% per year.

	$x = 1\%$	...	7%	8%	9%	10%	...
$n = 1$	0.990		0.935	0.926	0.917	0.909	
2	1.970		1.808	1.783	1.759	1.736	
3	2.941		2.624	.	.	2.487	
4	3.902		3.387	.	.	.	
5	4.853		4.100	.	.	.	
6	5.795		4.767	.	.	.	
⋮							

10.6.4 Fixed interest stock: market price

Another application of present value analysis is the calculation of the market price of fixed interest stock.

Consider the following example: Stock has 5 years to run to maturity and has a nominal value of £250. It pays a dividend of 5 per cent and the current market rate of interest is 7 per cent. What is its present market value?

Holding this stock would yield £12.50 per year (5% × £250). The present value of these dividends is found as follows:

$$P_1 = \frac{12.5[1-(1.07)^{-5}]}{0.07} = 12.5 \times 4.1002 = £51.25$$

In addition the stockholder will receive the nominal value of the stock, £250, in five years' time. The present value of this is:

$$P_2 = 250(1.07)^{-5} = 250 \times 0.7129861 = £178.25$$

Hence, the total worth of the stock (present market value) is

$$£229.50 \ (P_1 + P_2)$$

Result P_1 could have been found using cumulative present value tables (4.100 is the appropriate factor: see table 10.3). P_2 could have been calculated using present value tables (0.7130 is the appropriate factor: see table 10.1).

10.7 Investment appraisal

Business activities involve satisfying market needs, whether it be in the provision of services or manufacture of goods. Investment is one step towards satisfying such needs. The investment might take the form of building new premises or purchasing specialist machinery, or investing in a project which may include elements of either or both of these. The idea behind such investment is to postpone consumption now in the expectation that investment will provide more to consume in the future.

Investment decisions are based on, amongst other considerations, certain time factors: is it better to invest and consequently gain at some time in the future or is it preferable to use available resources today?

This section is concerned with the criteria that influence financial analysts. They are the people most likely to be faced with the problems inherent in providing information for top management. They in their turn will have to take responsibility for investment-related decisions.

Investment considerations include some or all of the following:

1 **scale**–large sums of money are often involved;

2 **uncertainty**–estimates of interest rates, cash flows and future costs have to be made. It is impossible to guarantee any of these;

3 **irreversibility**–once a commitment has been made to, for instance, building a factory, it is difficult or impossible to change one's mind without dire and expensive consequences.

Investment appraisal involves comparison of expected returns and required investment. A number of techniques with which you will be expected to be familiar will be considered. The following example will be used to illustrate and indicate the advantages and disadvantages of each method:

The planning team of Holden Investment Company is considering four projects and wishes to choose the most effective:

Project A: This has an initial capital outlay of £50 000 and a life expectancy of 8 years. The estimated cash flows are £12 000 in years 1, 2 and 3; £15 000 in year 4; £16 000 in year 5; and £18 000 in years 6, 7 and 8.

Project B: This has an initial capital cost of £20 000 and a life expectancy of five years. The estimated cash flows are £10 000 during each of the first three years and £3000 during years 4 and 5. Additionally there is a scrap value attached to some of the purchased machinery of £2000, collectable at the end of the project life.

Project C: This has an initial capital cost of £30 000 and a life expectancy of five years. The estimated cash flows are £10 000 per annum.

Project D: This involves an initial investment of £20 000 and a further sum of £28 600 to be paid at the end of year 2. The life expectancy of the project is two years. There is one inflow of cash, £48 000, at the end of the first year.

The cost of capital for each project is 10 per cent.

The following is a summary of the project details:

Project:	**A**	**B**	**C**	**D**
Year	**Cash flow (£)**	**Cash flow (£)**	**Cash flow (£)**	**Cash flow (£)**
0	(50 000)	(20 000)	(30 000)	(20 000)
1	12 000	10 000	10 000	48 000
2	12 000	10 000	10 000	(28 600)
3	12 000	10 000	10 000	
4	15 000	3 000	10 000	
5	16 000	3 000	10 000	
6	18 000			
7	18 000			
8	18 000			

Consideration of the different decision criteria follows.

10.7.1 Payback period

The payback period is the time taken for a project's 'cash flows' to equal the project's 'outflows'. If this is used as a decision criteria, the usual decision is to choose the project with the shortest payback period.

Project B is judged to be the most favourable, giving a return of funds in two years. (See table 10.4; details are shown for the calculations for project A.)

The advantages of this method are:

1 It is simple to understand and to calculate.

2 It uses actual cash flows, rather than profits (as defined in accounting terms, which could be misleading; see section 10.7.2).

3 By definition, it favours quick return projects which may encourage faster growth, enhancing liquidity. This could prove to be advantageous to a company. Perhaps this is central to the widespread use of this method.

4 It will tend to minimize time-related risk.

The disadvantages are:

1 Cash flows after the payback period are not taken into account. Project B could fail to have any further cash flow, but this would be undetected by this method.
2 Only a crude indication of cash flow is provided. For instance, a zero return in year 1 followed by a return of £20 000 in year 2 would not change the payback period for B, but such a cash flow would be less favourable than the one shown here.
3 Essentially, payback is not a measure of the overall worth of a project, but a measure of liquidity.

Table 10.4 Payback period for projects A, B, C and D

Project	Cost (£)	Payback period	Rank
A	50 000	3.933 years*	3
B	20 000	2 years	1
C	30 000	3 years	2
D	20 000 (+ 28 600 at the end of year 2)	**	

***Calculation of payback period for project A**

Year	Cash flow (£)	Cumulative cash flow (£)	
0	(50 000)	(50 000)	Payback period =
1	12 000	(38 000)	3 years + 14/15
2	12 000	(26 000)	= 3.933
3	12 000	(14 000)†	
4	15 000	1 000‡	
5	16 000	17 000	
6	18 000	etc	
7	18 000		
8	18 000		

† £36 000 total inflow of cash, leaving £14 000 (36 000 – 50 000) to come before the break even point is reached.

‡ £51 000 total inflow of cash, consequently a credit situation has been reached (this ignores the time value of money).

** This project is non-conventional (it does not follow the pattern of an initial outflow followed by a series of inflows), the capital costs occur at different periods during the life of the project. It is not possible to calculate a unique payback period as overall the project does not pay back the total investment.

10.7.2 Accounting rate of return (ARR)

The ARR is the ratio of average profits, after depreciation, to the capital invested. A single definition does not exist, which limits its application possibilities. It is one example of a number of interpretations of average annual rate of return techniques. Another is the **gross average annual rate of return** which is defined as the average proceeds per year over the life of the project, expressed as a percentage of the initial capital outlay; that is, the total cash flow divided by the life of the project, expressed as a percentage of the capital investment.

For Project A this would be 30.25%:

$$\frac{121\,000/8}{50\,000} = 0.3025 = 30.25\%$$

Project B, 36%; project C, 33.3%; project D, 49.4%.

The advantages of this method are:

1 It is easy to calculate.
2 It takes into account the revenue over the life of the project.
3 It is a concept which is easily understood.

The disadvantages are:

1 It does not allow for the timing of cash flows.
2 It does not indicate the time span of the return.
3 The more sophisticated versions, those which calculate net cash flows (profits, say) relative to some accounting criteria, make it difficult to interpret the measure.

10.7.3 Net present value (NPV)

We saw in section 10.5.1 how to calculate the present value of future cash flows. The technique is a central one in investment appraisal, giving due attention to the time value of money. Money is more efficiently utilized if it is received sooner rather than later and its value is eroded the longer one has to wait for it.

The NPV method involves calculating the present values of expected inflows and outflows of cash, giving NPV = PV − intial outlay. The outline methodology and main assumptions are:

1 Net incremental cash flows are known or can be estimated. It is assumed that these are certain. Examination questions tend to assume year-end cash flows.
2 An appropriate discount rate must be established. The cost of capital is usually the basis for the choice of discount rate, effectively ignoring inflation. In practice, establishing the cost of capital is a complex business depending on many factors too complicated to consider within this text. Typical examination questions assume that the discount rate, based on the cost of capital, is known.
3 It is assumed that a perfect capital market exists. That is, there is no restriction on available funds.
4 Discount factors can be calculated but are more usually read directly from discount tables.
5 Having established the net cash flows and the discount rate to be used, the cash flows are discounted.
6 A project is acceptable if there is a positive NPV.

Table 10.5 summarizes the NPV appraisal for the four projects. Project A has the superior NPV.

Table 10.5 Net present values of projects A, B, C and D

Project A:

Year	Cash flow	10% discount factor $(1+0.1)^{-n}$	Present value (£)
0	(50 000)		
1	12 000	0.9091	10 909.09
2	12 000	0.8264	9 917.36*
3	12 000	0.7513	9 015.78
4	15 000	0.6830	10 245.20
5	16 000	0.6209	9 934.74
6	18 000	0.5645	10 160.53
7	18 000	0.5132	9 236.85
8	18 000	0.4665	8 397.13
			77 816.68

NPV (Project A) = £77 816.68 − £50 000 = £27 816.68

The present values have been calculated using the discount factors $(1.1)^{-n}$. In using the tabulated values rounding error is introduced into the calculation, e.g. * is calculated using $(1.1)^{-2} \times$ £12 000 = £9917.36; using the tabulated discount factor gives £9916.80, an error of approximately 0.006%.

Project B:

Details left as an exercise: NPV = £30 022.17 − £20 000 = £10 022.17

Note: End of year 5 inflow + scrap value = £3000 + £2000 = £5000

Project C:

Details left as an exercise: NPV = £7907.86

Project D:

Year	Cash flow	10% discount factor $(1+0.1)^{-n}$	Present value (£)
0	(20 000)		
1	48 000	0.9091	43 636.36
2	(28 600)	0.8116	(23 636.36)
			20 000.00

NPV = £20 000 − £20 000 = 0

Project	NPV (£)	Rank
A	27 816.68	1
B	10 022.17	2
C	7 907.86	3
D	0.00	4

Advantages of this method are:

1 It takes full account of the time value of money. Note: It is versatile enough to accommodate a change in capital cost. That is, if after year 3, say, the cost of capital increases, one needs merely to change the discount factor for that period.

2 It provides a clear and relatively easy way to compare competing projects.

Disadvantages are:

1 The NPV criterion relies on the choice of discount rate. A change in discount rate (see advantage **1** above) can be accommodated but such an occurrence might result in committing oneself to a project, only to find it was not the best choice after all.

2 While the scales of return of competing projects are effectively established, one is not told the efficiency of the capital employment (see the next section).

10.7.4 Profitability index

The profitability index is defined as the ratio of the net present value to the capital cost:

$$\textbf{Profitability index} = \frac{\text{NPV}}{\text{Capital cost}}$$

It is a measure of the efficiency of the employed capital. That is, it establishes the return per £ of capital employed.

Table 10.6 Profitability indices for projects A, B, C and D

Project	Cost (£)	NPV (£)	Index (= NPV/Cost)	Rank
A	50 000	27 816.68	0.556	1
B	20 000	10 022.17	0.501	2
C	30 000	7 907.86	0.264	3
D	20 000 (+ 28 600 at the end of year 2)			*

* Because this project is non-conventional (see table 10.4), it is difficult to calculate the index and more difficult to interpret it.

So far we have considered various aspects of investments, namely:

1 the contribution to profits, using the NPV criterion;

2 the contribution of each £ invested, using the profitability index; and

3 the level of profits as a percentage of capital cost, using a rate of return method.

The next section looks at the internal rate of return which provides an analyst with information about how profitable an investment proposal is, not just the information that it is profitable.

10.7.5 Internal rate of return (IRR)

The IRR is particularly useful if we have a number of investment proposals to consider but can only afford to finance one.

The IRR of a project can be defined as that rate of interest which results in a zero net present value; i.e. it enables us to find the rate of interest which will equate the future cash flows with the present cost of capital.

Equally important, is the fact that we can use the IRR if we do not have a value for the rate of interest. In such a case it is possible to compare the IRR of a project with some predetermined target. When a project produces an IRR in excess of the target rate of return, then the project is acceptable.

Calculating the IRR at this level is essentially a matter of trial and error. However, there is a way to shorten the process.

Consider project C: we need firstly to calculate the average annual profits as a percentage. The steps are as follows:

1 Calculate the total cash flow less the capital cost (= profit).

$$£(50\,000 - 30\,000) = £20\,000$$

2 Calculate this amount as a percentage of the capital.

$$\frac{20\,000}{30\,000} \times 100 = 66.67\%$$

3 Calculate the average annual percentage profit.

$$\frac{66.75}{5} = 13.3\%$$

13.3% is our first guess. It tends to be an underestimate, so usually it needs to be increased by somewhere between a third and a half of its value. That is, in this case, to somewhere between 17.8 and 20 per cent. We will choose 19 per cent as our starting point. The NPV is now calculated. The NPV for project C with a 19 per cent discount rate is seen to be +£576.35 (see table 10.7 and fig. 10.1).

Table 10.7 Internal rate of return of projects A, B, C and D

Project A:
Details left as an exercise.
NPV at 22% = +£799.17
NPV at 23% = −£766.86

$$\text{IRR} = 22 + \frac{799.17}{799.17 + 766.86} = 22.5\%$$

Project B:
Details left as an exercise.
NPV at 29% = +£342.65
NPV at 30% = −£707.69

$$\text{IRR} = 29 + \frac{342.65}{342.65 + 707.69} = 29.3\%$$

Project C:

End of year	Cash flow (£)	Present values 19%	Present values 20%
0	(30 000)		
1	10 000	8 403.36	8 333.33
2	10 000	7 061.65	6 944.44
3	10 000	5 934.16	5 787.04
4	10 000	4 986.69	4 822.53
5	10 000	4 190.49	4 018.78
		30 576.35	29 906.12
Net present values are:		+£576.35	−93.88

NPV at 19% = +£576.35
NPV at 20% = −£93.88

$$\text{IRR} = 19 + \frac{576.35}{93.88 + 576.35} = 19.9\%$$

Project D:
NPV = 0 for interest rates of 10 and 30%, i.e. project D does not have a unique IRR.

Project	IRR	Rank
A	22.5%	2
B	29.3%	3
C	19.9%	1
D	10 and 30%	–

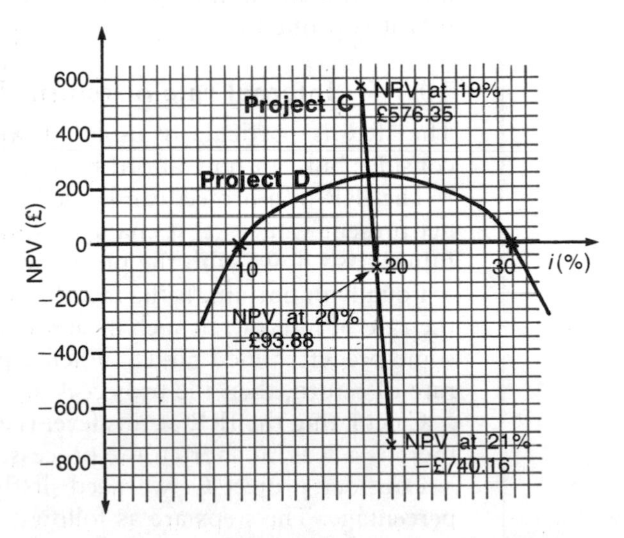

Fig. 10.1 Present value profiles of projects C and D

To establish the discount rate which will give a zero NPV it is necessary to find a rate which gives a positive NPV value and an interest rate which gives a negative NPV. Again refer to table 10.7 and fig. 10.1.

An advantage of the IRR is that it provides the rate of return of the project itself. Consequently, if the target rate changes it is not necessary to re-calculate the IRR of a project to establish if it is still acceptable. As long as the IRR is greater than the target rate of interest, the project remains acceptable. In such a case, if the NPV method had been applied, a new NPV would need to be found at every change in the interest rate.

A disadvantage is that the IRR gives a relative value. It does not indicate the scale of a project, as does the NPV method. Also, as can be seen from table 10.7 and fig. 10.1, there are instances when it is impossible to get a unique value for the IRR. The NPV for project D is zero for two values, 10 and 30 per cent. This is because project D is non-conventional (see table 10.4**). However, even when a unique value for the IRR does not exist it is always possible to find the NPV for a given value of r. In such cases, the NPV is considered to be the superior method to employ.

10.7.6 Summary

Overall, the type of investment appraisal used depends on the needs of the company or individual.

In the example considered here, project A provides the best and most efficient returns on capital cost. It also commits the investor to the project for eight years, unlike project B which has almost as good a return (judged on the profitability index) but is only five years long and demands less in the way of initial cost.

Sample questions

1 In the near future a company has to make a decision about its computer C which has a market value of £15 000. There are three possibilities:

(i) sell C and buy a new computer costing £75 000;
(ii) overhaul and upgrade C;
(iii) continue with C as at present.

Relevant data on these decisions are given below:

Decision	Initial outlay	Economic life	Re-sale value after 5 years	Annual service contract plus operating cost (payable annually in advance)
	£	years	£	£
(i)	75 000	5	10 000	20 000
(ii)	25 000	5	10 000	27 000
(iii)	0	5	0	32 000

Assume the appropriate rate of interest to be 12% and ignore taxation.

You are required, using the concept of net present value, to find which decision would be in the best financial interest of the company, stating why, and including any reservations or assumptions.

CIMA; Foundation Stage, Section B; Mathematics and Statistics; May 1986.

Solution/hints

1 Assumptions:

(a) Rate of interest remains constant.
(b) Estimate of outlays are reliable.
(c) Estimates of economic life of competing choices is accurate.

2 NPV computations:

Year	Decision (i) outlay*	Present value outlay	Decision (ii) outlay	Present value outlay	Decision (iii) outlay	Present value outlay
0	−80 000	−80 000	−52 000	−52 000	−32 000	−32 000
1	−20 000		−27 000		−32 000	
2	−20 000		−27 000		−32 000	
3	−20 000	−60 747	−27 000	−82 008	−32 000	−97 195
4	−20 000		−27 000		−32 000	
5	+10 000	+5 674	+10 000	+5 674	0	
NPV =		−£135 073		−£128 334		−129 195

***Details of calculations of decision** (i):

Year 0 = beginning of first year:

−£80 000 = initial outlay + revenue from sale of old computer + cost of maintenance (paid 1yr in advance)
= −£75 000 + £15 000 + (−£20 000)

Years 1 – 4 (end of year in each case):
−£20 000 = maintenance costs (paid 1yr in advance)
Year 5: +£10 000 = re-sale value
Present values: −£60 747 = £20 000 × (sum of discount values for yrs 1 – 4)
= £20 000 × (0.8929 + 0.7972 + 0.7118 + 0.6355)

where $0.8929 = \frac{1}{1+0.12}$, $0.7972 = \frac{1}{(1+0.12)^2}$, etc.

Note: There is some 'rounding' error in the totals. For example, −£135 073, the decision (i) NPV, is actually:

$$-80\,000 + (-60\,746.987) + 5674.2686 = -135\,072.72$$

On the basis of the NPV analysis one would advise the company to make decision (ii), namely to overhaul and update the existing computer.

3 Reservations

(a) Concern about the rate of interest: the three NPVs are relatively similar. If the rate of interest turns out to be different to the one used in the analysis the decision-making could be problematical.
(b) Overhauling the existing machine is less than £1000 cheaper in 'real terms' than doing nothing. Is the saving worth the disruption involved?
(c) Has the social cost been taken into account? That is, will the employees be unsettled by the disruption and will the resulting costs outstrip the 'saving'?

2 A company's cricket club sets up a reserve fund so that its equipment can be renewed in exactly five years from now at an expected cost, then, of £1000. How much (to the nearest 1 decimal place) should be put into the fund at each of the five year-starts if compound interest at 9% per annum can be obtained?

A £153.30 **B** £167.10 **C** £183.50 **D** £218.00 **E** £917.40

CIMA; Foundation Stage, Section B; Mathematics and Statistics; November 1986.

Solution/hints

This is a 'sinking fund' problem. Be careful to use the formula for payments at year-*starts* not year-*ends* (see section 10.4.4)

Answer: **A**

3(**a**) What do you understand by the phrase 'discounting in real terms' in association with capital expenditure decisions? Explain in terms that a non-accountant would understand.

(**b**) JKL plc, a well-established print and photocopy business, is considering the purchase of an additional copier to augment its existing service. Two suitable copiers are available, with information as follows:

	M	N
Purchase cost	£4500	£7500
Expected life	3 years	3 years
Residual value	Nil	Nil
Annual net receipts		
Year 1	£1800	£3000
Year 2	£2250	£3750
Year 3	£1640	£3132

You are required to advise JKL's management on:
(i) which of the two models should be purchased using a 10% discount rate;
(ii) the effect of using alternative discount rates, as a basis for the calculation.

CIPFA; Diploma in Public Sector Audit and Accounting; Quantitative analysis; March 1986.

Solution/hints

Part (a)

When embarking on an investment scheme one wishes to know how much one is gaining as a result of investing one's money. Two main areas of concern would be:
1 Will the money earn as much as it might if it were invested elsewhere?
2 How will inflation influence the value of any returns?

If one merely subtracted initial capital costs from the monetary returns achieved at some future date, then one would be ignoring both these concerns. To avoid this, accountants try to calculate the value of any returns taking into account **1** and **2**. That is, they discount, in real terms, the influence of **1** and **2**.

Part (b)

(i) Discounting at 10% you will find that: copier M has a NPV of £228.02;
copier N has a NPV of £679.56.

When advising JKL consider initial outlay and whether other factors might be relevant, for example pay-back period.

(ii) Other discount rates which could be used include

1 the market rate of interest;
2 a rate of return comparable with returns from other company projects;
3 a target rate set by the board; and so on.

Whatever the rationale, if the rate of interest chosen is higher than 10%, then the NPV will be a lower figure in each case. It may indicate a loss, which would presumably preclude the company from purchasing an extra copier.

4 (**a**) A department store charges its credit customers interest at a rate of 2% per month. Calculate the rate of interest at an annual rate assuming that interest is compounded monthly.

(**b**) A man pays £10 000 into an account on 1st January 1987. If the annual rate of interest is constant and is equal to 10%, find the amount that will be in the account on 1st January 1993

(**i**) if he makes no withdrawals; and

(**ii**) if he withdraws £2000 on 31st December of each year.

(**c**) A project involves an investment of £60 000. The net cash flows for the life of the investment are as follows:

Year	Cash flows (£)
1	−4000
2	20 000
3	25 000
4	45 000
5	30 000

Assuming a discount rate of 10%, calculate the net present value. Comment on your result.

CIS; Part 1; Quantitative studies; December 1986.

Solution/hints

Part (a)

See section 10.2.1

Answer: APR = 26.8%

Part (b)

(i) Compound interest (see section 10.2) *Answer*: £17 715.61

(ii) See section 10.3: equation (4), $a = -£2000$. *Answer*: £2284.39

Part (c)

Answer: NPV = £81 038.68

5(**a**) A project has a positive net present value (NPV) of £80 when discounted at 10%. Explain clearly what this means.

(**b**) The following data is available to a company faced with a mutually exclusive investment decision:

Project A		Project B		Project C		Project D	
Year	Net cash flow (£)	Year	Net cash flow (£)	Year	Net cash flow (£)	Year	Net cash flow (£)
0	−100	0	−100	0	−100	0	−100
1	+50	1	+20	1	+50	1	+100
2	+50	2	+40	2	+100	2	−20
3	+50	3	+80	3	−30		
		4	+160	4	+100		

(i) Using the payback method, indicate which project the company should choose and comment on the result obtained.

(ii) State the assumptions of the approach used in (i), and indicate under which circumstances the payback method would provide useful investment decision information.

(iii) Calculate the NPV and internal rate of return (IRR) of the following project at a discount rate of 10%.

Assume cash flows occur at the year end.

Year	Cash flow (£)
0	−100
1	+ 20
2	+ 40
3	+ 80

CIPFA; Diploma in Public Sector Audit and Accounting; Quantitative analysis; May 1987.

Solution/hints

Part (a)

See section 10.6.3.

Part (b)

(i) Payback periods: **A,** 2 years; **B,** 2.5 years; **C,** 1.5 years; **D,** 1 year.

The answers suggest that project D should be chosen, as the investment is 'paid back' sooner than for any of the other projects. Further observation indicates that project D goes on to make a loss. In general, it is unwise to rely on this method in isolation as it fails to take into account the nature of the project after the capital has been regained and the time value of money.

(ii) Left as an exercise. See section 10.6.3.

(iii) *Answers:* NPV = £11.34; IRR = 15.12%

6 (a) An organization is considering investing in capital equipment which will cost £300 000. The expected life of this equipment is six years. The revenue derived from this equipment is expected to be £200 000 in the first year and is expected to increase at the rate of 10% SIMPLE interest. The cost of labour, materials, repairs, maintenance, etc. is expected to be £100 000 in the first year and is expected to increase at an annual rate of 20% COMPOUND interest.

Find for each of the six years of the life of the machine:

(i) revenue;

(ii) costs; and

(iii) net revenue (i.e. revenue less costs).

(b) Assuming that the net revenues found in part (a)(iii) are received at the end of each year, calculate the net present value assuming a discount rate of 15%. Comment on your result.

(c) Compare and contrast the net present value and the internal rate of return methods of investment appraisal.

CIS; Part 1; Quantitative studies; June 1987.

Solution/hints

Part (a)

	(i) Revenues (£)	(ii) Costs (£)	(iii) Net revenues (£)
End of year 1	200 000	100 000	100 000
2	220 000	120 000	100 000
3	240 000	144 000	96 000
4	260 000	172 800	87 200
5	280 000	207 360	72 649
6	300 000	248 832	51 168
		$[100\,000(1+0.2)^5]$	

Part (b)

	Present values (15%) (£)
year 0	−300 000.00
1	86 956.52
2	75 614.37
3	63 121.56
4	49 856.88
5	36 114.92
6	22 121.34
NPV =	£33 785.59

7 (**a**) In two years from now some machinery in your company will need replacing. It is estimated that £500 000 will then be required. To provide for this, £X is to be allocated now, invested at 12 per cent per year, with interest compounded quarterly. What should £X be?

(**b**) Assume that the money required in (a) is not immediately available, but that the company can put £30 000 into a reserve fund every quarter, starting now (i.e. 9 quarterly amounts can be set aside in two years). Annual interest of 12 per cent is payable, compounded quarterly.

(**i**) How much short of the target figure of £500 000 will be the reserve fund in exactly two years from now?

(**ii**) What quarterly amount should be put into the reserve fund, starting now, to ensure that £500 000 is available in exactly two years time?

$$S = A(R^n - 1)/(R - 1)$$

CIMA; Stage 1; Quantitative methods; June 1987.

Solution/hints

Part (a)

We require the present value of an outlay of £500 000 discounted at 12 per cent per annum (or, more precisely, at 3 per cent per quarter).

Using the formula for compound interest:

$$£500\,000 = £X \times \left(1 + \frac{0.12}{4}\right)^{4\times2} = £X \times (1.03)^8$$

Therefore, $£X = 500\,000 \times (1.03)^{-8} = £394\,704.62$

Part (b)

(i) Using the formula given, where $R = 1 + r$ and n is the number of payments made (in this case, $n = 9$), we have

$$S = \frac{30\,000\,(1.03^9 - 1)}{0.03} = £304\,773.18$$

i.e. there is a shortfall of £195 226.82 (500 000 − 304 773.18).

(ii) Amount, a, required to achieve £500 000 at the end of two years:

$$a = \frac{500\,000 \times 0.03}{1.03^9 - 1} = £49\,216.93$$

Recommended reading

Curwin, J. and Slater, R., *Quantitative Methods for Business Decisions* (Van Nostrand Reinhold, 1986), chapter 8.

Lucey, T., *Quantitative Techniques* (D.P. Publications, 1980), chapters 24–26.

Owen, F. and Jones, R., *Modern Analytical Techniques* (Pitman, 1988), chapter 15.

11 Linear Programming

11.1 Introduction

When faced with a problem, it is common for business people to want to use efficiently the resources available to them. If it is possible to describe such a business problem numerically, linear programming can on occasions provide a valuable problem-solving structure. The inherent techniques allow one to allocate scarce resources, while optimizing an objective function.

For example, suppose a company produces two computers for the home-user market. The standard model contributes £500 to profits for each unit sold. The deluxe model contributes £800 to profits for each unit sold. A company production analysis has shown that, for each model, one hour of labour, including all necessary checks, is required to produce one unit. Also, the standard model contains two X15 chips and three X225 chips, while the deluxe model has three and six of these chips, respectively. Each week there are available, 90 hours of labour, 200 X15 chips and 360 X225 chips. Finally the sales department has secured a contract to supply 25 of the standard model to a retail outlet. The aim of the company is to maximize profits subject to contractual obligations and the limits on available resources.

Linear programming techniques can be applied to this problem because:

1 it has been possible to state the problem numerically;

2 there exist constraints – e.g. only 90 hours of labour are available;

3 it is possible to derive an objective function – in this instance a profit function, $£(500x + 800y)$ – which the company wishes to maximize (x is the number of standard computers produced, y the number of deluxe computers produced);

4 there are a number of possible outcomes – the profit function may be maximized by manufacturing only the deluxe model or a production mix of both models.

In addition to the above points, it is also necessary to make assumptions in all linear programming problems. The following is the most important: All resources or inputs (labour, raw materials, etc) are assumed to be linearly related to the outputs produced. For instance, in the example above, one hour of labour is needed to produce one standard model computer and one hour of labour is needed to produce one deluxe model computer. The profit contributions for these inputs are £500 and £800, respectively. Hence, to produce 20 standard computers and 15 deluxe computers would require the necessary hours to increase proportionally (20 for the standard computer and 15 for the deluxe computer), resulting in a total of 35 hours and a total profit of £22 000 ($20 \times £500 + 15 \times £800$).

In practice, many resources are not linearly related to the outputs but they can be assumed to have an approximately linear relationship over the range of values being considered. Most linear programming problems you are likely to encounter at this level involve a number of stages of analysis:

1 The objective function and the constraint equations or inequalities have to be formulated from the question. This is an important aspect of linear programming problems and a substantial proportion of marks is usually attached to carrying out this stage successfully (see section 11.2).

2 The objective function and the constraints often have to be presented in graphical or matrix form. This depends on the method of solution required. We will only be concerned with the graphical method.

3 The optimum solution has to be found by graphical methods or Simplex analysis. (The latter is beyond the scope of this book.)

4 Comments and/or further analysis relative to some minor change in the constraints and/or objective function are often required. This is the basis of sensitivity analysis.

11.2 Formulation of the linear programme

Consider the computer sales problem described above. As already mentioned, the objective function can be expressed as:

$$\text{Profit} = 500x + 800y$$

where x = number of standard model computers produced
and y = number of deluxe model computers produced.

We need to formulate the constraints, which in this example are concerned with available labour, available components (X15 and X255 chips) and contracted supply. Consider the available labour first: Each unit of production (of both models) requires 1 hour of labour input. Consequently production of x units of the standard model will require x hours of labour input and production of y units of the deluxe model will require y hours of labour input. Hence, the total hours of labour required is $(x+y)$ hours. We are told that the weekly availability of labour is 90 hours, therefore the total hours required to produce x standard and y deluxe computers cannot exceed 90 hours. This can be written as an inequality

$$x + y \leqslant 90$$

which is referred to as the 'labour constraint'.

Inequalities can be formulated for the other constraints in a similar way: A production output of x standard models and y deluxe models requires $2x$ and $3y$ X15 chips, respectively – a total of $(2x+3y)$ X15 chips. There is a weekly limit of 200 of these chips, which gives the X15 chip constraint:

$$2x + 3y \leqslant 200$$

All the relevant constraints can be represented, enabling us to state the problem in **standard linear programming format:**

Maximize the objective function, Profit = £$(500x+800y)$, subject to

$x + y \leqslant 90$ (labour constraint);
$2x + 3y \leqslant 200$ (X15 chip constraint);
$3x + 6y \leqslant 360$ (X255 chip constraint)
(Simplified, this constraint becomes $x + 2y \leqslant 120$);
$x \geqslant 25$ (contract constraint);
$y \geqslant 0$ (non-negativity).

All variables should be non-negative (0 or positive). The **non-negativity constraints** should always be explicitly stated, not assumed. In this case it is unnecessary to state $x \geqslant 0$ because it is ensured by constraint $x \geqslant 25$.

The next stage would be to represent this information graphically (or to prepare the information for Simplex analysis). This aspect will be considered in the following sections. Before moving on, we will consider two variations of the type of problem you are likely to encounter.

11.2.1 A minimization problem

The example considered above is a maximization problem – the aim is to maximize profits. You will notice that most of the constraints are less-than-or-equal-to constraints, which is typical for maximization problems. Now consider the following minimization example. You will notice that most of the constraints are of the greater-than-or-equal-to type.

A road transport company specializes in the transportation of food. Fresh and frozen foods are transported in sectioned carriers at temperatures of 0 degrees centigrade (cool compartment) and −25 degrees centigrade (cold compartment), respectively. The company uses two types of carriers: Type *A* has a cool compartment providing 10 cubic metres of space and a cold department providing 15 cubic metres; type *B* has a cool department providing 15 cubic metres of space and a cold department providing 12 cubic metres of space.

The company is contracted to carry 1000 cubic metres of fresh food and 1200 cubic metres of frozen food per week. The average cost per journey for type *A* carriers is £700 and for type *B* carriers, £500. The aim is to minimize costs with respect to the carrier constraints.

Formulated as a linear programme, this becomes:
Minimize £$(700a+b)$ subject to:

$$10a+15b \geqslant 1000 \text{ (fresh food constraint)}$$
$$15a+12b \geqslant 1200 \text{ (frozen food constraint)}$$
$$\left.\begin{array}{r} a \geqslant 0 \\ b \geqslant 0 \end{array}\right\} \text{(non-negativity constraints)}$$

11.2.2 A minimization problem: product mix

A keen gardener is planning to market a new fertilizer, which he can produce by using two ingredients, F and G. His analysis of the ingredients produced the following results:

	Content (kg component per kg fertilizer)				
	Ash	Nitrogen	Phosphoric acid	Potash	Calcium
F	0.18	0.01	0.001	0.0005	0.008
G	0.10	0.02	0.02	0.001	0.004

The ingredients F and G cost £0.20 and £0.40 per kg, respectively.

The fertilizer is to be sold in 25kg bags (minimum weight). The gardener has calculated that each 25kg bag of the new fertilizer must contain:

1 a minimum of 3.6 kg ash;

2 a minimum of 0.3 kg nitrogen;

3 a minimum of 0.01 kg potash.

(*Note*: Only three elements are constrained.)

He wishes to minimize costs subject to meeting the above requirements. Formulated as a linear programme this becomes:

Minimize $0.20f+0.40g$ subject to:

$$f+g \geqslant 25 \text{ (weight constraint);}$$
$$0.18f+0.1g \geqslant 3.6$$
simplified, this is $18f+10g \geqslant 360$ (ash constraint);
$$0.01f+0.02g \geqslant 0.3$$
simplified, this is $f+2g \geqslant 30$ (nitrogen constraint);
$$0.0005f+0.001g \geqslant 0.01$$
simplified, this is $f+2g \geqslant 20$ (potash constraint).

11.3 Graphical solution of linear programmes

Having formulated the problem as a linear programme, you will probably be required to present and solve the problem graphically. We will now solve, using graphical techniques, each of the examples outlined in section 11.2.

The computer example will be considered first. The first step was to represent the problem in a standard form. The equation and inequalities, in their simplest form, are reproduced for reference:

Maximize £$(500x+800y)$

$$\begin{array}{rll} \text{subject to: } x+y \leqslant & 90 & (1) \\ 2x+3y \leqslant & 200 & (2) \\ 3x+6y \leqslant & 360 & \text{which simplifies to} \\ x+2y \leqslant & 120 & (3) \\ x \geqslant & 25 & (4) \\ y \geqslant & 0 & (5) \end{array}$$

11.3.1 Graphing the constraints

The second step involves drawing lines on a graph to represent the constraints. In this instance the variables (the number of standard and deluxe computers, respectively) are labelled x and y.

The constraints are represented by inequalities, not equations. However, to represent the constraints graphically, we draw the corresponding equation. Constraint (1), reflecting the hours of labour available, is illustrated in fig. 11.1.

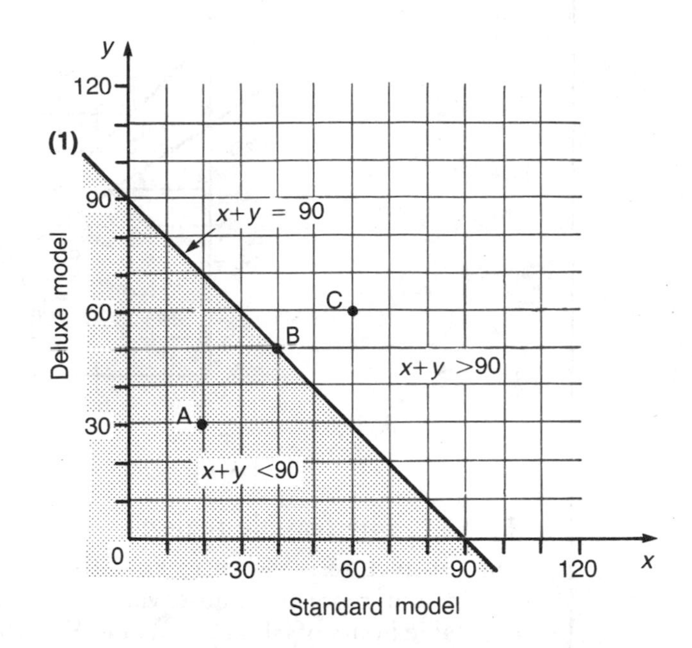

Fig. 11.1 Limitation of available labour (hours), represented by $x+y \leqslant 90$ (1)

The line, $x+y = 90$, represents the limit of available labour. The area on and below that line is within the limits of available labour. This can be confirmed by considering the situation at typical points A, B and C:

Point A has coordinates (20, 30). When these x and y values are substituted into the expression $x+y$, we find

$$x+y = 20+30 = 50 \leqslant 90$$

which satisfies the labour constraint.

Point B has coordinates (40, 50) which, in a similar manner, results in

$$x+y = 40+50 = 90$$

which also satisfies the constraints.

Point C has coordinates (60, 60), resulting in

$$x+y = 60+60 = 120 > 90$$

which contravenes the constraints.

The shaded region in fig. 11.1 represents the values of x and y (the production levels of the two types of computer) which can be produced if (1) were the only constraint.

Constraint (2) can be illustrated in the same way (see fig. 11.2). However, the problem demands that all the constraints are satisfied simultaneously. Figure 11.3 (overleaf) shows the region which satisfies constraints (1) and (2). The values of x and y of points in regions A, B and C contravene one or both constraints and would consequently represent unacceptable production targets.

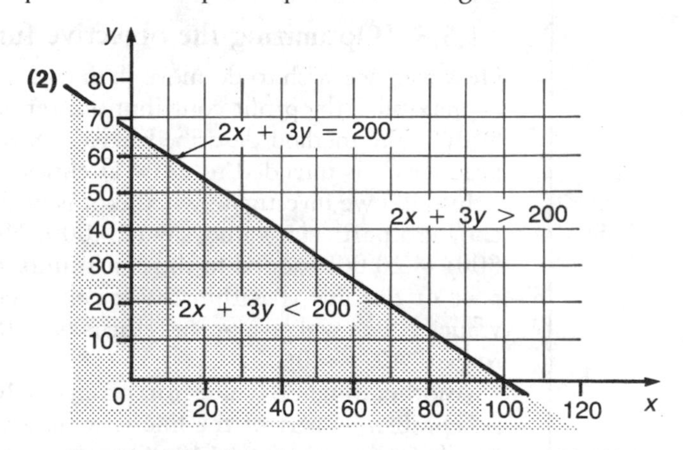

Fig. 11.2 The X15 chip constraint: $2x+3y \leqslant 200$ (2)

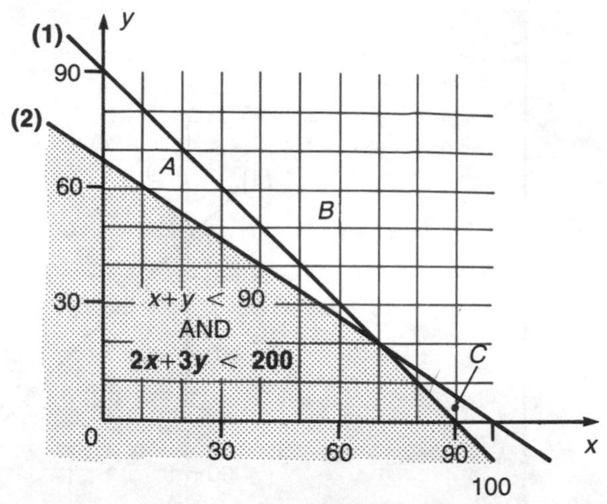

Fig. 11.3 The region which satisfies constraints (1) and (2)

11.3.2 The feasible region

The full picture is shown in fig. 11.4. The shaded region is known as the feasible region and represents those values of x and y which satisfy all the constraints (available hours of labour, available X15 chips, available X255 chips, non-negativity and contractual constraints) simultaneously. Production levels represented by points anywhere outside the feasible region would not be acceptable because one or more of the constraints would be contravened.

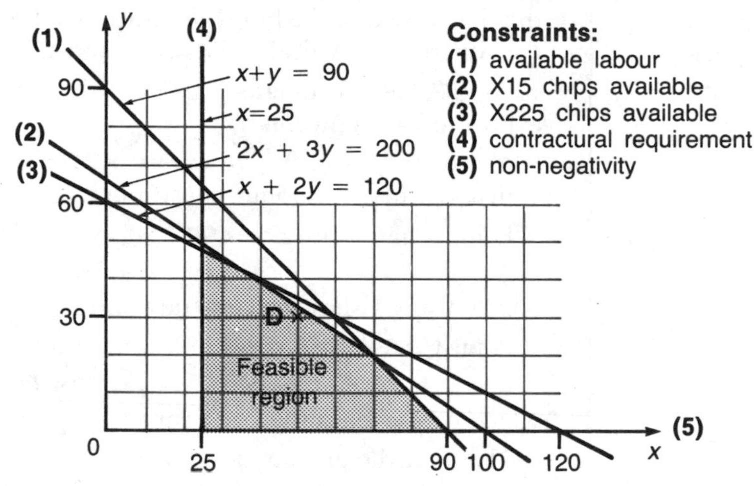

Fig. 11.4 The feasible region

We are now able to give possible production levels for the two computers which are feasible, that is, values of x and y which do not break any of the constraints. Point D (fig. 11.4) represents one such pair of production levels: 50 standard models and 30 deluxe models.

11.3.3 Optimizing the objective function

However, we wish to do more than find a possible production level; the stated aim is to maximize the profit contribution, represented by the objective function £$(500x+800y)$. The method described here is a little more detailed than you will require in practice. It is intended to draw attention to the reasons for the various stages.

Initially we investigate whether it is possible to achieve a particular profit level. We choose a particular value, say £20,000. The objective function for that value ($500x+800y = 20\ 000$) added to the constraints graph, (line (**A**), fig. 11.5). Notice that only part of the line is within the feasible region, representing the feasible production values which will give profit returns of £20,000. These range from ($x = 25, y = 9.375$) to ($x = 40, y = 0$).

Another (higher) value for profits, £40,000, (line (**B**), fig. 11.5) would also be acceptable, as part of the line is within the feasible region. The feasible production levels giving a profit of £40,000 range from ($x = 25, y = 34.375$) to ($x = 80, y = 0$).

It should be clear that as the objective function moves away from the origin ($x = 0$, $y = 0$), the profit contribution increases. Consequently the profit contribution is maximized at that point (or those points) furthest from the origin but still within the feasible region.

In this case, line (C) touches the feasible region at one point, C, where $x = 40$ and $y = 40$. This production combination will result in a profit contribution of £52 000.

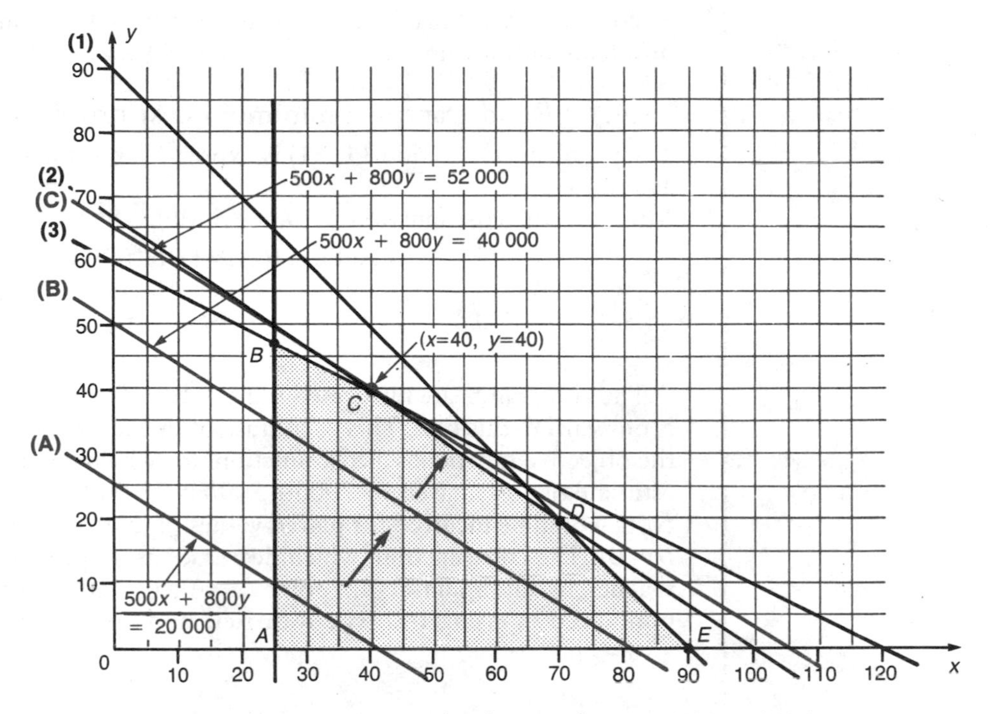

Fig. 11.5 Optimizing the objective function ($500x + 800y$)

11.4 Graphical solution: summary

Notes:

1 The different examples of the profit function have the same slope, that is, they are represented by parallel lines.

2 As the profit function moves further from the origin the value of profits increases.

3 Point C (fig. 11.5), the optimum production mix, is the point of intersection of two constraint lines (2 and 3). One of the lines has a gradient less than, and the other has a gradient greater than, that of the objective function line:

Line (3) gradient $= -0.5$;
'Objective' line gradient $= -0.625$;
Line (2) gradient $= -0.667$ (3 d.p.).

4 Once the optimum point(s) have been identified, their exact coordinates can be calculated by solving simultaneously the equations of the intersecting lines.

5 The optimum values are not always whole numbers. If you get a fractional answer, check that it makes sense. For instance, it is not possible to produce 9.375 deluxe model computers. In such a case you would have to look for the nearest whole number production value which satisfies the constraints.

6 Only **two** variables can be dealt with in any linear programming analysis e.g. standard and deluxe models, two types of carrier, two ingredients.

7 There can be any number of constraints. However, as the number of constraints increases it will become increasingly difficult to read the graphical information.

8 The feasible region is that area that satisfies **all** the constraints. Sometimes such a region may not exist (see section 11.5).

9 The optimum solution is that point (or points; see section 11.4.2) in the feasible region that maximizes (or minimizes; see section 11.4.1) the objective function. This is usually a corner of the feasible region, though it can be a line joining two corners (see section 11.3.2).

In general, the steps involved in solving a linear programming problem are:

1 Formulate the problem in standard linear programming format.
2 Graph the constraints and identify the feasible region.
3 Find the point or points within the feasible region which will optimize the objective function. The optimum point is found on the edge of the feasible region, usually at one of the corners.

We will now briefly look at the graphical solutions of the two minimization problems outlined in sections 11.2.1 and 11.2.2.

11.4.1 Road transport minimum cost problem

The problem (see section 11.2.1) is expressed in standard linear programming format as follows:

Minimize the cost function £$(700a + 500b)$,

$$\text{subject to: } 2a + 3b \geqslant 200 \quad (1)$$
$$5a + 4b \geqslant 400 \quad (2)$$
$$a \geqslant 0 \quad (3)$$
$$b \geqslant 0 \quad (4)$$

The constraints are illustrated in fig. 11.6. You should note that the feasible region is outward in this case. This is because of the nature of the inequalities. Also note that the objective function is a cost function, for which we require the minimum. An initial estimate of costs, given by $700a + 500b = 78\ 000$, is shown in fig. 11.6. This level of costs would be incurred for paired values (a, b) from $(a = 0, b = 156)$ to $(a = 111.43, b = 0)$. The cost function, which decreases as it approaches the origin $(a = 0, b = 0)$, reaches its minimum at the point $(a = 0, b = 100)$, that is, the transport costs are minimized by using 100 type B carriers and no type A carriers.

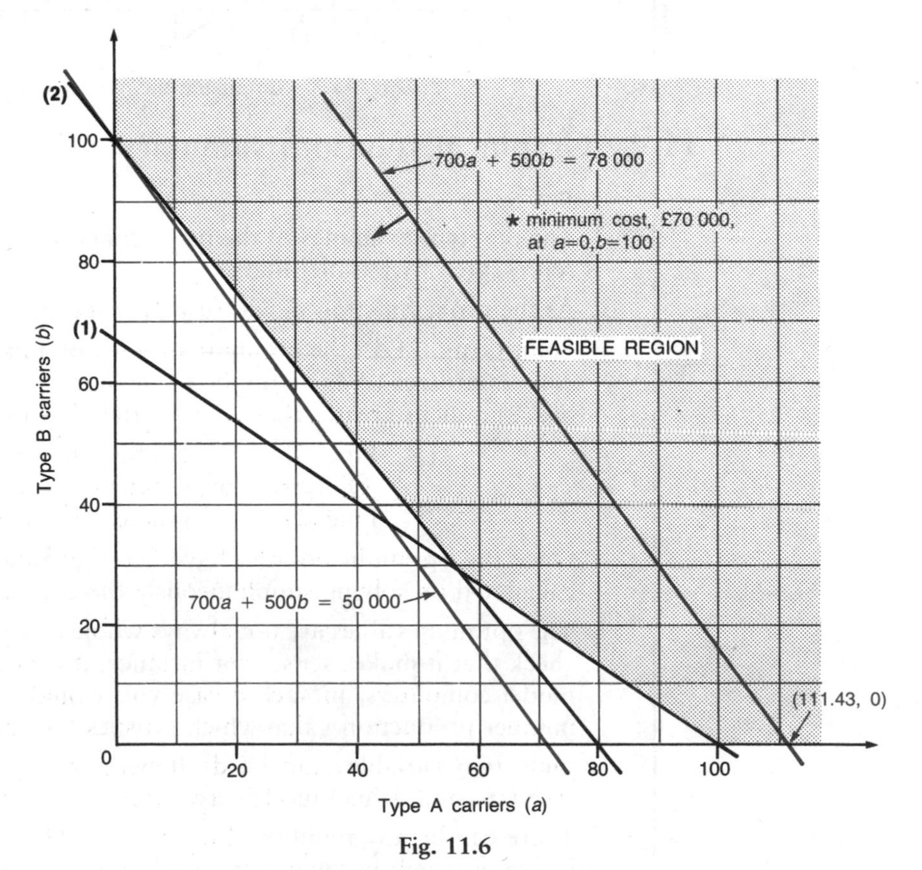

Fig. 11.6

11.4.2 Minimum cost: multiple solutions

The third example is also a minimization problem. It involves mixing two items to produce a third. In this type of problem the most difficult step is often the formulation of the linear programme in standard format. You should practice this aspect of linear programming with as many different types of problem as possible (see section 11.2.2).

The formulation for this example is:
Minimize the cost, $0.20x + 0.40y$,

$$\text{subject to:} \quad f+g \geqslant 25 \quad (1)$$
$$0.18f + 0.1g \geqslant 3.6$$
$$\text{which simplifies to } 18f + 10g \geqslant 360 \quad (2)$$
$$0.01f + 0.02g \geqslant 0.3$$
$$\text{which simplifies to } f + 2g \geqslant 30 \quad (3)$$
$$0.0005f + 0.001g \geqslant 0.01$$
$$\text{which simplifies to } f + 2g \geqslant 20 \quad (4)$$

The problem is illustrated in fig. 11.7. The method employed to find the solution is the same as in the previous example, but in this case there is no unique solution. The reason is that the slope of line (3), which represents the nitrogen constraint, is equal to the slope of the objective function. Therefore, the minimum cost (£6) can be achieved by combinations of the two ingredients ranging from ($f = 20$, $g = 5$) to ($f = 30$, $g = 0$).

It is also worth noting that the potash constraint has a degree of 'slack', that is, the amount of potash could be raised significantly above the minimum requirement (represented by line (4)) before it influences the optimum solution. (In fact it would need to increase from its present value of 0.01 kg to over 0.015 kg.)

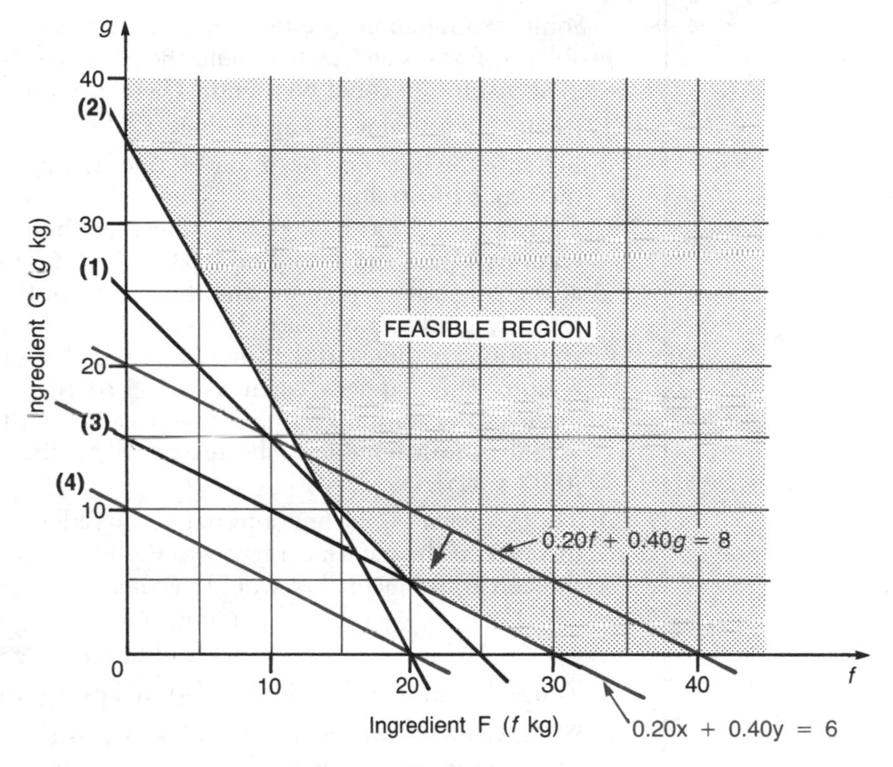

Fig. 11.7

11.5 Non-standard problems

11.5.1 Mutually exclusive constraints

Suppose the keen gardener suddenly realized that in stating that the potash content should have a minimum value of 0.01 kg, he had made a mistake: he should have said that potash content should be *not more than* 0.01 kg (i.e. a maximum value of 0.01). The linear programme would now be:
Minimize the cost, $0.10f + 0.20g$,

subject to:	$f+g \geqslant 25$	(1)	(as above)
	$18f + 10g \geqslant 360$	(2)	(as above)
	$f + 2g \geqslant 30$	(3)	(as above)
	$f + 2g \leqslant 20$	(4′)	(constraint reversed)

This situation is illustrated in fig. 11.8 (overleaf). Clearly there is no area which is simultaneously feasible for all the constraints. Therefore, there is no solution.

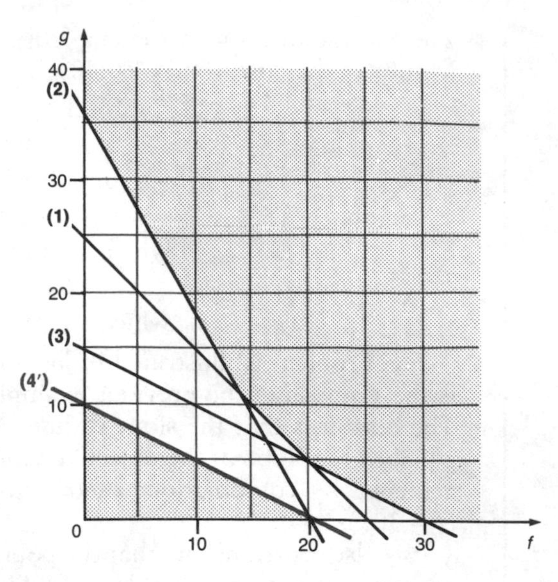

Figure 11.8

11.5.2 Sensitivity analysis

Some examination questions require you to suggest remedies for non-standard problems. For instance, what could the gardener do in the situation described above, so that a solution could be found? There are a number of possibilities:

1 Remove constraint (4′) altogether.

2 Increase the limit on constraint (4′), so that it coincides with the feasible region of the other constraints.

This would mean increasing the top limit of potash content from 0.01 (see sections 11.2.2 and 11.4.2) to 0.015. How has this figure been arrived at? Looking at the constraint equations and the graphs of these equations in fig. 11.8, one sees that the gradients of lines (3) and (4′) are the same (−0.5). To obtain a feasible solution it is sufficient that line (4′) coincides with line (3), giving a linear feasible region. This happens when $x+2y \leqslant 20$ (line (4′)) moves to $x+2y \leqslant 30$. This involves the constant term, 20, increasing by a factor of 1.5, to 30. Therefore the actual potash limit must be increased by the same factor, from 0.01 to 0.015 (0.01×1.5).

3 The gardener could find alternative ingredients with lower potash contents. For instance, if the potash content was 0.0001 kg per kg for both ingredients *F* and *G*, the constraint inequality would become

$$0.0001f + 0.0001g \leqslant 0.01,$$

which simplifies to $f+g \leqslant 100$

You will find that this change also makes it possible to solve the problem.

We have merely touched on the idea of sensitivity analysis here. Sensitivity analysis looks at situations, such as changes in constraint values and/or objective function coefficient, which influence the optimum solution.

Sample questions

1 A company makes insulating foam for injection into the wall cavities of buildings. The foam is made in batches from a mixture of water and two chemical agents, X and Y. Strict quality and safety regulations specify the minimum quantities of agent X and agent Y that must be used in each standard batch of foam.

Details of the mixture are as follows:

Standard batch quantity: 500 gallons.

Minimum quantities specified per standard batch:

120 gallons agent X; 120 gallons agent Y.

Agents X and Y are available from two products, AB which costs £100 per ten gallon drum, and CD which costs £60 per ten gallon drum. Each drum of AB contains three gallons of agent X, six gallons of agent Y, the rest is water; each drum of CD contains four gallons of agent X, three gallons of agent Y, the rest is water.

You are required to:

(a) formulate the above information as a linear programming problem, solve the problem by graph and therefore advise the company on the least expensive mixture of the two chemical agents consistent with the quality and safety regulations;

(b) indicate how this problem would change, if, as a result of a health scare, the company were advised to use no more than 50 gallons of AB in each standard batch.

CIPFA; Diploma in Public Sector Audit and Accounting; Quantitative Analysis; March 1986.

Solution/hints

Part (a)

Problem in linear programming format:

It is helpful to first set out the details about the two ingredients AB and CD as follows:

	X	Y	Water	Price (10 gallons)
AB	3	6	1	£100
CD	4	3	3	£60

We are also given $X \geqslant 120$ and $Y \geqslant 120$, the minimum quantities required per batch.

Let the quantity of AB and CD in each batch of foam be a and c, respectively. The problem can now be stated as follows:

Minimize $£(100a + 60c)$

subject to: $3a + 4c \geqslant 120$ (minimum X allowed) (1)

$6a + 3c \geqslant 120$ (minimum Y allowed) (2)

$a, c \geqslant 0$ (non-negativity constraints)

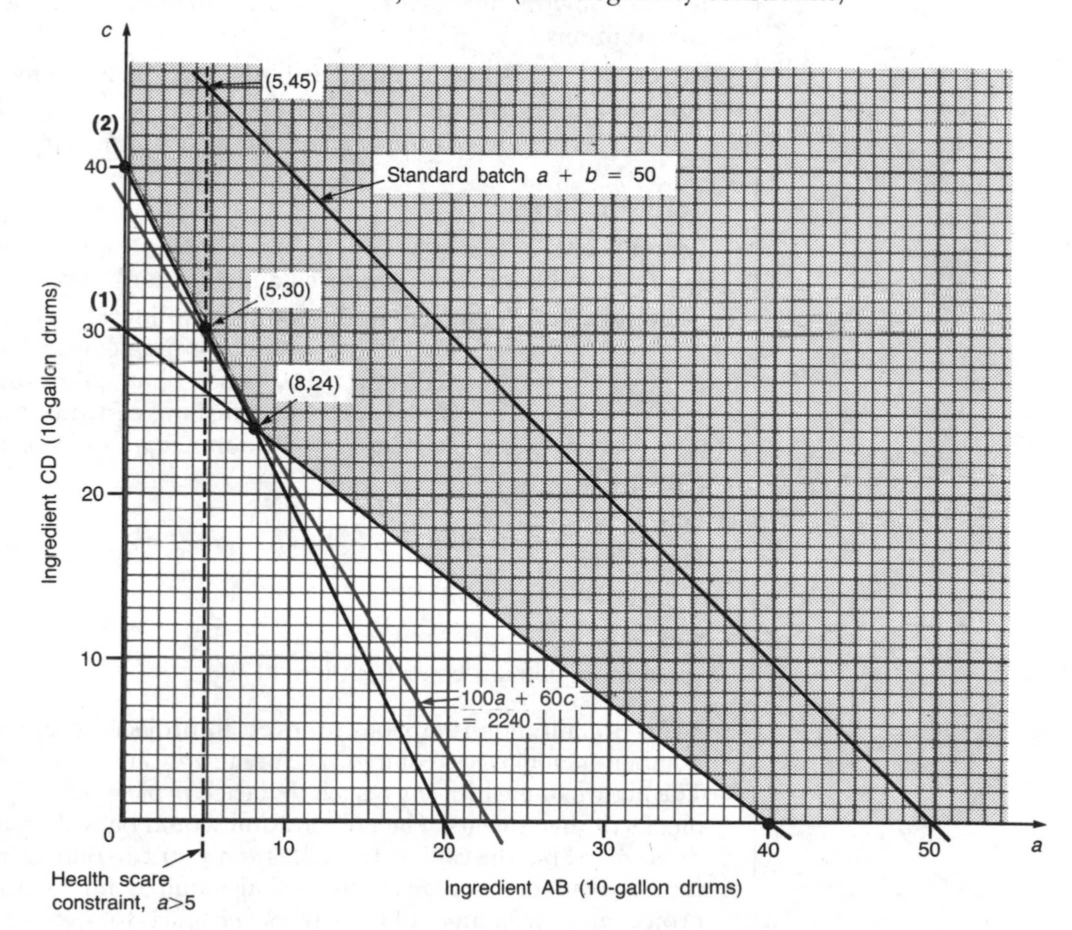

If the standard batch quantity can be varied, the optimum solution is achieved when $a = 8$ and $c = 24$, that is, a mixture of 8 drums of agent AB (80 gallons) and 24 drums of agent CD (240 gallons), a batch size of 320 gallons. Such a batch would cost £2240 ($100 \times 8 + 60 \times 24$).

If the batch quantity is rigid, the constraint $a + c = 50$ is introduced. The feasible region becomes that portion of the line $a + c = 50$ which lies within in the feasible region. The minimum cost will be incurred by a batch using all agent CD (50 drums; £3000).

Part (b)

Answers:

$a = 5$, $c = 30$, cost £2300, cost increase £760, if batch size can vary;
$a = 5$, $c = 45$, cost £3200, cost increase £200, if batch size must be 500 gallons.

2 A small furniture manufacturer makes two speciality products, tables and chairs. Three stages in the manufacturing process are required – machining, assembling and finishing. The number of minutes required for each unit is shown below:

	Machining	Assembling	Finishing
Table	4	10	12
Chair	12	5	12

Each day the machining equipment is available for six hours (360 minutes), the assembly shop for six hours and the finishing equipment for eight hours. The contribution is £3 per table and £5 per chair. All equipment can be used for the production of either tables or chairs at all times it is available.

You are required to:

(a) state all the equations/inequalities (constraints) which describe the production process;

(b) graph these constraints;

(c) find how many tables and chairs the manufacturer should produce to maximize contributions;

(d) calculate this maximum contribution and include any comments or reservations you have about this analysis generally.

CIMA; Foundation stage, Section B; Mathematics and Statistics; May 1986.

Solution/hints

Part (a)

Let t = number of tables produced and c = number of chairs produced.

Maximize $3t + 5c$, subject to the production limitations:

$$4t + 12c \leqslant 360 \text{ (machining constraint);}$$
$$10t + 5c \leqslant 360 \text{ (assembling constraint);}$$
$$12t + 12c \leqslant 480 \text{ (finishing constraint);}$$
$$t, c \geqslant 0 \text{ (non-negativity constraints)}$$

Part (b)

Left as an exercise.

Part (c)

Answer: $t = 15$, $c = 25$

Part (d)

Answer: maximum contribution = £170

3 An organization is anxious to invest in projects where the aim is to maximize the number of employees required. The organization is considering two projects, A and B. The nature of the projects is such that each project may be made up of any number of blocks of investments. The organization would be willing to inject capital for each of three years, but the capital available for a year can only be used in that year. The total capital available each year, the capital required for each investment block for each project and the number of employees per block for each project are given in the table below:

	Project A	Project B	Capital available (£1000)
Year 1	1	3	120
Year 2	2	1	100
Year 3	1	1	60
Number of employees needed per block	4	3	

(a) Formulate as a linear programming problem.

(b) Represent graphically and shade feasible region.

(c) Advise the organization on the number of blocks to invest in each project to maximize the total number of employees, assuming that the number employed on each project is constant over the three years. What is the maximum number of employees?

(d) If the organization decides to accept the advice you have given in part (c), which year will some of the capital remain unused?

CIS; Part 1; Quantitative studies; December 1986.

Solution/hints

Part (a)

Maximize $4a+3b$, subject to:

$$\begin{aligned} a+3b &\leqslant 120 \text{ (year 1)} \quad (1)\\ 2a+b &\leqslant 100 \text{ (year 2)} \quad (2)\\ a+b &\leqslant 60 \text{ (year 3)} \quad (3)\\ a, b &\geqslant 0 \text{ (non-negativity constraints)} \end{aligned}$$

Part (b)

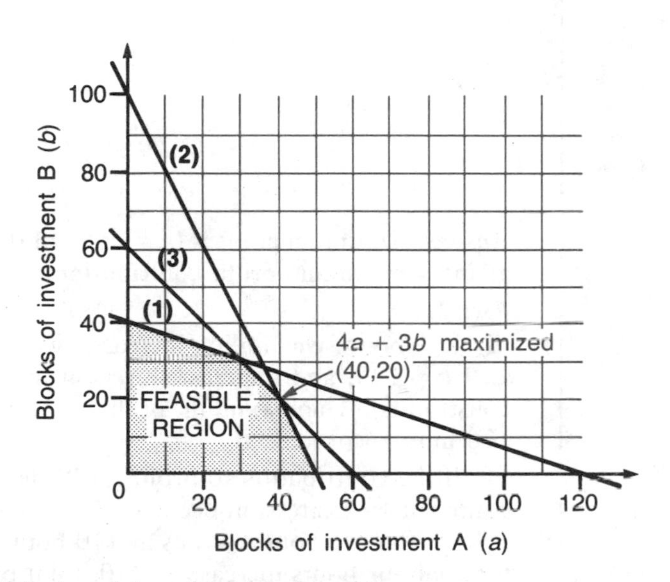

Answers: 40 blocks of A and 20 blocks of B;

Part (c)

Maximum no. of employees $= 4\times 40+3\times 20=220$

Part (d)

Answer: Year 1

At (40, 20), constraint (1) is not fully utilized: $40+3\times 20=100$ (<120); capital available = £120 000; therefore £20 000 will be unused.

4(a) Provide a graphical solution to the following problem of profit maximization: A company manufactures two products A and B. Contribution per unit from A = £8 and from B = £6. Total daily labour hours available to the company = 110 hours. Labour required per unit of output: A, 2 hours and B, 1 hour. Daily production capacity limitations are: 70 units of A

: 150 units of B.

(b) Owing to uncertainty, it is possible that the information given in (a) is subject to variability.

You are requested to ascertain the implications for the company of:

(i) a 10 hour increase in the total labour hours available each day;

and

(ii) a £1 decrease in the contribution per unit of B.

(*Note*: Both possibilities are independent).

CIPFA; Diploma in Public Sector Audit and Accounting; Quantitative analysis; May 1987.

Solution/hints

Part (a)

Maximize $8a+6b$ subject to:

$2a+b \leqslant 110$ (available labour hours)
$a \leqslant 70$ (production capacity)
$b \leqslant 150$ (production capacity)
$a,\ b \geqslant 0$ (non-negativity constraints)

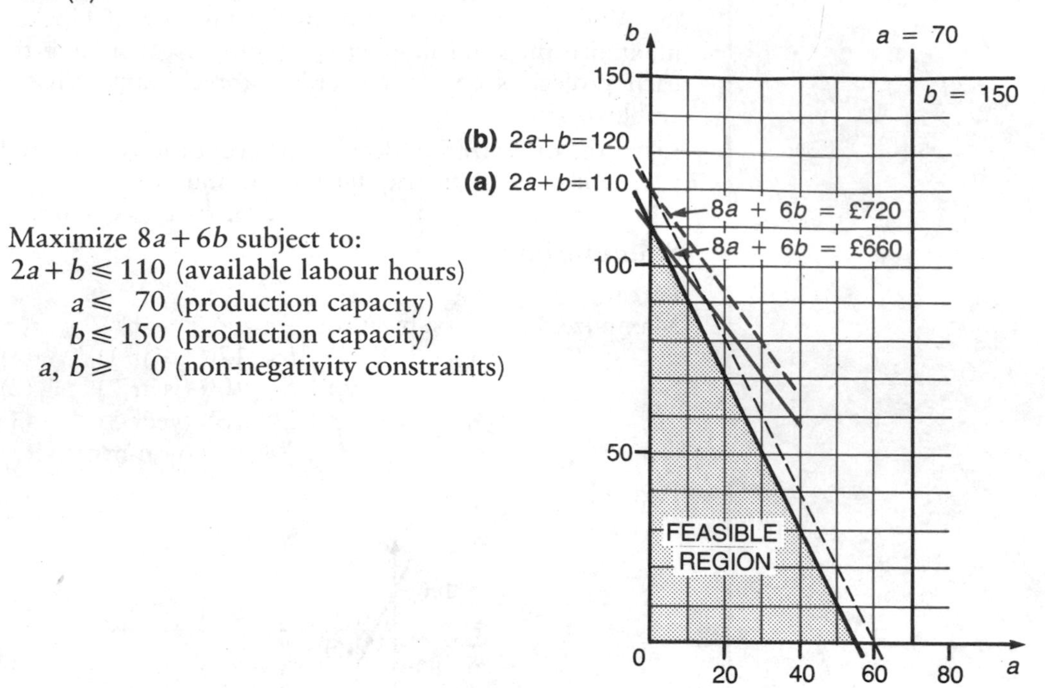

Answer: Optimum point is ($a = 0$, $b = 110$) i.e. produce 110 units of product B only, giving a maximum profit contribution of £660.

Part (b)

(i) If labour availability increases by 10 hours the labour constraint becomes $2a+b = 120$, and the feasible region increases accordingly (still within production constraints). The maximum profit contribution (£720) is now gained by producing 120 units of product B.

(ii) If the contribution to profit (of B) decreases by £1 per unit, then the total profit contribution expression becomes $8a \times 5b$. Optimum points are unchanged.

If available labour remains at 110 hours, total profit contribution will be £500; if total labour hours increase to 120, total profit contribution becomes £600.

5 Your company has to buy immediately two types of table for its canteens. A maximum sum of £24 000 is available for this purpose. A type X table costs £40 and seats four people. A type Y table costs £30 and seats two people. Seating for at least 1800 people is required. There must be at least as many type Y tables as type X because the tables are to be used for a variety of functions in the canteens. For reasons of maintenance, storage, etc, the company wishes to buy the smallest total number of tables to meet its requirements.

You are required to

(**a**) state the company's objective function;

(**b**) state all the constraints (equations/inequalities);

(**c**) draw a graph of these constraints, shading any unwanted regions;

(**d**) recommend the number of each type of table the company should buy; justify your answer.

CIMA; Stage 1; Quantitative methods; May 1987.

Solution/hints

Part (a)

Minimize $x+y$.

Part (b)

$x \leqslant y$	(at least as many type Y as type X)
$4x+2y \geqslant 1800$	($2x+y \geqslant 900$) (people seated)
$40x+30y \leqslant 24\ 000$	($4x+3y \leqslant 2400$) (cost)
$x,\ y \geqslant 0$	(non-negativity constraints)

Part (c)
Note: you are asked to shade **unwanted** regions.

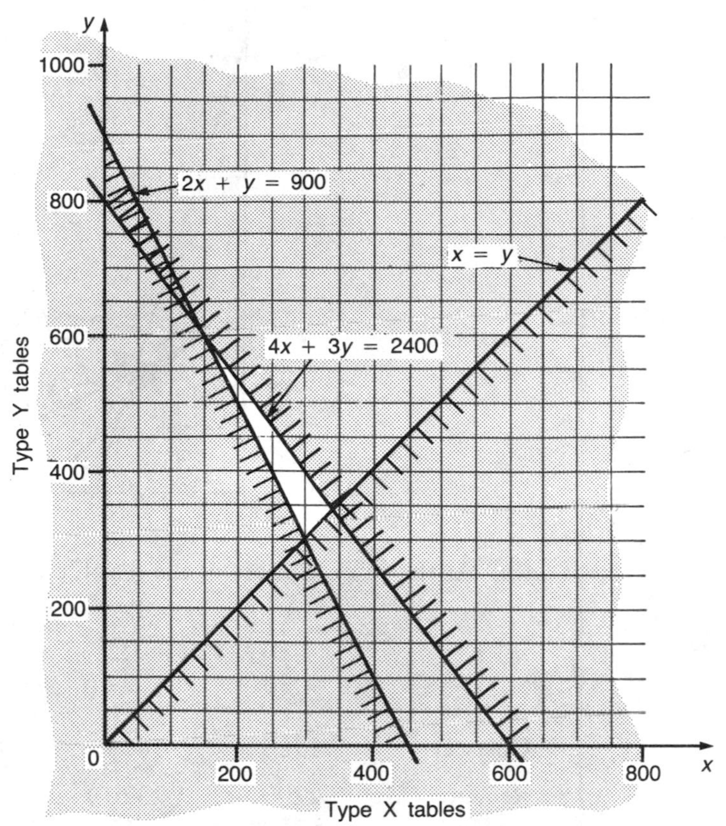

Part (d)
Recommend 300 type X, 300 type Y. Justification left as an exercise.

6 A company manufactures and sells two products X and Y. Each product is processed in two departments, electrical and assembly. Each unit of product X requires 30 minutes in the electrical department and 10 minutes in the assembly department. The corresponding times for product Y are 10 minutes and 20 minutes, respectively. Each day, 900 minutes are available in the electrical department and 700 minutes are available in the assembly department. The sales manager wishes to have daily output of at least 40 products, irrespective of whether they are X or Y.

The costs of manufacture of each unit of X and each unit of Y are £32 and £40, respectively. The selling prices are £40 and £60, respectively.

The company wishes to know the output levels required in each of the following situations:

(a) to maximize sales revenue;

(b) to minimize manufacturing costs;

(c) to maximize profits.

(i) Formulate as a linear programming problem.

(ii) Represent graphically and shade feasible region.

(iii) Obtain the output levels in each of the situations (a), (b) and (c).

CIS; Part 1; Quantitative Studies; June 1987.

Solution/hints

Part (a)

(i) (a) Maximize revenue: $£(40x+60y)$

(b) Minimize costs: $£(32x+40y)$

(c) Maximize profits: $£(8x+20y)$

Subject to:

$30x+10y \leqslant 900$ $\quad(3x+y \leqslant 90)$ (dept. constraint)

$10x+20y \leqslant 700$ $\quad(x+2y \leqslant 70)$ (assembly dept. constraint)

$x+y \geqslant 40$ (Sales manager's product quota)

$x, y \geqslant 0$ (non-negativity constraints)

(ii)

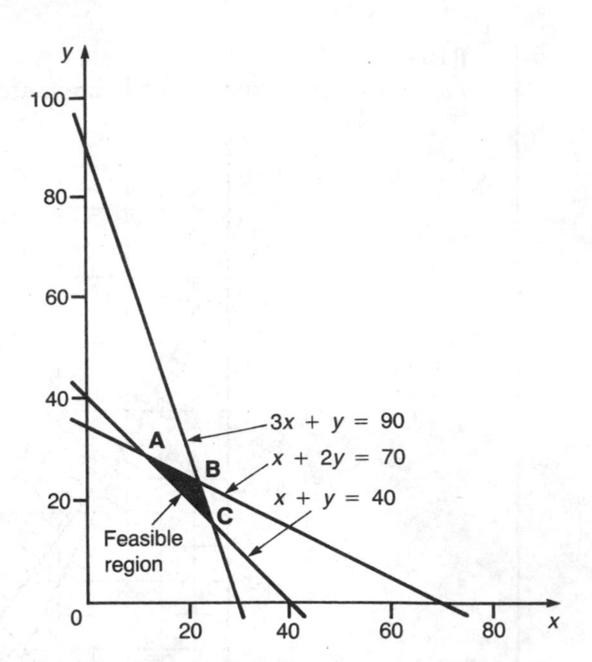

(iii) At the three corners of the feasible region, **A**, **B** and **C**:

	A	B	C	
x =	10	22	25	
y =	30	24	15	
Cost (£)	1520	1664	1400	(25, 15) minimizes cost;
Revenue (£)	2200	2320	1900	(22, 24) maximizes revenue;
Profit (£)	680	656	500	(10, 30) maximizes profit.

7 A tailor has 40 metres of silk, 60 metres of cotton and 64 metres of wool in stock. He makes two styles of garment, the first requiring 2 metres, 2 metres and 4 metres of these materials, respectively, and the second requiring 2 metres, 5 metres and 2 metres, respectively.

The tailor sells the first garment for £8 and the second for £10.

(**a**) Using a graphical linear programming approach, calculate how many of each garment the tailor should make and sell, to maximize his revenue.

(**b**) What are the implications of this decision as far as the use of his stocks are concerned?

(**c**) Suggest how, using the same level of stocks, the tailor could increase his revenue without changing his selling prices. Perform any further calculations which substantiate your suggestion.

The Polytechnic of North London; Accounting Foundation Course; Quantitative methods; December 1987.

Solution/hints

Part (a)

Maximize £$(8a+10b)$ (revenue),

subject to: $2a+2b\leqslant 40$ $(a+b\leqslant 20)$ (silk constraints);
$2a+5b\leqslant 60$ (cotton constraint);
$4a+2b\leqslant 64$ $(2a+b\leqslant 32)$ (wool constraint);
$a, b\geqslant 0$ (non-negativity constraints).

Graph left as an exercise.

Answer: To maximize revenue the tailor should produce 12 of the first garment and 7 of the second.

Part (b)

Calculate surpluses of various materials for the optimum production levels found in part (**a**).

Part (c)

Mention possible different composition of garments.

8 Twig and Leaf Ltd are a small company who are producing Bonsai trees for sale in the local supermarkets and off-licences. They are having some slight problems on the cost side of their business and have come to you for advice. Currently, with limited space, they can only have about 2500 pots to grow trees in at any one time. Hence, for their current crop, they estimate that they will require 1600 milligrams of postassium, 1200 grams of nitrates, and 3000 milligrams of phosphates.

These nutrients can be obtained from two different types of fertilizer:

Each kilo pack of Growfat fertilizer costs £6 and yields 4 milligrams of potassium, 2 grams of nitrates and 6 milligrams of phosphates.
Each kilo pack of Growbig fertilizer costs £8 and yields 2 milligrams of potassium, 3 grams of nitrates, and 5 milligrams of phosphates.

These fertilizers are then mixed with water and distributed evenly amongst the pots over the growing season.

Required:

(**a**) Formulate the above as a linear programme. Solve graphically (or otherwise) to determine the minimum cost at which they can obtain the required nutrients. State any assumptions that you are making.

(**b**) According to the suppliers of the fertilizers, the actual cost of Growfat per kilo is uncertain (due to uncertainty regarding supplies of guano droppings – its main ingredient). There is a small chance that its cost may fall. If this is so, by how much can its cost per kilo fall before your current solution becomes no longer optimal?

(**c**) It is more likely that the cost per kilo of Growfat will increase. If this is the case, Twig and Leaf wish to know two things:

(**i**) By how much can its cost per kilo rise before your current solution becomes no longer optimal?

(**ii**) By how much can its cost per kilo rise before it is no longer worth buying any Growfat fertilizer at all?

The Polytechnic of North London; ACCA internal course; Level 2, part II; Quantitative analysis; January 1988.

Solution/hints

Part (a)

Let x = kilos of Growfat
y = kilos of Growbig

Minimize costs, $C = 6x + 8y$

subject to: $4x + 2y \geqslant 1600$	(potassium constraint; mg)	(1)
$2x + 3y \geqslant 1200$	(nitrate constraint; g)	(2)
$6x + 5y \geqslant 3000$	(phosphate constraint; mg)	(3)
$x, y \geqslant 0$	(non-negativity constraints)	

Assumptions: linearity; no other constraints; equal absorption of the two fertilizers.

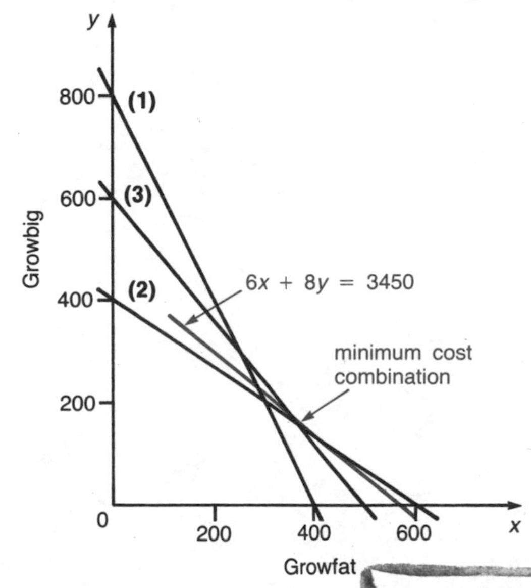

Minimum costs are given by the intersection of lines (2) and (3). Solving these two equations simultaneously:

$$\begin{aligned} 2x+3y &= 1200 && (1) \\ 6x+5y &= 3000 && (2) \\ 6x+9y &= 3600 && (4) \quad (3\times(2)) \\ 4y &= 600 && (4)-(2) \end{aligned}$$

Therefore, $y=150$, $x=375$ and $C=£3450$, i.e. to obtain the required nutrients the company should buy 375 kilo packs of Growfat and 150 kilo packs of Growbig, at a minimum total cost of £3450.

Part (b)

Suppose $C=(6+?)x+8y$, where ? represents the uncertain change in the cost of Growfat. The gradient of the line C is $-(6+?)/8$.

If the cost of Growfat falls, C will reach a point when it is parallel to constraint (2) (gradient -2).

C has a gradient equal to (2) when $-\dfrac{(6+?)}{8}=-\dfrac{2}{3}$

$$6+?=\frac{16}{3}$$

$$?=\frac{16}{3}-6=\frac{16-18}{3}=-\frac{2}{3}$$

Therefore the current solution is acceptable until the cost of Growfat falls below **£5.33** $(6-\frac{2}{3})$ per kilo pack.

Part (c)

Suppose the cost of Growfat increases:

(i) Current solution is acceptable until gradient of C is equal to gradient of (3), i.e. when

$$\frac{-(6+?)}{8}=\frac{-6}{5},\ ?=3.6$$

i.e. until cost of Growfat is **£9.60** per kilo pack.

(ii) The lowest cost combination in the feasible region with no Growfat at all is ($x=0$, $y=800$), on line (1); gradient of C is equal to gradient of (1) when

$$\frac{-(6+?)}{8}=-2,\ ?=10$$

Therefore the company should no longer purchase any Growfat once the cost exceeds **£16** per kilo pack.

Recommended reading

Curwin, J. and Slater, R., *Quantitative Methods for Business Decisions* (Van Nostrand Reinhold, 1986), chapter 20.

Lucey, T., *Quantitative Techniques* (D.P. Publications, 1980), chapters 12–15.

Owen, F. and Jones, R., *Modern Analytical Techniques* (Pitman, 1988), chapter 6.

12 Stock control

12.1 Introduction

An organization's stock is often a major element of its working capital. It can be defined as usable but idle resources and takes a number of forms: raw materials, work-in-progress (WIP) or finished goods waiting for despatch.

Stock and its control are important for a number of reasons:

1 Ample raw materials have to be available for production.

2 Finished goods need to be ready for despatch to customers.

3 It must be possible to control the amount of capital tied up in stock.

4 On-going work needs to be valued.

5 Wastage and pilferage of stock need to be monitored.

Some or all of the above may figure in the need to control stock. In exercising such control, these are two factors of fundamental importance: the costs involved in maintaining a particular level of stock and the costs involved in processing orders for stock. Almost all questions you are likely to encounter will involve these factors and, will lead to the associated problems of 'how much stock to order' and 'how often to place an order'.

This chapter will consider the mathematical models of stock control which you will need to be familiar with. You should also be aware that models of this type are limited in that they do not take into account the various social, environmental and other costs which may be present.

12.2 Simple stock control: assumptions

The models to be considered are based on a number of assumptions, many of which will seem unrealistic. In spite of this, simple stock control models are very useful and have been found to give reasonable solutions to practical problems. The assumptions for the most basic model are that:

1 the rate of demand is constant and known, i.e. withdrawals from stock are at a linear rate;

2 there is zero lead time (lead time is the time between placing an order and receiving the goods);

3 ordering costs are constant, irrespective of order quantity – ordering costs include administrative costs, telephone costs, postage costs, travel costs, etc;

4 holding costs are constant, either in terms of cost per item or as a proportion of item value – holding costs include costs of capital, handling, storage, taxes, insurance, etc;

5 no discounts are available for larger orders.

A number of these assumptions will be discarded as the basic model is developed.

12.3 The EOQ or EBQ stock control model

The model we will consider initially is the basic economic order quantity (EOQ) model, referring to the vendor/purchaser situation, or the economic batch quantity (EBQ) model, referring to items produced *in situ*. All the assumptions (**1** to **5**) listed above apply to this model.

The following notation and information is used in the derivation of this model:

1 The **annual demand,** D, is usually given;

2 The **order quantity,** Q, is usually unknown;

3 The **cost for each order** (**or batch**), c, is usually given and is fixed;

4 The **holding cost per year, *h***, is either given explicitly or is found by reference to the **value of each item, *p***, and the **cost of capital, *i***, resulting in a holding cost of $i \times p$. In this text we will assume the identity $h = i \times p$, where appropriate.

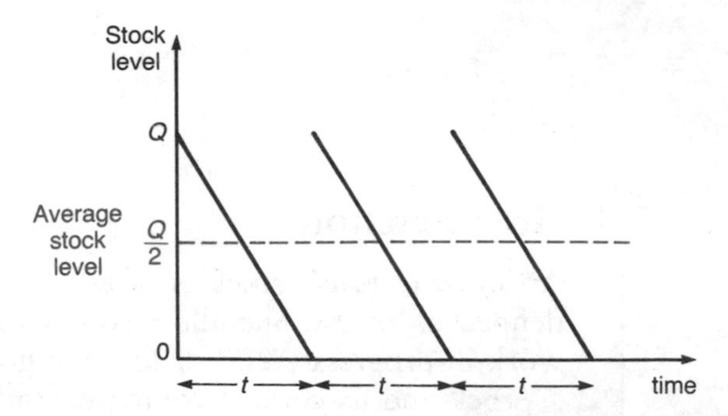

Fig. 12.1 The basic stock control model

Figure 12.1 illustrates this model for stock control. Note:

1 Uniform demand, illustrated by a linear decline in level of stock from Q to zero.
2 Instantaneous replenishment, illustrated by the immediate replacement of stock from a level of zero to a level of Q, at the end of each time period, t.
3 No 'stock-outs', i.e. stock level never falls below zero.
4 Average level of stock is $Q/2$.

As previously mentioned, two factors are of prime importance: ordering costs and holding costs:

12.3.1 Ordering costs

The annual demand is D and the size of an individual order is Q. Hence the total number of orders per annum is given by D/Q. As the cost of one order is c, the annual ordering cost is given by the expression

$$C_o = cD/Q \qquad (1)$$

12.3.2 Holding costs

The total annual stock-holding costs are found by multiplying the cost of keeping one item in stock for one year (h or $i \times p$) by the stock level. However, as the stock level is constantly changing we need to consider the average stock level, $Q/2$. Consequently, the total stock-holding cost, based on the average stock level, is given by the expression

$$C_h = h \times Q/2 \qquad (2)$$

$$\text{or } C_h = i \times p \times Q/2 \qquad (2')$$

The order quantity, Q, is usually the unknown factor in (1) and (2). The relationship between the ordering costs and Q, and also between the holding costs and Q, is illustrated in fig. 12.2.

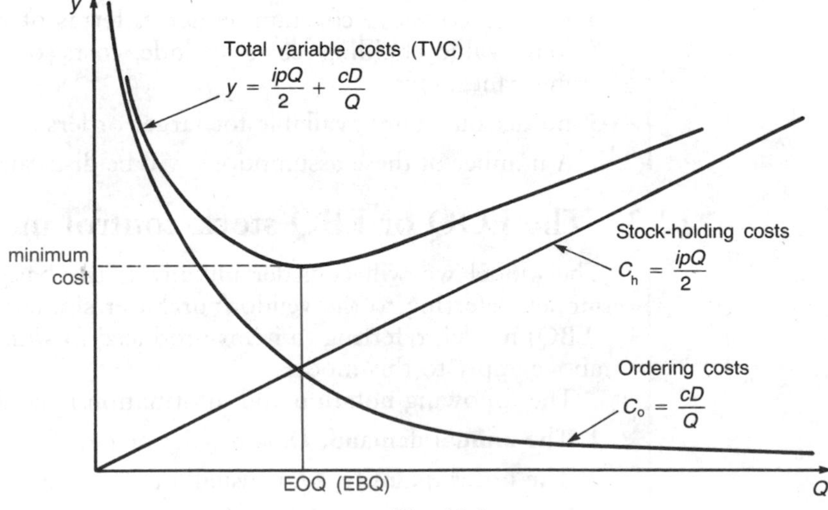

Fig. 12.2

12.3.3 Total variable costs (TVC)

The annual ordering costs, C_o (1), and the annual holding costs, C_h (2), combined give total variable costs (TVC) for an organization. Any fixed costs, such as purchasing costs, are extra and do not influence the ordering or holding costs. Total variable costs are given by the expression

$$\text{TVC} = C_o + C_h = cD/Q + hQ/2$$

12.3.4 Economic order (batch) quantity (EOQ or EBQ)

The economic order quantity, EOQ, is the order quantity which minimizes the total variable costs. We can find the minimum using calculus (refer back to section 9.7 if necessary):

$$\text{TVC} = cD/Q + hQ/2$$

$$\text{Therefore } \frac{\text{d (TVC)}}{\text{d}Q} = \frac{-cD}{Q^2} + \frac{h}{2}$$

$$\frac{\text{d (TVC)}}{\text{d}Q} = 0 \text{ gives } \frac{h}{2} = \frac{cD}{Q^2}$$

which rearranges to give $Q^2 = \frac{2cD}{h}$, therefore $Q = \sqrt{\frac{2cD}{h}}$.

It can be confirmed that this is a minimum. Hence, the economic batch quantity is given by the expression

$$\text{EOQ (or EBQ)} = \sqrt{\frac{2cD}{h}}$$

For this value of Q, the annual ordering cost, cD/Q, is identically equal to the annual holding cost, $hQ/2$ (see fig. 12.2).

By substituting this value of Q back into the original expression for TVC, it can also be shown that the minimum TVC equals $\sqrt{2cDh}$.

12.3.5 An example

E. G. Holden Company has to work a 50-week year and satisfies an annual demand for 10 000 building components. The components may be obtained from one of two sources, a subsidiary of the company, A. S. Supplies, or an outside supplier, Woodstock Builders. The details of the terms of supply are given in table 12.1.

Table 12.1 Costs of ordering and storing building supplies for E. G. Holden Company

	A. S. Supplies	Woodstock Builders
Purchase price per unit (£)	10	9.75
Storage, carrying costs, capital costs per unit per annum (£)	2	2
Cost of placing an order (£)	10	21
Cost of receiving an order (£)	15	15

We wish to find three factors relating to the company's stock control:

1 the EOQ;

2 the ordering pattern; and

3 the TVC.

With this information, and taking note of fixed costs, we wish to find which supplier E. G. Holden Company should deal with.

1 A. S. Supplies

The annual demand is given, $D = 10\ 000$.

The ordering costs per order are given in two parts – the cost of placing an order and the cost of receiving an order:

$$c = £10 + £15 = £25$$

The holding costs are given: $h = £2$/unit/annum.

We can now calculate the EOQ:

$$\text{EOQ} = \sqrt{2cD/h} = \sqrt{(2 \times 25 \times 10\ 000)/2} = \sqrt{250\ 000}$$
$$= 500$$

The ordering pattern can be found by reference to the order quantity, 500, and the annual demand, 10 000:

$$\text{number of orders per year} = D/Q = \frac{10\ 000}{500} = 20$$

$$\text{period of orders} = \frac{50}{20} = 2.5 \text{ weeks}$$

The total variable costs are:

$$\begin{aligned} \text{TVC} = C_h + C_o &= \frac{hQ}{2} + \frac{cD}{Q} \\ &= \frac{2 \times 500}{2} + \frac{25 \times 10\ 000}{500} \\ &= 500 + 500 \\ &= £1000 \end{aligned}$$

The fixed costs are found by multiplying the annual demand by the purchase price per unit:

$$\text{fixed costs} = 10\ 000 \times 10 = £100\ 000$$

Therefore, using A. S. Supplies would result in total costs to E. G. Holden Company of £101 000 (TVC + fixed costs).

2 Woodstock Builders

EOQ can be calculated as above:

$c = £36$, $h = £2$, $D = 10\ 000$: EOQ $= \sqrt{(2 \times 36 \times 10\ 000)/2} = 600$

The ordering pattern works out as an order of 600, 16.67 times a year, that is, an order of 600 every three weeks.

$$\text{TVC} = \frac{2 \times 600}{2} + \frac{36 \times 10\ 000}{600} = 600 + 600 = £1200$$

Fixed costs are £97 500 (10 000 × £9.75).

Hence, the total cost to the company, using Woodstock Builders, is £98 700 (£97 500 + £1200).

If E. G. Holden Company use Woodstock Builders rather than their own subsidiary company, A. S. Supplies, they stand to save £2300 per annum.

Examination questions frequently require candidates to indicate any non-mathematical factors which may be present, such as social or environmental issuses. Clearly, in this case, on financial grounds alone the best choice is Woodstock Builders. However, the damage done to the subsidiary company, A. S. Supplies, in not winning a contract, may be much more costly to the parent company in the long term, than the £2300 per annum saving. Any advice given should bear such effects in mind.

The EBQ stock control model has a number of variants. These will be considered in the following sections.

12.4 Stock control with discounts: fixed holding costs

We now consider a case in which assumption 5, that no discounts are given for large orders, no longer holds.

Consider an example: PJ Cold Storage operates for 50 weeks of the year and has to satisfy an average weekly demand for goods of 100 units. The cost per order is fixed at £9 per order. Stockholding costs are fixed at £4 per unit per annum. The company's supplier suggests a discount scheme, to encourage larger orders, the details of which are shown in table 12.2. What should the company decide?

Table 12.2 Discounts on variable order levels

Price range	Price per unit (£)	Order quantity, Q
A	£9	< 100
B	£8	100 but < 200
C	£7.50	$\geqslant 200$

Initially, we need to calculate the EOQ and the TVC:

$$\text{EOQ} = \sqrt{(2\times 9\times 5000)/4} = 150$$

$$\text{TVC} = \frac{4\times 150}{2} + \frac{9\times 5000}{150}$$

$$= 300 + 300 = £600$$

This order quantity is in price range B. Consequently, the total cost to the company per annum (TVC plus total fixed costs) is £600 + (5000 × £8) = £40 600.

It is not worth considering a smaller batch quantity, because:

1 the price would be more per unit (an increase of £1 per unit); and

2 the TVC would also be larger, as we have already established that 150 is the best order quantity.

We need to consider the savings (if any) resulting from ordering in batches of 200 or more units. If $Q = 200$, then

$$\text{TVC} = \frac{4\times 200}{2} + \frac{9\times 5000}{200}$$

$$= 400 + 225 = £625$$

and total purchase costs = 5000 × £7.50
= £37 500

resulting in total annual costs of £38 125, a saving of £2475 over the result for batches of 150.

PJ Cold Storage should therefore increase their order quantity from 150 (total cost, £40 600) to 200 (total cost, £38 125) to save £2475. This is assuming that the company will experience no problems in accommodating extra stock. Be aware of the assumptions you are making when answering examination questions. As much as 20 per cent of the marks for a question can be allotted to responding correctly to the request to 'state your assumptions'.

12.5 Stock control with discounts: variable holding costs

As it is reasonable to expect a proportion of holding costs to include capital costs, any change in purchase costs must influence holding costs. We will therefore dispense with assumption **4**, that holding costs are constant.

For instance, suppose the £4 attributed to the holding costs of PJ Cold Storage in the previous section is split between a number of sub-costs:

Warehouse staff costs	£2.00 per unit per annum
Security/insurance costs	£0.75 per unit per annum
Capital costs (15% of £8 (the purchase price))	£1.20 per unit per annum
Other costs	£0.05 per unit per annum

Any drop in purchase costs would result in a drop in capital costs, which would affect part of the holding costs, see section 12.4.1. PJ Cold Storage's holding costs would vary as follows:

Purchase price per unit		Holding costs part-costs	Holding costs per unit per annum
A	£9	2 + 0.75 + 1.35 + 0.05	£4.15
B	£8	2 + 0.75 + 1.20 + 0.05	£4.00
C	£7.50	2 + 0.75 + 1.13 + 0.05	£3.93

12.5.1 Capital costs

If a company commits capital to the purchase of goods or the production of components, then that capital can no longer be in the bank earning interest or be used for other investment opportunities. Consequently there is a cost attributed to holding stock that is directly related to the lost opportunity of the capital involved. This is called capital cost and is usually presented as a proportion (or percentage) of the price of a unit of stock, $i\times p$.

The data shown above confirm that the capital cost portion of holding costs drops as the value of the goods drops, resulting in a net fall in holding costs. If the order quantity is increased to 200, the holding costs per unit drops from £4 to £3.93. The TVC will then be

$$\text{TVC} = \frac{3.93 \times 200}{2} + \frac{9 \times 5000}{200}$$

$$= 393 + 225 = £618$$

giving a total cost, including fixed costs of £37 500, of £38 118, a saving of £2482 over the EOQ costs.

12.6 Continuous replenishment

We next consider a further variant of the basic model in which the assumption of instantaneous replenishment (assumption 2) is abandoned.

For example, suppose a company produces components for use in a manufacturing process. It would not be reasonable to expect a batch of components to be available at a moment's notice. Consequently, when a new batch is ordered there will be a period when stock is both coming into storage and being used. Figure 12.3 illustrates this model:

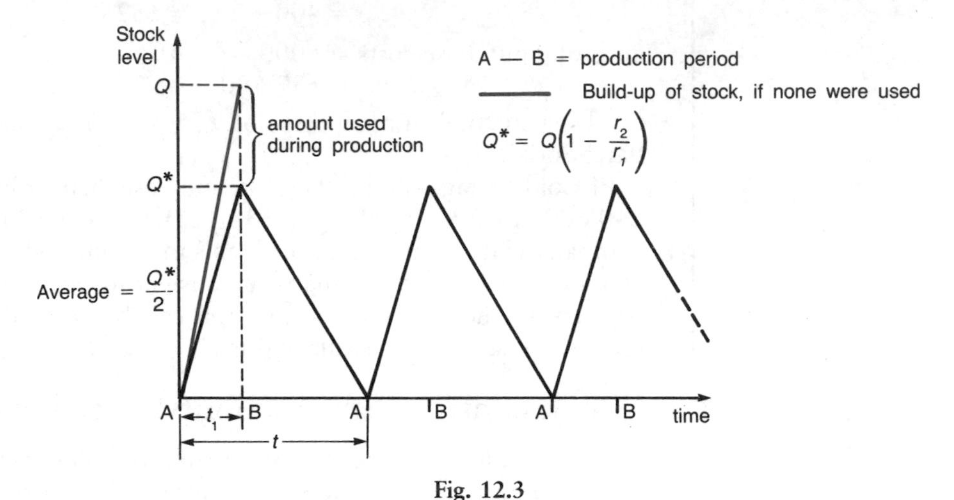

Fig. 12.3

Note:

1 The company can produce components at a rate r_1 (per week, say);

2 The weekly rate of demand for components can be calculated using $r_2 = D/T$, where T is the number of working weeks per annum (Some questions quote a rate directly, making it unnecessary to carry out any further calculations);

3 To be able to meet demand, the rate of production must be greater than or equal to the rate of demand, i.e. $r_1 \geqslant r_2$;

4 Q is the quantity produced each production run;

5 $Q^* = Q(1 - r_2/r_1)$ is the maximum stock level, therefore the average stock level is $Q^*/2$;

6 the difference $Q - Q^*$ is the amount used during production.

Consider an example: Storit Company Ltd produces components and keeps them in stock until required as input for a manufacturing process. The average rate at which the company can produce the components is 266.67 per week. The annual demand for these components, over a working year of 50 weeks, is 10 000 or 200 per week. The cost of setting up a production run is £36 (equivalent to ordering costs in previous examples) and the holding costs are 20 per cent of component costs per annum. The cost of producing one component is £10. We wish to find the company's total annual costs with respect to this component.

12.6.1 The maximum stock level, Q^*

Because stock is produced over a period of time, t_1, and the demand on stock is assumed to be constant, a proportion of the incoming stock will be used before production is complete. Consequently, the stock level never reaches Q, the batch order quantity. The amount of stock used during this production period is $r_2/r_1 \times Q$.

For example, in the example outlined above the rate of demand is 75 per cent of the rate of production. Hence, during the production period 75 per cent of the components produced will be used and the maximum stock level reached is only 25 per cent of the batch order quantity, $0.25Q$. In general,

$$Q^* = Q - (r_2/r_1 \times Q) = Q(1 - r_2/r_1)$$

Here $Q^* = Q(1 - 200/266.67) = Q(1 - 0.75) = 0.25Q$

Using Q^*, the average stock level can be calculated as before:

$$Q^*/2 = 0.25Q/2 = 0.125Q$$

12.6.2 Solution to the continuous replenishment problem

We are now in a position to solve the above problem. The relevant factors are similar to those already encountered:

Production set-up costs, $C_o = cD/Q$ (1)
Holding costs, $C_h = hQ(1 - r_2/r_1)/2$ (2)
Economic batch quantity, $\text{EBQ} = \sqrt{2cD/h(1 - r_2/r_1)}$
Total variable costs, $\text{TVC} = C_o + C_h$
$= (1) + (2)$

In this case, C_o = production set-up costs and $Q(1 - r_2/r_1)/2$ = average stock level. The method is essentially the same as before:

1 What is the EBQ?

$$\text{EBQ} = \sqrt{2 \times 36 \times 10\,000/2(1 - 0.75)}$$
$$= \sqrt{720\,000/0.5} = \sqrt{1\,440\,000} = 1200$$

2 How often does the company need to produce components?
The annual demand, D, is 10 000. Hence, there will need to be 10 000/1200 = 8.33 production runs per year, i.e. a production run every 6 weeks (50/8.33).

3 How long will a production run last?
If the company can produce 267.67 per week (on average), it will take 4.5 weeks (1200/267.67) to produce a batch of 1200.

4 What will be the maximum stock level?

$$Q^* = Q(1 - r_2/r_1) = 1200(1 - 0.75) = 300$$

5 What are the TVC?

$$\text{TVC} = C_o + C_h$$
$$= \frac{36 \times 10\,000}{1200} + \frac{(20\% \times 10) \times 1200(1 - 0.75)}{2}$$
$$= 300 + 300 = £600$$

6 What is the total cost (including production costs)?
Components cost £10 each to produce, therefore

$$\text{Total cost} = (10 \times 10\,000) + 600 = £100\,600$$

The pattern is illustrated in fig. 12.4.

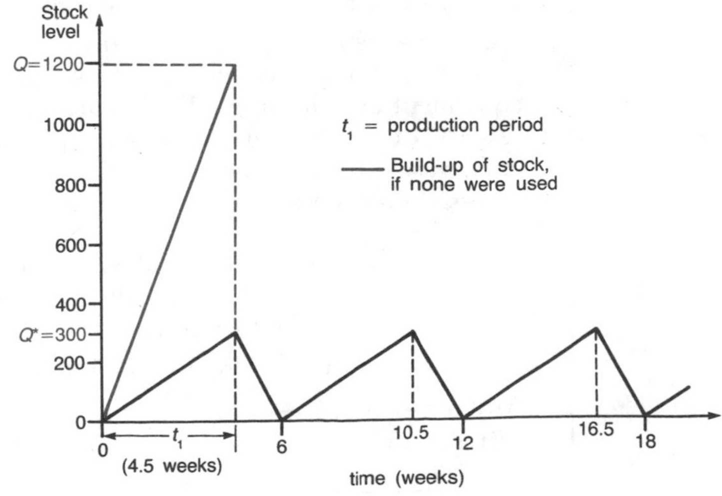

Fig. 12.4

12.7 Sensitivity analysis

Often estimates of the various parameters, annual demand, D, holding costs, C_h, and ordering costs (or production set-up costs), c, include an element of error. If this is so, it is important to know how this error affects the TVC.

12.7.1 Set-up cost error

Suppose at the end of 12 months Storit Company Ltd discover that they underestimated the setup cost of production by 25 per cent, i.e. $c = £45$ (not £36). One consequence is that the TVC will be higher than expected:

$$\text{TVC} = \frac{45 \times 10\,000}{1200} + \frac{(20\% \times 10) \times 1200(1 - 0.75)}{2}$$
$$= 375 + 300 = £675 \quad (*)$$

This is an increase of £75 (12.5%) over the expected total variable costs.

The model seems sensitive to errors in ordering cost estimates. But wait! We must look again. If the company had known that ordering costs would be £45 (not £36), the analysis would have produced a different EBQ, TVC, and so on:

If $c = £45$, EBQ $= 1341.64 \approx 1342$ and TVC $= £670.82$, only £4.18 less than (*).

Thus the higher TVC value resulting from the underestimation (*) is only 0.6% above what it should be, resulting from a 25 per cent error in ordering costs. It can be seen that the stock control model is actually quite *in*sensitive to changes in the value of the set-up cost.

12.7.2 Holding cost error

Now suppose that it was discovered that the holding costs should have been 30 per cent of component costs (not 20 per cent). This is an error of 50 per cent (50% of 20% = 10%).

The company would find that:

$$\text{TVC} = \frac{36 \times 10\,000}{1200} + \frac{(30\% \times 10) \times 1200(1 - 0.75)}{2}$$
$$= 300 + 450 = 750 \quad (**)$$

If the holding cost had been known when the original analysis was made they would have found EBQ = 980 and TVC = £734.85. Comparing this with the result (**) we see that extra cost incurred because of holding cost error is just over 2% higher than the costs the company should have paid.

As an exercise you should carry out a sensitivity analysis relating to an error in the estimation of the annual demand, D. You should find that your result confirms what we have seen above, that is, that the EBQ model is quite insensitive to change in the value of the various parameters.

Sample questions

1(a) Why is a system of inventory control necessary?

(b) A General Hospital has its own central stores department responsible for the purchase, storage and delivery to wards of small items of medical and surgical equipment and dressings. The department operates an inventory control system in respect of these items. One such product, disposable syringes, is kept under continuous review, and each syringe costs £10 to buy. On average, 100 such syringes are used each week, and each order for syringes incurs the following costs:

	£
Supplies Officer's time	5.00
Telephone calls, postage, etc	2.00
Other costs	1.12

Annual unit stock-holding cost is estimated to be $\frac{1}{2}$% of the purchase price of stock.

You are required to:

(i) determine the economic order quantity and the number of orders per year for this item;

(ii) derive the total cost of purchasing and holding this item per year;

(iii) state what assumptions you have made in the calculations of (i) and (ii).

CIPFA; Diploma in Public Sector Audit and Accounting; Quantitative analysis; March 1986.

Solution/hints

Part (a)

See the introduction to this chapter.

Part (b)

Purchase price = £10 per syringe
Ordering costs = £(5 + 2 + 1.12) = £8.12 per order
Holding costs = 0.5% of purchase price per unit per annum
= 0.005 × £10 = £0.05 per unit per annum

(i) $\text{EOQ} = \sqrt{2cD/h}$
$= \sqrt{(2 \times 8.12 \times 5200)/0.05} = \sqrt{1\,688\,960}$
$= 1299.60$
≈ 1300 syringes

Number of orders per year $= \dfrac{5200}{1300} = 4$

(ii) $\text{TVC} = C_o + C_h = cD/Q + hQ/2$
$= (8.12 \times 4) + (0.05 \times 1300)/2$
$= 32.48 + 32.50 = £64.98$

Total purchase cost = 5200 × £10 = £52 000

Therefore the total cost of purchasing and holding this item is £52 064.98.

(iii) Assumptions: Demand is known and accurate; 52-week working year; adequate storage space; no stockouts allowed; uniform demand throughout the year.

2 Your organization uses 1000 packets of paper each year of 48 working weeks. The variable costs of placing an order, progressing delivery and payment have been estimated at £12 per order.

Storage and interest costs have been estimated at £0.50 per packet per annum based on the average annual stock.

The price from the usual supplier is £7.50 per packet for any quantity.

The usual supplier requires four weeks between order and delivery.

A potential supplier has offered the following schedule of prices and quantities:

£7.25 per packet for a minimum quantity of 500 at any one time;

£7.00 per packet for a minimum quantity of 750 at any one time;

£6.94 per packet for a minimum quantity of 1000 at any one time.

If more than 450 packets are received at the same time an additional fixed storage cost of £250 will be payable for the use of additional space for the year.

Assume certainty of demand, lead time and costs.

You are required to:

(a) calculate and state the EBQ from the existing supplier;

(b) calculate and state the stock level at which the orders will be placed;

(c) calculate and state the total minimum cost for the year from the existing supplier;

(d) calculate and state the total minimum cost if you change to the new supplier.

CIMA; Professional Stage, part 1; Quantitative analysis; May 1986.

Solution/hints

Part (a)

EBQ, existing supplier = 219

Part (b)

Lead time = 4 weeks, i.e. it takes 4 weeks for an order to be processed and delivered. We solve the problem of when to re-order in three stages:

1 find the number of orders per annum;

2 find the period of time between orders;

3 find the stock level with 4 weeks' of stock left.

If the quantity ordered is 219, the number of orders needed per year is (on average) 1000/219 = 4.566. Therefore the period between orders is 48/4.566 = 10.5 weeks.

The stock level at which orders should be placed is

$$\frac{4}{10.5} \times 219 = 83 \text{ (see fig. 12.5)}$$

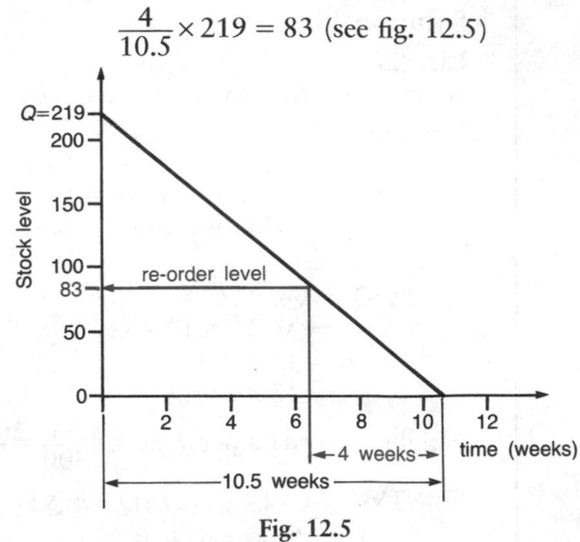

Fig. 12.5

Part (c)
TVC, existing supplier = £109.54
Total purchasing cost, existing supplier = 100 × 7.50 = £7500
Total minimum annual cost, existing supplier = £7609.54 (*)

Part (d)
Costs relating to potential new supplier:

We need to consider all three minimum order quantities, 500, 750 and 1000. Remember that there is an additional storage charge of £250 for orders of 450 or more.

1 $Q = 500$

$$\text{TVC} = cD/Q + hQ/2 + 250 = \frac{12 \times 1000}{500} + \frac{0.5 \times 500}{2} + 250$$

$$= £399$$

Total purchasing cost = 7.25 × 1000 = £7250
Total annual cost = £7649

2 $Q = 750$
TVC = £435.50; Total purchasing cost = £7000
Total annual cost = £7435.50

3 $Q = 1000$
TVC = £512; Total purchasing cost = £6940
Total annual cost = £7452

The best deal from the new supplier is obtained with an order quantity of 750 (total annual cost, £7435.50). (You could note that this is also better than the total cost using the existing supplier, *.)

3 Your organization uses 5000 units of part F each year and 10 000 units of part G each year of 48 working weeks.

From the regular supplier, part F costs £25 per unit.

From a different regular supplier, part G costs £32 per unit.

Because of the technical characteristics of the product the cost of ordering, receiving and testing each batch of either F or G has been estimated at the relatively high amount of £150.

A third supplier has recently offered to supply parts F and G on a regular monthly basis, i.e. twelve times per annum. He quotes the same price (per unit) for both F and G, and your organization has estimated the cost of ordering, receiving and testing of each monthly batch at £250 per month.

Whichever supplier is used, the storage costs are £2 per unit per annum based on the average annual stock.

Assume certainty of demand, lead time and costs.

You are required to:

(a) calculate and state the EOQ for part F and for part G from the present supplier;

(b) calculate and state the total minimum cost for part F and for part G for the year from the present supplier;

(c) calculate and state the price for the part F and for part G if a change to the new supplier is to be worthwhile.

CIMA; Professional stage, part 1; Quantitative techniques; November 1986.

Solution/hints

Details left as an exercise.

Part (a)
Part F: EOQ = 866
Part G: EOQ = 1225

Part (b)
Total minimum costs = TVC + purchase costs

$$\text{Part F: Total minimum cost} = 1732.05 + 125\,000 = £126\,732.05 \quad (1)$$

$$\text{Part G: Total minimum cost} = 2449.49 + 320\,000 = £322\,449.49 \quad (2)$$

Part (c) New supplier

$$\text{TVC(F)} = (250 \times 12) + \frac{2 \times (5000/12)}{2}$$

$$= 3000 + 416.67 = £3416.67$$

Let the price per unit be p. Total costs can then be expressed as £3416.67 + 5000p. For it to be worthwhile for the organization to change to this new supplier, total costs must be less than (1), that is:

$$£3416.67 + 5000p < £126\,732.05$$

Therefore, the price of F must be less than £24.66.

Using the same method for G (remembering that, for G, $D = 10\,000$), TVC = £3833.33 and the price of G must be less than £31.87.

4 A small business requires 400 boxes of paper each year. The purchase price is £20 per box. The cost to the business to place an order with the supplier is £40, which includes telephone calls, clerical work and management time. The holding costs are estimated to be 20% of the purchase price of stock.

Assume the lead time is zero and:

(a) determine the economic order quantity in respect of the boxes;

(b) determine the frequency of placing an order;

(c) determine the total holding, ordering and purchase costs in a year of the policy recommended in (a);

(d) state any assumptions made in the calculations undertaken in (a), (b) or (c).

CIPFA; Diploma in Public Sector Audit and Accounting; Quantitative analysis; May 1987.

Solution/hints

This is quite straightforward and answers only will be provided:

(a) EOQ = 89

(b) Order frequency: 4.5 times a year.

(c) Total annual costs = £8357.78

(d) Assumptions: uniform demand; accuracy of estimates; zero lead time; etc (see section 12.1).

Recommended reading

Owen, F. and Jones, R., *Modern Analytical Techniques* (Pitman, 1988), chapter 4.

Curwin, J. and Slater, R., *Quantitative Methods for Business Decisions* (Van Nostrand Reinhold, 1986), chapter 22.

Lucey, T., *Quantitative Techniques* (D.P. Publications, 1980), chapters 6–8.

Appendices

1 Random numbers

03 47 43 73 86	36 96 47 36 61	46 98 63 71 62	33 26 16 80 45	60 11 14 10 95	53 74 23 99 67	61 32 28 69 84	94 62 67 86 24	98 33 41 19 95	47 53 53 38 09
97 74 24 67 62	42 81 14 57 20	42 53 32 37 32	27 07 36 07 51	24 51 79 89 73	63 38 06 86 54	99 00 65 26 94	02 82 90 23 07	79 62 67 80 60	75 91 12 81 19
16 76 62 27 66	56 50 26 71 07	32 90 79 78 53	13 55 38 58 59	88 97 54 14 10	35 30 58 21 46	06 72 17 10 94	25 21 31 75 96	49 28 24 00 49	55 65 79 78 07
12 56 85 99 26	96 96 68 27 31	05 03 72 93 15	57 12 10 14 21	88 26 49 81 76	63 43 36 82 69	65 51 18 37 88	61 38 44 12 45	32 92 85 88 65	54 34 81 85 35
55 59 56 35 64	38 54 82 46 22	31 62 43 09 90	06 18 44 32 53	23 83 01 30 30	98 25 37 55 26	01 91 82 81 46	74 71 12 94 97	24 02 71 37 07	03 92 18 66 75
16 22 77 94 39	49 54 43 54 82	17 37 93 23 78	87 35 20 96 43	84 26 34 91 64	02 63 21 17 69	71 50 80 89 56	38 15 70 11 48	43 40 45 86 98	00 83 26 91 03
84 42 17 53 31	57 24 55 06 88	77 04 74 47 67	21 76 33 50 25	83 92 12 06 76	64 55 22 21 82	48 22 28 06 00	61 54 13 43 91	82 78 12 23 29	06 66 24 12 27
63 01 63 78 59	16 95 55 67 19	98 10 50 71 75	12 86 73 58 07	44 39 52 38 79	85 07 26 13 89	01 10 07 82 04	59 63 69 36 03	69 11 15 83 80	13 29 54 19 28
33 21 12 34 29	78 64 56 07 82	52 42 07 44 38	15 51 00 13 42	99 66 02 79 54	58 54 16 24 15	51 54 44 82 00	62 61 65 04 69	38 18 65 18 97	85 72 13 49 21
57 60 86 32 44	09 47 27 96 54	49 17 46 09 62	90 52 84 77 27	08 02 73 43 28	34 85 27 84 87	61 48 64 56 26	90 18 48 13 26	37 70 15 42 57	65 65 80 39 07
18 18 07 92 46	44 17 16 58 09	79 83 86 19 62	06 76 50 03 10	55 23 64 05 05	03 92 18 27 46	57 99 16 96 56	30 33 72 85 22	84 64 38 56 98	99 01 30 98 64
26 62 38 97 75	84 16 07 44 99	83 11 46 32 24	20 14 85 88 45	10 93 72 88 71	62 95 30 27 59	37 75 41 66 48	86 97 80 61 45	23 53 04 01 63	45 76 08 64 27
23 42 40 64 74	82 97 77 77 81	07 45 32 14 08	32 98 94 07 72	93 85 79 10 75	08 45 93 15 22	60 21 75 46 91	98 77 27 85 42	28 88 61 08 84	69 62 03 42 73
52 36 28 19 95	50 92 26 11 97	00 56 76 31 38	80 22 02 53 53	86 60 42 04 53	07 08 55 18 40	45 44 75 13 90	24 94 96 61 02	57 55 66 83 15	73 42 37 11 61
37 85 94 35 12	83 39 50 08 30	42 34 07 96 88	54 42 06 87 98	35 85 29 48 39	01 85 89 95 66	51 10 19 34 88	15 84 97 19 75	12 76 39 43 78	64 63 91 08 25
70 29 17 12 13	40 33 20 38 26	13 89 51 03 74	17 76 37 13 04	07 74 21 19 30	72 84 71 14 35	19 11 58 49 26	50 11 17 17 76	86 31 57 20 18	95 60 78 46 75
56 62 18 37 35	96 83 50 87 75	97 12 25 93 47	70 33 24 03 54	97 77 46 44 80	88 78 28 16 84	13 52 53 94 53	75 45 69 30 96	73 89 65 70 31	99 17 43 48 76
99 49 57 22 77	88 42 95 45 72	16 64 36 16 00	04 43 18 66 79	94 77 24 21 90	45 17 75 65 57	28 40 19 72 12	25 12 74 75 67	60 40 60 81 19	24 62 01 61 16
16 08 15 04 72	33 27 14 34 09	45 59 34 68 49	12 72 07 34 45	99 27 72 95 14	96 76 28 12 54	22 01 11 94 25	71 96 16 16 88	68 64 36 74 45	19 59 50 88 92
31 16 93 32 43	50 27 89 87 19	20 15 37 00 49	52 85 66 60 44	38 68 88 11 80	43 31 67 72 30	24 02 94 08 63	38 32 36 66 02	69 36 38 25 39	48 03 45 15 22
68 34 30 13 70	55 74 30 77 40	44 22 78 84 26	04 33 46 09 52	68 07 97 06 57	50 44 66 44 21	66 06 58 05 62	68 15 54 35 02	42 35 48 06 32	14 52 41 52 48
74 57 25 65 76	59 29 97 68 60	71 91 38 67 54	13 58 18 24 76	15 54 55 95 52	22 66 22 15 86	26 63 75 41 99	58 42 36 72 24	58 37 52 18 51	03 37 18 39 11
27 42 37 86 53	48 55 90 65 72	96 57 69 36 10	96 46 92 42 45	97 60 49 04 91	96 24 40 14 51	23 22 30 88 57	95 67 47 29 83	94 69 40 06 07	18 16 36 78 86
00 39 68 29 61	66 37 32 20 30	77 84 57 03 29	10 45 65 04 26	11 04 96 67 24	31 73 91 61 19	60 20 72 93 48	98 57 07 23 69	65 95 39 69 58	56 80 30 19 44
29 94 98 94 24	68 49 69 10 82	53 75 91 93 30	34 25 20 57 27	40 48 73 51 92	78 60 73 99 84	43 89 94 36 45	56 69 47 07 41	90 22 91 07 12	78 35 34 08 72
16 90 82 66 59	83 62 64 11 12	67 19 00 71 74	60 47 21 29 68	02 02 37 03 31	84 37 90 61 56	70 10 23 98 05	85 11 34 76 60	76 48 45 34 60	01 64 18 39 96
11 27 94 75 06	06 09 19 74 66	02 94 37 34 02	76 70 90 30 86	38 45 94 30 38	36 67 10 08 23	98 93 35 08 86	99 29 76 29 81	33 34 91 58 93	63 14 52 32 52
35 24 10 16 20	33 32 51 26 38	79 78 45 04 91	16 92 53 56 16	02 75 50 95 98	07 28 59 07 48	89 64 58 80 75	03 05 02 27 89	30 14 78 56 27	86 63 59 80 02
38 23 16 86 38	42 38 97 01 50	87 75 00 81 41	40 01 74 91 62	48 51 84 08 32	10 15 83 87 60	79 24 31 66 56	21 48 24 06 93	91 98 94 05 49	01 47 59 38 00
31 96 25 91 47	96 44 33 49 13	34 86 82 53 91	00 52 43 48 85	27 55 26 89 62	55 19 68 97 65	03 73 52 16 56	00 53 55 90 27	33 42 29 38 87	22 13 88 83 34
66 67 40 67 14	64 05 71 95 86	11 05 65 09 68	76 83 20 37 90	57 16 00 11 66	53 81 29 13 39	35 01 20 71 34	62 33 74 82 14	53 73 19 09 03	56 54 29 56 93
14 90 84 45 11	75 73 88 05 90	52 27 41 14 86	22 98 12 22 08	07 52 74 95 80	51 86 32 68 92	33 98 74 66 99	40 14 71 94 58	45 94 19 38 81	14 44 99 81 07
68 05 51 18 00	33 96 02 75 19	07 60 62 93 55	59 33 82 43 90	49 37 38 44 59	35 91 70 29 13	80 03 54 07 27	96 94 78 32 66	50 95 52 74 33	13 80 55 62 54
20 46 78 73 90	97 51 40 14 02	04 02 33 31 08	39 54 16 49 36	47 95 93 13 30	37 71 67 95 13	20 02 44 95 94	64 85 04 05 72	01 32 90 76 14	53 89 74 60 41
64 19 58 97 79	15 06 15 93 20	01 90 10 75 06	40 78 78 89 62	02 67 74 17 33	93 66 13 83 27	92 79 64 64 72	28 54 96 53 84	48 14 52 98 94	56 07 93 89 30
05 26 93 70 60	22 35 85 15 13	92 03 51 59 77	59 56 78 06 83	52 91 05 70 74	02 96 08 45 65	13 05 00 41 84	93 07 54 72 59	21 45 57 09 77	19 48 56 27 44
07 97 10 88 23	09 98 42 99 64	61 71 62 99 15	06 51 29 16 93	58 05 77 09 51	49 83 43 48 35	82 88 33 69 96	72 36 04 19 76	47 45 15 18 60	82 11 08 95 97
68 71 86 85 85	54 87 66 47 54	73 32 08 11 12	44 95 92 63 16	29 56 24 29 48	84 60 71 62 46	40 80 81 30 37	34 39 23 05 38	25 15 35 71 30	88 12 57 21 77
26 99 61 65 53	58 37 78 80 70	42 10 50 67 42	32 17 55 85 74	94 44 67 16 94	18 17 30 88 71	44 91 14 88 47	89 23 30 63 15	56 34 20 47 89	99 82 93 24 98
14 65 52 68 75	87 59 36 22 41	26 78 63 06 55	13 08 27 01 50	15 29 39 39 43	79 69 10 61 78	71 32 76 95 62	87 00 22 58 40	92 54 01 75 25	43 11 71 99 31
17 53 77 58 71	71 41 61 50 72	12 41 94 96 26	44 95 27 36 99	02 96 74 30 83	75 93 36 57 83	56 20 14 82 11	74 21 97 90 65	96 42 68 63 86	74 54 13 26 94
90 26 59 21 19	23 52 23 33 12	96 93 02 18 39	07 02 18 36 07	25 99 32 70 23	38 30 92 29 03	06 28 81 39 38	62 25 06 84 63	61 29 08 93 67	04 32 92 08 09
41 23 52 55 99	31 04 49 69 96	10 47 48 45 88	13 41 43 89 20	97 17 14 49 17	51 29 50 10 34	31 57 75 95 80	51 97 02 74 77	76 15 48 49 44	18 55 63 77 09
60 20 50 81 69	31 99 73 68 68	35 81 33 03 76	24 30 12 48 60	18 99 10 72 34	21 31 38 86 24	37 79 81 53 74	73 24 16 10 33	52 83 90 94 76	70 47 14 54 36
91 25 38 05 90	94 58 28 41 36	45 37 59 03 09	90 35 57 29 12	82 62 54 65 60	29 01 23 87 88	58 02 39 37 67	42 10 14 20 92	16 55 23 42 45	54 96 09 11 06
34 50 57 74 37	98 80 33 00 91	09 77 93 19 82	74 94 80 04 04	45 07 31 66 49	95 33 95 22 00	18 74 72 00 18	38 79 58 69 32	81 76 80 26 92	82 80 84 25 39
85 22 04 39 43	73 81 53 94 79	33 62 46 86 28	08 31 54 46 31	53 94 13 38 47	90 84 60 79 80	24 36 59 87 38	82 07 53 89 35	96 35 23 79 18	05 98 90 07 35
09 79 13 77 48	73 82 97 22 21	05 03 27 24 83	72 89 44 05 60	35 80 39 94 88	46 40 62 98 82	54 97 20 56 95	15 74 80 08 32	16 46 70 50 80	67 72 16 42 79
88 75 80 18 14	22 95 75 42 49	39 32 82 22 49	02 48 07 70 37	16 04 61 67 87	20 31 89 03 43	38 46 82 68 72	32 14 82 99 70	80 60 47 18 97	63 49 30 21 30
90 96 23 70 00	39 00 03 06 90	55 85 78 38 36	94 37 30 69 32	90 89 00 76 33	71 59 73 05 50	08 22 23 71 77	91 01 93 20 49	82 96 59 26 94	66 39 67 98 60
22 17 68 65 84	68 95 23 92 35	87 02 22 57 51	61 09 43 95 06	58 24 82 03 47	10 27 53 96 23	71 50 54 36 23	54 31 04 82 98	04 14 12 15 09	26 78 25 47 47
19 36 27 59 46	13 79 93 37 55	39 77 32 77 09	85 52 05 30 62	47 83 51 62 74	28 41 50 61 88	64 85 27 20 18	83 36 36 05 56	39 71 65 09 62	94 76 62 11 89
16 77 23 02 77	09 61 87 25 21	28 06 24 25 93	16 71 13 59 78	23 05 47 47 25	34 21 42 57 02	59 19 18 97 48	80 30 03 30 98	05 24 67 70 07	84 97 50 87 46
78 43 76 71 61	20 44 90 32 64	97 67 63 99 61	46 38 03 93 22	69 81 21 99 21	61 81 77 23 23	82 82 11 54 08	53 28 70 58 96	44 07 39 55 43	42 34 43 39 28
03 28 28 26 08	73 37 32 04 05	69 30 16 09 05	88 69 58 28 99	35 07 44 75 47	61 15 18 13 54	16 86 20 26 88	90 74 80 55 09	14 53 90 51 17	52 01 63 01 59
93 22 53 64 39	07 10 63 76 35	87 03 04 79 88	08 13 13 85 51	55 34 57 72 69	91 76 21 64 64	44 91 13 32 97	75 31 62 66 54	84 80 32 75 77	56 08 25 70 29
78 76 58 54 74	92 38 70 96 92	52 06 79 79 45	82 63 18 27 44	69 66 92 19 09	00 97 79 08 06	37 30 28 59 85	53 56 68 53 40	01 74 39 59 73	30 19 99 85 48
23 68 35 26 00	99 53 93 61 28	52 70 05 48 34	56 65 05 61 86	90 92 10 70 80	36 46 18 34 94	75 20 80 27 77	78 91 69 16 00	08 43 18 73 68	67 69 61 34 25
15 39 25 70 99	93 86 52 77 65	15 33 59 05 28	22 87 26 07 47	86 96 98 29 06	88 98 99 60 50	65 95 79 42 94	93 62 40 89 96	43 56 47 71 66	46 76 29 67 02
58 71 96 30 24	18 46 23 34 27	85 13 99 24 44	49 18 09 79 49	74 16 32 23 02	04 37 59 87 21	05 02 03 24 17	47 97 81 56 51	92 34 86 01 82	55 51 33 12 91
57 35 27 33 72	24 53 63 94 09	41 10 76 47 91	44 04 95 49 66	39 60 04 59 81	63 62 06 34 41	94 21 78 55 09	72 76 45 16 94	29 95 81 83 83	79 88 01 97 30
48 50 86 54 48	22 06 34 72 52	82 21 15 65 20	33 29 94 71 11	15 91 29 12 03	78 47 23 53 90	34 41 92 45 71	09 23 70 70 07	12 38 92 79 43	14 85 11 47 23
61 96 48 95 03	07 16 39 33 66	98 56 10 56 79	77 21 30 27 12	90 49 22 23 62	87 68 62 15 43	53 14 36 59 25	54 47 33 70 15	59 24 48 40 35	50 03 42 99 36
36 93 89 41 26	29 70 83 63 51	99 74 20 52 36	87 09 41 15 09	98 60 16 03 03	47 60 92 10 77	88 59 53 11 52	66 25 69 07 04	48 68 64 71 06	61 65 70 22 12
18 87 00 42 31	57 90 12 02 07	23 47 37 17 31	54 08 01 88 63	39 41 88 92 10	56 88 87 59 41	65 28 04 67 53	95 79 88 37 31	50 41 06 94 76	81 83 17 16 33

2 Cumulative binomial probabilities

p = probability of success in a single trial; n = number of trials. The table gives the probability of obtaining ***r* or more** successes in n independent trials, i.e.

$$\sum_{x=r}^{n} {}^{n}C_x p^x (1-p)^{n-x}$$

When there is no entry for a particular pair of values of r and p, this indicates that the appropriate probability is less than 0.00005. Similarly, except for the case $r = 0$, when the entry is exact, a tabulated value of 1.0000 represents a probability greater than 0.99995.

	p=	0.01	0.02	0.03	0.04	0.05	0.06	0.07	0.08	0.09
n=2	r=0	1.0000	1.0000	1.0000	1.0000	1.0000	1.0000	1.0000	1.0000	1.0000
	1	.0199	.0396	.0591	.0784	.0975	.1164	.1351	.1536	.1719
	2	.0001	.0004	.0009	.0016	.0025	.0036	.0049	.0064	.0081
n=5	r=0	1.0000	1.0000	1.0000	1.0000	1.0000	1.0000	1.0000	1.0000	1.0000
	1	.0490	.0961	.1413	.1846	.2262	.2661	.3043	.3409	.3760
	2	.0010	.0038	.0085	.0148	.0226	.0319	.0425	.0544	.0674
	3		.0001	.0003	.0006	.0012	.0020	.0031	.0045	.0063
	4						.0001	.0001	.0002	.0003
n=10	r=0	1.0000	1.0000	1.0000	1.0000	1.0000	1.0000	1.0000	1.0000	1.0000
	1	.0956	.1829	.2626	.3352	.4013	.4614	.5160	.5656	.6106
	2	.0043	.0162	.0345	.0582	.0861	.1176	.1517	.1879	.2254
	3	.0001	.0009	.0028	.0062	.0115	.0188	.0283	.0401	.0540
	4			.0001	.0004	.0010	.0020	.0036	.0058	.0088
	5					.0001	.0002	.0003	.0006	.0010
	6									.0001
n=20	r=0	1.0000	1.0000	1.0000	1.0000	1.0000	1.0000	1.0000	1.0000	1.0000
	1	.1821	.3324	.4562	.5580	.6415	.7099	.7658	.8113	.8484
	2	.0169	.0599	.1198	.1897	.2642	.3395	.4131	.4831	.5484
	3	.0010	.0071	.0210	.0439	.0755	.1150	.1610	.2121	.2666
	4		.0006	.0027	.0074	.0159	.0290	.0471	.0706	.0993
	5			.0003	.0010	.0026	.0056	.0107	.0183	.0290
	6				.0001	.0003	.0009	.0019	.0038	.0068
	7						.0001	.0003	.0006	.0013
	8								.0001	.0002
n=50	r=0	1.0000	1.0000	1.0000	1.0000	1.0000	1.0000	1.0000	1.0000	1.0000
	1	.3950	.6358	.7819	.8701	.9231	.9547	.9734	.9845	.9910
	2	.0894	.2642	.4447	.5995	.7206	.8100	.8735	.9173	.9468
	3	.0138	.0784	.1892	.3233	.4595	.5838	.6892	.7740	.8395
	4	.0016	.0178	.0628	.1391	.2396	.3527	.4673	.5747	.6697
	5	.0001	.0032	.0168	.0490	.1036	.1794	.2710	.3710	.4723
	6		.0005	.0037	.0144	.0378	.0776	.1350	.2081	.2928
	7		.0001	.0007	.0036	.0118	.0289	.0583	.1019	.1596
	8			.0001	.0008	.0032	.0094	.0220	.0438	.0768
	9				.0001	.0008	.0027	.0073	.0167	.0328
	10					.0002	.0007	.0022	.0056	.0125
	11						.0002	.0006	.0017	.0043
	12							.0001	.0005	.0013
	13								.0001	.0004
	14									.0001

	p=	0.01	0.02	0.03	0.04	0.05	0.06	0.07	0.08	0.09
n=100	r=0	1.0000	1.0000	1.0000	1.0000	1.0000	1.0000	1.0000	1.0000	1.0000
	1	.6340	.8674	.9524	.9831	.9941	.9979	.9993	.9998	.9999
	2	.2642	.5967	.8054	.9128	.9629	.9848	.9940	.9977	.9991
	3	.0794	.3233	.5802	.7679	.8817	.9434	.9742	.9887	.9952
	4	.0184	.1410	.3528	.5705	.7422	.8570	.9256	.9633	.9827
	5	.0034	.0508	.1821	.3711	.5640	.7232	.8368	.9097	.9526
	6	.0005	.0155	.0808	.2116	.3840	.5593	.7086	.8201	.8955
	7	.0001	.0041	.0312	.1064	.2340	.3936	.5557	.6968	.8060
	8		.0009	.0106	.0475	.1280	.2517	.4012	.5529	.6872
	9		.0002	.0032	.0190	.0631	.1463	.2660	.4074	.5506
	10			.0009	.0068	.0282	.0775	.1620	.2780	.4125
	11			.0002	.0022	.0115	.0376	.0908	.1757	.2882
	12				.0007	.0043	.0168	.0469	.1028	.1876
	13				.0002	.0015	.0069	.0224	.0559	.1138
	14					.0005	.0026	.0099	.0282	.0645
	15					.0001	.0009	.0041	.0133	.0341
	16						.0003	.0016	.0058	.0169
	17						.0001	.0006	.0024	.0078
	18							.0002	.0009	.0034
	19							.0001	.0003	.0014
	20								.0001	.0005
	21									.0002
	22									.0001

	p=	0.10	0.15	0.20	0.25	0.30	0.35	0.40	0.45	0.50
n=2	r=0	1.0000	1.0000	1.0000	1.0000	1.0000	1.0000	1.0000	1.0000	1.0000
	1	.1900	.2775	.3600	.4375	.5100	.5775	.6400	.6975	.7500
	2	.0100	.0225	.0400	.0625	.0900	.1225	.1600	.2025	.2500
n=5	r=0	1.0000	1.0000	1.0000	1.0000	1.0000	1.0000	1.0000	1.0000	1.0000
	1	.4095	.5563	.6723	.7627	.8319	.8840	.9222	.9497	.9688
	2	.0815	.1648	.2627	.3672	.4718	.5716	.6630	.7438	.8125
	3	.0086	.0266	.0579	.1035	.1631	.2352	.3174	.4069	.5000
	4	.0005	.0022	.0067	.0156	.0308	.0540	.0870	.1312	.1875
	5		.0001	.0003	.0010	.0024	.0053	.0102	.0185	.0313
n=10	r=0	1.0000	1.0000	1.0000	1.0000	1.0000	1.0000	1.0000	1.0000	1.0000
	1	.6513	.8031	.8926	.9437	.9718	.9865	.9940	.9975	.9990
	2	.2639	.4557	.6242	.7560	.8507	.9140	.9536	.9767	.9893
	3	.0702	.1798	.3222	.4744	.6172	.7384	.8327	.9004	.9453
	4	.0128	.0500	.1209	.2241	.3504	.4862	.6177	.7430	.8281
	5	.0016	.0099	.0328	.0781	.1503	.2485	.3669	.4956	.6230
	6	.0001	.0014	.0064	.0197	.0473	.0949	.1662	.2616	.3770
	7		.0001	.0009	.0035	.0106	.0260	.0548	.1020	.1719
	8			.0001	.0004	.0016	.0048	.0123	.0274	.0547
	9					.0001	.0005	.0017	.0045	.0107
	10							.0001	.0003	.0010
n=20	r=0	1.0000	1.0000	1.0000	1.0000	1.0000	1.0000	1.0000	1.0000	1.0000
	1	.8784	.9612	.9885	.9968	.9992	.9998	1.0000	1.0000	1.0000
	2	.6083	.8244	.9308	.9757	.9924	.9979	.9995	.9999	1.0000
	3	.3231	.5951	.7939	.9087	.9645	.9879	.9964	.9991	.9998
	4	.1330	.3523	.5886	.7748	.8929	.9556	.9840	.9951	.9987
	5	.0432	.1702	.3704	.5852	.7625	.8818	.9490	.9811	.9941
	6	.0113	.0673	.1958	.3828	.5836	.7546	.8744	.9447	.9793
	7	.0024	.0219	.0867	.2142	.3920	.5834	.7500	.8701	.9423
	8	.0004	.0059	.0321	.1018	.2277	.3990	.5841	.7480	.8684
	9	.0001	.0013	.0100	.0409	.1133	.2376	.4044	.5857	.7483
	10		.0002	.0026	.0139	.0480	.1218	.2447	.4086	.5881
	11			.0006	.0039	.0171	.0532	.1275	.2493	.4119
	12			.0001	.0009	.0051	.0196	.0565	.1308	.2517
	13				.0002	.0013	.0060	.0210	.0580	.1316
	14					.0003	.0015	.0065	.0214	.0577
	15						.0003	.0016	.0064	.0207
	16							.0003	.0015	.0059
	17								.0003	.0013
	18									.0002

	p=	0.10	0.15	0.20	0.25	0.30	0.35	0.40	0.45	0.50
n=50	r=0	1.0000	1.0000	1.0000	1.0000	1.0000	1.0000	1.0000	1.0000	1.0000
	1	.9948	.9997	1.0000	1.0000	1.0000	1.0000	1.0000	1.0000	1.0000
	2	.9662	.9971	.9998	1.0000	1.0000	1.0000	1.0000	1.0000	1.0000
	3	.8883	.9858	.9987	.9999	1.0000	1.0000	1.0000	1.0000	1.0000
	4	.7497	.9540	.9943	.9995	1.0000	1.0000	1.0000	1.0000	1.0000
	5	.5688	.8879	.9815	.9979	.9998	1.0000	1.0000	1.0000	1.0000
	6	.3839	.7806	.9520	.9930	.9993	.9999	1.0000	1.0000	1.0000
	7	.2298	.6387	.8966	.9806	.9975	.9998	1.0000	1.0000	1.0000
	8	.1221	.4812	.8096	.9547	.9927	.9992	.9999	1.0000	1.0000
	9	.0579	.3319	.6927	.9084	.9817	.9975	.9998	1.0000	1.0000
	10	.0245	.2089	.5563	.8363	.9598	.9933	.9992	.9999	1.0000
	11	.0094	.1199	.4164	.7378	.9211	.9840	.9978	.9998	1.0000
	12	.0032	.0628	.2893	.6184	.8610	.9658	.9943	.9994	1.0000
	13	.0010	.0301	.1861	.4890	.7771	.9339	.9867	.9982	.9998
	14	.0003	.0132	.1106	.3630	.6721	.8837	.9720	.9955	.9995
	15	.0001	.0053	.0607	.2519	.5532	.8122	.9460	.9896	.9987
	16		.0019	.0308	.1631	.4308	.7199	.9045	.9780	.9967
	17		.0007	.0144	.0983	.3161	.6111	.8439	.9573	.9923
	18		.0002	.0063	.0551	.2178	.4940	.7631	.9235	.9836
	19		.0001	.0025	.0287	.1406	.3784	.6644	.8727	.9675
	20			.0009	.0139	.0848	.2736	.5535	.8026	.9405
	21			.0003	.0063	.0478	.1861	.4390	.7138	.8987
	22			.0001	.0026	.0251	.1187	.3299	.6100	.8389
	23				.0010	.0123	.0710	.2340	.4981	.7601
	24				.0004	.0056	.0396	.1562	.3866	.6641
	25				.0001	.0024	.0207	.0978	.2840	.5561
	26					.0009	.0100	.0573	.1966	.4439
	27					.0003	.0045	.0314	.1279	.3359
	28					.0001	.0019	.0160	.0780	.2399
	29						.0007	.0076	.0444	.1611
	30						.0003	.0034	.0235	.1013
	31						.0001	.0014	.0116	.0595
	32							.0005	.0053	.0325
	33							.0002	.0022	.0164
	34							.0001	.0009	.0077
	35								.0003	.0033
	36								.0001	.0013
	37									.0005
	38									.0002

This table gives binomial probalities only for a limited range of values of n and p since, in practice, either the more compact tabulation of the Poisson distribution (appendix 3) or that of the normal distribution (appendix 4) can usually be used to give an adequate approximation.

As a reasonable working rule,

1 use the Poisson approximation if $p<0.05$ and $np \geqslant 5$, putting $m = np$.

2 use the normal approximation if $0.1 \leqslant p \leqslant 0.9$ and $np > 5$, putting $\mu = np$ and $\sigma = \sqrt{np(1-p)}$.

	p=	0.10	0.15	0.20	0.25	0.30	0.35	0.40	0.45	0.50
n=100	r=0	1.0000	1.0000	1.0000	1.0000	1.0000	1.0000	1.0000	1.0000	1.0000
	1	1.0000	1.0000	1.0000	1.0000	1.0000	1.0000	1.0000	1.0000	1.0000
	2	.9997	1.0000	1.0000	1.0000	1.0000	1.0000	1.0000	1.0000	1.0000
	3	.9981	1.0000	1.0000	1.0000	1.0000	1.0000	1.0000	1.0000	1.0000
	4	.9922	.9999	1.0000	1.0000	1.0000	1.0000	1.0000	1.0000	1.0000
	5	.9763	.9996	1.0000	1.0000	1.0000	1.0000	1.0000	1.0000	1.0000
	6	.9424	.9984	1.0000	1.0000	1.0000	1.0000	1.0000	1.0000	1.0000
	7	.8828	.9953	.9999	1.0000	1.0000	1.0000	1.0000	1.0000	1.0000
	8	.7939	.9878	.9997	1.0000	1.0000	1.0000	1.0000	1.0000	1.0000
	9	.6791	.9725	.9991	1.0000	1.0000	1.0000	1.0000	1.0000	1.0000
	10	.5487	.9449	.9977	1.0000	1.0000	1.0000	1.0000	1.0000	1.0000
	11	.4168	.9006	.9943	.9999	1.0000	1.0000	1.0000	1.0000	1.0000
	12	.2970	.8365	.9874	.9996	1.0000	1.0000	1.0000	1.0000	1.0000
	13	.1982	.7527	.9747	.9990	1.0000	1.0000	1.0000	1.0000	1.0000
	14	.1239	.6526	.9531	.9975	.9999	1.0000	1.0000	1.0000	1.0000
	15	.0726	.5428	.9196	.9946	.9998	1.0000	1.0000	1.0000	1.0000
	16	.0399	.4317	.8715	.9889	.9996	1.0000	1.0000	1.0000	1.0000
	17	.0206	.3275	.8077	.9789	.9990	1.0000	1.0000	1.0000	1.0000
	18	.0100	.2367	.7288	.9624	.9978	.9999	1.0000	1.0000	1.0000
	19	.0046	.1628	.6379	.9370	.9955	.9999	1.0000	1.0000	1.0000
	20	.0020	.1065	.5398	.9005	.9911	.9997	1.0000	1.0000	1.0000
	21	.0008	.0663	.4405	.8512	.9835	.9992	1.0000	1.0000	1.0000
	22	.0003	.0393	.3460	.7886	.9712	.9983	1.0000	1.0000	1.0000
	23	.0001	.0221	.2611	.7136	.9521	.9966	.9999	1.0000	1.0000
	24		.0119	.1891	.6289	.9245	.9934	.9997	1.0000	1.0000
	25		.0061	.1314	.5383	.8864	.9879	.9994	1.0000	1.0000
	26		.0030	.0875	.4465	.8369	.9789	.9988	1.0000	1.0000
	27		.0014	.0558	.3583	.7756	.9649	.9976	.9999	1.0000
	28		.0006	.0342	.2776	.7036	.9442	.9954	.9998	1.0000
	29		.0003	.0200	.2075	.6232	.9152	.9916	.9996	1.0000
	30		.0001	.0112	.1495	.5377	.8764	.9852	.9992	1.0000
	31			.0061	.1038	.4509	.8270	.9752	.9985	1.0000
	32			.0031	.0693	.3669	.7669	.9602	.9970	.9999
	33			.0016	.0446	.2893	.6971	.9385	.9945	.9998
	34			.0007	.0276	.2207	.6197	.9087	.9902	.9996
	35			.0003	.0164	.1629	.5376	.8697	.9834	.9991
	36			.0001	.0094	.1161	.4542	.8205	.9728	.9982
	37			.0001	.0052	.0799	.3731	.7614	.9571	.9967
	38				.0027	.0530	.2976	.6932	.9349	.9940
	39				.0014	.0340	.2301	.6178	.9049	.9895
	40				.0007	.0210	.1724	.5379	.8657	.9824
	41				.0003	.0125	.1250	.4567	.8169	.9716
	42				.0001	.0072	.0877	.3775	.7585	.9557
	43				.0001	.0040	.0594	.3033	.6913	.9334
	44					.0021	.0389	.2365	.6172	.9033
	45					.0011	.0246	.1789	.5387	.8644
	46					.0005	.0150	.1311	.4587	.8159
	47					.0003	.0088	.0930	.3804	.7579
	48					.0001	.0050	.0638	.3069	.6914
	49					.0001	.0027	.0423	.2404	.6178
	50						.0015	.0271	.1827	.5398
	51						.0007	.0168	.1346	.4602
	52						.0004	.0100	.0960	.3822
	53						.0002	.0058	.0662	.3086
	54						.0001	.0032	.0441	.2421
	55							.0017	.0284	.1841
	56							.0009	.0176	.1356
	57							.0004	.0106	.0967
	58							.0002	.0061	.0666
	59							.0001	.0034	.0443
	60								.0018	.0284
	61								.0009	.0176
	62								.0005	.0105
	63								.0002	.0060
	64								.0001	.0033
	65									.0018
	66									.0009
	67									.0004
	68									.0002
	69									.0001

3 Cumulative Poisson probabilities

The table gives the probability that ***r* or more** random events are contained in an interval when the average number of such events per interval is *m*, i.e.

$$\sum_{x=r}^{\infty} e^{-m}\frac{m^x}{x!}$$

Where there is no entry for a particular pair of values of *r* and *m*, this indicates that the appropriate probability is less than 0.00005. Similarly, except for the case $r=0$ when the entry is exact, a tabulated value of 1.0000 represents a probability greater than 0.99995.

m =	0.1	0.2	0.3	0.4	0.5	0.6	0.7	0.8	0.9	1.0
r = 0	1.0000	1.0000	1.0000	1.0000	1.0000	1.0000	1.0000	1.0000	1.0000	1.0000
1	.0952	.1813	.2592	.3297	.3935	.4512	.5034	.5507	.5934	.6321
2	.0047	.0175	.0369	.0616	.0902	.1219	.1558	.1912	.2275	.2642
3	.0002	.0011	.0036	.0079	.0144	.0231	.0341	.0474	.0629	.0803
4		.0001	.0003	.0008	.0018	.0034	.0058	.0091	.0135	.0190
5				.0001	.0002	.0004	.0008	.0014	.0023	.0037
6							.0001	.0002	.0003	.0006
7										.0001

m =	1.1	1.2	1.3	1.4	1.5	1.6	1.7	1.8	1.9	2.0
r = 0	1.0000	1.0000	1.0000	1.0000	1.0000	1.0000	1.0000	1.0000	1.0000	1.0000
1	.6671	.6988	.7275	.7534	.7769	.7981	.8173	.8347	.8504	.8647
2	.3010	.3374	.3732	.4082	.4422	.4751	.5068	.5372	.5663	.5940
3	.0996	.1205	.1429	.1665	.1912	.2166	.2428	.2694	.2963	.3233
4	.0257	.0338	.0431	.0537	.0656	.0788	.0932	.1087	.1253	.1429
5	.0054	.0077	.0107	.0143	.0186	.0237	.0296	.0364	.0441	.0527
6	.0010	.0015	.0022	.0032	.0045	.0060	.0080	.0104	.0132	.0166
7	.0001	.0003	.0004	.0006	.0009	.0013	.0019	.0026	.0034	.0045
8			.0001	.0001	.0002	.0003	.0004	.0006	.0008	.0011
9							.0001	.0001	.0002	.0002

m =	2.1	2.2	2.3	2.4	2.5	2.6	2.7	2.8	2.9	3.0
r = 0	1.0000	1.0000	1.0000	1.0000	1.0000	1.0000	1.0000	1.0000	1.0000	1.0000
1	.8775	.8892	.8997	.9093	.9179	.9257	.9328	.9392	.9450	.9502
2	.6204	.6454	.6691	.6916	.7127	.7326	.7513	.7689	.7854	.8009
3	.3504	.3773	.4040	.4303	.4562	.4816	.5064	.5305	.5540	.5768
4	.1614	.1806	.2007	.2213	.2424	.2640	.2859	.3081	.3304	.3528
5	.0621	.0725	.0838	.0959	.1088	.1226	.1371	.1523	.1682	.1847
6	.0204	.0249	.0300	.0357	.0420	.0490	.0567	.0651	.0742	.0839
7	.0059	.0075	.0094	.0116	.0142	.0172	.0206	.0244	.0287	.0335
8	.0015	.0020	.0026	.0033	.0042	.0053	.0066	.0081	.0099	.0119
9	.0003	.0005	.0006	.0009	.0011	.0015	.0019	.0024	.0031	.0038
10	.0001	.0001	.0001	.0002	.0003	.0004	.0005	.0007	.0009	.0011
11					.0001	.0001	.0001	.0002	.0002	.0003
12									.0001	.0001

m =	3.1	3.2	3.3	3.4	3.5	3.6	3.7	3.8	3.9	4.0
r = 0	1.0000	1.0000	1.0000	1.0000	1.0000	1.0000	1.0000	1.0000	1.0000	1.0000
1	.9550	.9592	.9631	.9666	.9698	.9727	.9753	.9776	.9798	.9817
2	.8153	.8288	.8414	.8532	.8641	.8743	.8838	.8926	.9008	.9084
3	.5988	.6201	.6406	.6603	.6792	.6973	.7146	.7311	.7469	.7619
4	.3752	.3975	.4197	.4416	.4634	.4848	.5058	.5265	.5468	.5665
5	.2018	.2194	.2374	.2558	.2746	.2936	.3128	.3322	.3516	.3712
6	.0943	.1054	.1171	.1295	.1424	.1559	.1699	.1844	.1994	.2149
7	.0388	.0446	.0510	.0579	.0653	.0733	.0818	.0909	.1005	.1107
8	.0142	.0168	.0198	.0231	.0267	.0308	.0352	.0401	.0454	.0511
9	.0047	.0057	.0069	.0083	.0099	.0117	.0137	.0160	.0185	.0214
10	.0014	.0018	.0022	.0027	.0033	.0040	.0048	.0058	.0069	.0081
11	.0004	.0005	.0006	.0008	.0010	.0013	.0016	.0019	.0023	.0028
12	.0001	.0001	.0002	.0002	.0003	.0004	.0005	.0006	.0007	.0009
13				.0001	.0001	.0001	.0001	.0002	.0002	.0003
14									.0001	.0001

m =	4.1	4.2	4.3	4.4	4.5	4.6	4.7	4.8	4.9	5.0
r = 0	1.0000	1.0000	1.0000	1.0000	1.0000	1.0000	1.0000	1.0000	1.0000	1.0000
1	.9834	.9850	.9864	.9877	.9889	.9899	.9909	.9918	.9926	.9933
2	.9155	.9220	.9281	.9337	.9389	.9437	.9482	.9523	.9561	.9596
3	.7762	.7898	.8026	.8149	.8264	.8374	.8477	.8575	.8667	.8753
4	.5858	.6046	.6228	.6406	.6577	.6743	.6903	.7058	.7207	.7350
5	.3907	.4102	.4296	.4488	.4679	.4868	.5054	.5237	.5418	.5595
6	.2307	.2469	.2633	.2801	.2971	.3142	.3316	.3490	.3665	.3840
7	.1214	.1325	.1442	.1564	.1689	.1820	.1954	.2092	.2233	.2378
8	.0573	.0639	.0710	.0786	.0866	.0951	.1040	.1133	.1231	.1334
9	.0245	.0279	.0317	.0358	.0403	.0451	.0503	.0558	.0618	.0681
10	.0095	.0111	.0129	.0149	.0171	.0195	.0222	.0251	.0283	.0318
11	.0034	.0041	.0048	.0057	.0067	.0078	.0090	.0104	.0120	.0137
12	.0011	.0014	.0017	.0020	.0024	.0029	.0034	.0040	.0047	.0055
13	.0003	.0004	.0005	.0007	.0008	.0010	.0012	.0014	.0017	.0020
14	.0001	.0001	.0002	.0002	.0003	.0003	.0004	.0005	.0006	.0007
15				.0001	.0001	.0001	.0001	.0001	.0002	.0002
16									.0001	.0001

m =	5.2	5.4	5.6	5.8	6.0	6.2	6.4	6.6	6.8	7.0
r = 0	1.0000	1.0000	1.0000	1.0000	1.0000	1.0000	1.0000	1.0000	1.0000	1.0000
1	.9945	.9955	.9963	.9970	.9975	.9980	.9985	.9986	.9989	.9991
2	.9658	.9711	.9756	.9794	.9826	.9854	.9877	.9897	.9913	.9927
3	.8912	.9052	.9176	.9285	.9380	.9464	.9537	.9600	.9656	.9704
4	.7619	.7867	.8094	.8300	.8488	.8658	.8811	.8948	.9072	.9182
5	.5939	.6267	.6579	.6873	.7149	.7408	.7649	.7873	.8080	.8270
6	.4191	.4539	.4881	.5217	.5543	.5859	.6163	.6453	.6730	.6993
7	.2676	.2983	.3297	.3616	.3937	.4258	.4577	.4892	.5201	.5503
8	.1551	.1783	.2030	.2290	.2560	.2840	.3127	.3419	.3715	.4013
9	.0819	.0974	.1143	.1328	.1528	.1741	.1967	.2204	.2452	.2709
10	.0397	.0488	.0591	.0708	.0839	.0984	.1142	.1314	.1498	.1695
11	.0177	.0225	.0282	.0349	.0426	.0514	.0614	.0726	.0849	.0985
12	.0073	.0096	.0125	.0160	.0201	.0250	.0307	.0373	.0448	.0534
13	.0028	.0038	.0051	.0068	.0088	.0113	.0143	.0179	.0221	.0270
14	.0010	.0014	.0020	.0027	.0036	.0048	.0063	.0080	.0102	.0128
15	.0003	.0005	.0007	.0010	.0014	.0019	.0026	.0034	.0044	.0057
16	.0001	.0002	.0002	.0004	.0005	.0007	.0010	.0014	.0018	.0024
17		.0001	.0001	.0001	.0002	.0003	.0004	.0005	.0007	.0010
18					.0001	.0001	.0001	.0002	.0003	.0004
19								.0001	.0001	.0001

m =	7.2	7.4	7.6	7.8	8.0	8.2	8.4	8.6	8.8	9.0
r = 0	1.0000	1.0000	1.0000	1.0000	1.0000	1.0000	1.0000	1.0000	1.0000	1.0000
1	.9993	.9994	.9995	.9996	.9997	.9997	.9998	.9998	.9998	.9999
2	.9939	.9949	.9957	.9964	.9970	.9975	.9979	.9982	.9985	.9988
3	.9745	.9781	.9812	.9839	.9862	.9882	.9900	.9914	.9927	.9938
4	.9281	.9368	.9446	.9515	.9576	.9630	.9677	.9719	.9756	.9788
5	.8445	.8605	.8751	.8883	.9004	.9113	.9211	.9299	.9379	.9450
6	.7241	.7474	.7693	.7897	.8088	.8264	.8427	.8578	.8716	.8843
7	.5796	.6080	.6354	.6616	.6866	.7104	.7330	.7543	.7744	.7932
8	.4311	.4607	.4900	.5188	.5470	.5746	.6013	.6272	.6522	.6761
9	.2973	.3243	.3518	.3796	.4075	.4353	.4631	.4906	.5177	.5443
10	.1904	.2123	.2351	.2589	.2834	.3085	.3341	.3600	.3863	.4126
11	.1133	.1293	.1465	.1648	.1841	.2045	.2257	.2478	.2706	.2940
12	.0629	.0735	.0852	.0980	.1119	.1269	.1429	.1600	.1780	.1970
13	.0327	.0391	.0464	.0546	.0638	.0739	.0850	.0971	.1102	.1242
14	.0159	.0195	.0238	.0286	.0342	.0405	.0476	.0555	.0642	.0739
15	.0073	.0092	.0114	.0141	.0173	.0209	.0251	.0299	.0353	.0415
16	.0031	.0041	.0052	.0066	.0082	.0102	.0125	.0152	.0184	.0220
17	.0013	.0017	.0022	.0029	.0037	.0047	.0059	.0074	.0091	.0111
18	.0005	.0007	.0009	.0012	.0016	.0021	.0027	.0034	.0043	.0053
19	.0002	.0003	.0004	.0005	.0006	.0009	.0011	.0015	.0019	.0024
20	.0001	.0001	.0001	.0002	.0003	.0003	.0005	.0006	.0008	.0011
21				.0001	.0001	.0001	.0002	.0002	.0003	.0004
22							.0001	.0001	.0001	.0002
23										.0001

m =	9.2	9.4	9.6	9.8	10.0	11.0	12.0	13.0	14.0	15.0
r = 0	1.0000	1.0000	1.0000	1.0000	1.0000	1.0000	1.0000	1.0000	1.0000	1.0000
1	.9999	.9999	.9999	.9999	1.0000	1.0000	1.0000	1.0000	1.0000	1.0000
2	.9990	.9991	.9993	.9994	.9995	.9998	.9999	1.0000	1.0000	1.0000
3	.9947	.9955	.9962	.9967	.9972	.9988	.9995	.9998	.9999	1.0000
4	.9816	.9840	.9862	.9880	.9897	.9951	.9977	.9990	.9995	.9998
5	.9514	.9571	.9622	.9667	.9707	.9849	.9924	.9963	.9982	.9991
6	.8959	.9065	.9162	.9250	.9329	.9625	.9797	.9893	.9945	.9972
7	.8108	.8273	.8426	.8567	.8699	.9214	.9542	.9741	.9858	.9924
8	.6990	.7208	.7416	.7612	.7798	.8568	.9105	.9460	.9684	.9820
9	.5704	.5958	.6204	.6442	.6672	.7680	.8450	.9002	.9379	.9626
10	.4389	.4651	.4911	.5168	.5421	.6595	.7576	.8342	.8906	.9301
11	.3180	.3424	.3671	.3920	.4170	.5401	.6528	.7483	.8243	.8815
12	.2168	.2374	.2588	.2807	.3032	.4207	.5384	.6468	.7400	.8152
13	.1393	.1552	.1721	.1899	.2084	.3113	.4240	.5369	.6415	.7324
14	.0844	.0958	.1081	.1214	.1355	.2187	.3185	.4270	.5356	.6368
15	.0483	.0559	.0643	.0735	.0835	.1460	.2280	.3249	.4296	.5343
16	.0262	.0309	.0362	.0421	.0487	.0926	.1556	.2364	.3306	.4319
17	.0135	.0162	.0194	.0230	.0270	.0559	.1013	.1645	.2441	.3359
18	.0066	.0081	.0098	.0119	.0143	.0322	.0630	.1095	.1728	.2511
19	.0031	.0038	.0048	.0059	.0072	.0177	.0374	.0698	.1174	.1805
20	.0014	.0017	.0022	.0028	.0035	.0093	.0213	.0427	.0765	.1248
21	.0006	.0008	.0010	.0012	.0016	.0047	.0116	.0250	.0479	.0830
22	.0002	.0003	.0004	.0005	.0007	.0023	.0061	.0141	.0288	.0531
23	.0001	.0001	.0002	.0002	.0003	.0010	.0030	.0076	.0167	.0327
24			.0001	.0001	.0001	.0005	.0015	.0040	.0093	.0195
25						.0002	.0007	.0020	.0050	.0112
26						.0001	.0003	.0010	.0026	.0062
27							.0001	.0005	.0013	.0033
28							.0001	.0002	.0006	.0017
29								.0001	.0003	.0009
30									.0001	.0004
31									.0001	.0002
32										.0001

m =	16.0	17.0	18.0	19.0	20.0	21.0	22.0	23.0	24.0	25.0	26.0	27.0	28.0	29.0	30.0
r = 0	1.0000	1.0000	1.0000	1.0000	1.0000	1.0000	1.0000	1.0000	1.0000	1.0000	1.0000	1.0000	1.0000	1.0000	1.0000
1	1.0000	1.0000	1.0000	1.0000	1.0000	1.0000	1.0000	1.0000	1.0000	1.0000	1.0000	1.0000	1.0000	1.0000	1.0000
2	1.0000	1.0000	1.0000	1.0000	1.0000	1.0000	1.0000	1.0000	1.0000	1.0000	1.0000	1.0000	1.0000	1.0000	1.0000
3	1.0000	1.0000	1.0000	1.0000	1.0000	1.0000	1.0000	1.0000	1.0000	1.0000	1.0000	1.0000	1.0000	1.0000	1.0000
4	.9999	1.0000	1.0000	1.0000	1.0000	1.0000	1.0000	1.0000	1.0000	1.0000	1.0000	1.0000	1.0000	1.0000	1.0000
5	.9996	.9998	.9999	1.0000	1.0000	1.0000	1.0000	1.0000	1.0000	1.0000	1.0000	1.0000	1.0000	1.0000	1.0000
6	.9986	.9993	.9997	.9998	.9999	1.0000	1.0000	1.0000	1.0000	1.0000	1.0000	1.0000	1.0000	1.0000	1.0000
7	.9960	.9979	.9990	.9995	.9997	.9999	.9999	1.0000	1.0000	1.0000	1.0000	1.0000	1.0000	1.0000	1.0000
8	.9900	.9946	.9971	.9985	.9992	.9996	.9998	.9999	1.0000	1.0000	1.0000	1.0000	1.0000	1.0000	1.0000
9	.9780	.9874	.9929	.9961	.9979	.9989	.9994	.9997	.9998	.9999	1.0000	1.0000	1.0000	1.0000	1.0000
10	.9567	.9739	.9846	.9911	.9950	.9972	.9985	.9992	.9996	.9998	1.0000	1.0000	1.0000	1.0000	1.0000
11	.9226	.9509	.9696	.9817	.9892	.9937	.9965	.9980	.9989	.9994	.9999	1.0000	1.0000	1.0000	1.0000
12	.8730	.9153	.9451	.9653	.9786	.9871	.9924	.9956	.9975	.9986	.9997	.9998	.9999	1.0000	1.0000
13	.8069	.8650	.9083	.9394	.9610	.9755	.9849	.9909	.9946	.9969	.9992	.9996	.9998	.9999	.9999
14	.7255	.7991	.8574	.9016	.9339	.9566	.9722	.9826	.9893	.9935	.9982	.9990	.9994	.9997	.9998
15	.6325	.7192	.7919	.8503	.8951	.9284	.9523	.9689	.9802	.9876	.9962	.9978	.9987	.9993	.9996
16	.5333	.6285	.7133	.7852	.8435	.8889	.9231	.9480	.9656	.9777	.9924	.9954	.9973	.9984	.9991
17	.4340	.5323	.6249	.7080	.7789	.8371	.8830	.9179	.9437	.9623	.9858	.9912	.9946	.9967	.9981
18	.3407	.4360	.5314	.6216	.7030	.7730	.8310	.8772	.9129	.9395	.9752	.9840	.9899	.9937	.9961
19	.2577	.3450	.4378	.5305	.6186	.6983	.7675	.8252	.8717	.9080	.9580	.9726	.9821	.9885	.9927
20	.1878	.2637	.3491	.4394	.5297	.6157	.6940	.7623	.8197	.8664	.9354	.9555	.9700	.9801	.9871
21	.1318	.1945	.2693	.3528	.4409	.5290	.6131	.6899	.7574	.8145	.9032	.9313	.9522	.9674	.9781
22	.0892	.1385	.2009	.2745	.3563	.4423	.5284	.6106	.6861	.7527	.8613	.8985	.9273	.9489	.9647
23	.0582	.0953	.1449	.2069	.2794	.3595	.4436	.5277	.6083	.6825	.8095	.8564	.8940	.9233	.9456
24	.0367	.0633	.1011	.1510	.2125	.2840	.3626	.4449	.5272	.6061	.7483	.8048	.8517	.8896	.9194
25	.0223	.0406	.0683	.1067	.1568	.2178	.2883	.3654	.4460	.5266	.6791	.7441	.8002	.8471	.8854
26	.0131	.0252	.0446	.0731	.1122	.1623	.2229	.2923	.3681	.4471	.6041	.6758	.7401	.7958	.8428
27	.0075	.0152	.0282	.0486	.0779	.1174	.1673	.2277	.2962	.3706	.5261	.6021	.6728	.7363	.7916
28	.0041	.0088	.0173	.0313	.0525	.0825	.1225	.1726	.2323	.2998	.4481	.5256	.6003	.6699	.7327
29	.0022	.0050	.0103	.0195	.0343	.0564	.0871	.1274	.1775	.2366	.3730	.4491	.5251	.5986	.6671
30	.0011	.0027	.0059	.0118	.0218	.0374	.0602	.0915	.1321	.1821	.3033	.3753	.4500	.5247	.5969
31	.0006	.0014	.0033	.0070	.0135	.0242	.0405	.0640	.0958	.1367	.2407	.3065	.3774	.4508	.5243
32	.0003	.0007	.0018	.0040	.0081	.0152	.0265	.0436	.0678	.1001	.1866	.2447	.3097	.3794	.4516
33	.0001	.0004	.0010	.0022	.0047	.0093	.0169	.0289	.0467	.0715	.1411	.1908	.2485	.3126	.3814
34	.0001	.0002	.0005	.0012	.0027	.0055	.0105	.0187	.0314	.0498	.1042	.1454	.1949	.2521	.3155
35		.0001	.0002	.0006	.0015	.0032	.0064	.0118	.0206	.0338	.0751	.1082	.1495	.1989	.2556
36			.0001	.0003	.0008	.0018	.0038	.0073	.0132	.0225	.0528	.0787	.1121	.1535	.2027
37			.0001	.0002	.0004	.0010	.0022	.0044	.0082	.0146	.0363	.0559	.0822	.1159	.1574
38				.0001	.0002	.0005	.0012	.0026	.0050	.0092	.0244	.0388	.0589	.0856	.1196
39					.0001	.0003	.0007	.0015	.0030	.0057	.0160	.0263	.0413	.0619	.0890
40					.0001	.0001	.0004	.0008	.0017	.0034	.0103	.0175	.0283	.0438	.0648
41						.0001	.0002	.0004	.0010	.0020	.0064	.0113	.0190	.0303	.0463
42							.0001	.0002	.0005	.0012	.0039	.0072	.0125	.0205	.0323
43								.0001	.0003	.0007	.0024	.0045	.0080	.0136	.0221
44								.0001	.0002	.0004	.0014	.0027	.0050	.0089	.0148
45									.0001	.0002	.0008	.0016	.0031	.0056	.0097
46										.0001	.0004	.0009	.0019	.0035	.0063
47											.0002	.0005	.0011	.0022	.0040
48											.0001	.0003	.0006	.0013	.0025
49											.0001	.0002	.0004	.0008	.0015
												.0001	.0002	.0004	.0009

For values of m greater than 30, use the table of areas under the normal curve (appendix 4) to obtain approximate Poisson probabilities, putting $\mu = m$ and $\sigma = \sqrt{m}$.

4 Areas in tail of the normal distribution

The function tabluted is $1-\Phi(z)$ where $\Phi(z)$ is the cumulative distribution function of a standardized normal variable z. Thus $1-\Phi(z)$ is the probability that a standardized normal variable selected at random will be greater than a value of z $\left(=\frac{x-\mu}{\sigma}\right)$.

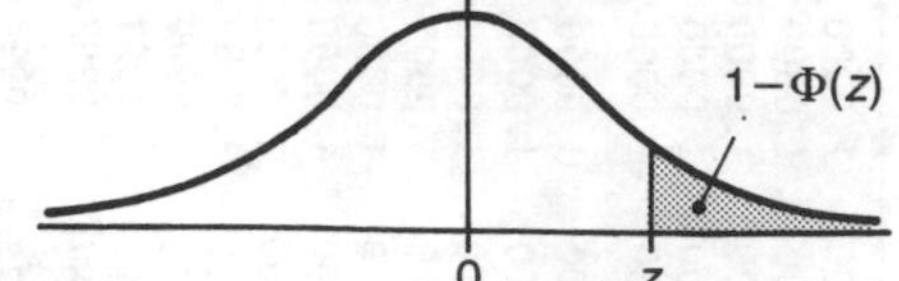

$\frac{(x-\mu)}{\sigma}$	.00	.01	.02	.03	.04	.05	.06	.07	.08	.09
0.0	.5000	.4960	.4920	.4880	.4840	.4801	.4761	.4721	.4681	.4641
0.1	.4602	.4562	.4522	.4483	.4443	.4404	.4364	.4325	.4286	.4247
0.2	.4207	.4168	.4129	.4090	.4052	.4013	.3974	.3936	.3897	.3859
0.3	.3821	.3783	.3745	.3707	.3669	.3632	.3594	.3557	.3520	.3483
0.4	.3446	.3409	.3372	.3336	.3300	.3264	.3228	.3192	.3156	.3121
0.5	.3085	.3050	.3015	.2981	.2946	.2912	.2877	.2843	.2810	.2776
0.6	.2743	.2709	.2676	.2643	.2611	.2578	.2546	.2514	.2483	.2451
0.7	.2420	.2389	.2358	.2327	.2296	.2266	.2236	.2206	.2177	.2148
0.8	.2119	.2090	.2061	.2033	.2005	.1977	.1949	.1922	.1894	.1867
0.9	.1841	.1814	.1788	.1762	.1736	.1711	.1685	.1660	.1635	.1611
1.0	.1587	.1562	.1539	.1515	.1492	.1469	.1446	.1423	.1401	.1379
1.1	.1357	.1335	.1314	.1292	.1271	.1251	.1230	.1210	.1190	.1170
1.2	.1151	.1131	.1112	.1093	.1075	.1056	.1038	.1020	.1003	.0985
1.3	.0968	.0951	.0934	.0918	.0901	.0885	.0869	.0853	.0838	.0823
1.4	.0808	.0793	.0778	.0764	.0749	.0735	.0721	.0708	.0694	.0681
1.5	.0668	.0655	.0643	.0630	.0618	.0606	.0594	.0582	.0571	.0559
1.6	.0548	.0537	.0526	.0516	.0505	.0495	.0485	.0475	.0465	.0455
1.7	.0446	.0436	.0427	.0418	.0409	.0401	.0392	.0384	.0375	.0367
1.8	.0359	.0351	.0344	.0336	.0329	.0322	.0314	.0307	.0301	.0294
1.9	.0287	.0281	.0274	.0268	.0262	.0256	.0250	.0244	.0239	.0233
2.0	.02275	.02222	.02169	.02118	.02068	.02018	.01970	.01923	.01876	.01831
2.1	.01786	.01743	.01700	.01659	.01618	.01578	.01539	.01500	.01463	.01426
2.2	.01390	.01355	.01321	.01287	.01255	.01222	.01191	.01160	.01130	.01101
2.3	.01072	.01044	.01017	.00990	.00964	.00939	.00914	.00889	.00866	.00842
2.4	.00820	.00798	.00776	.00755	.00734	.00714	.00695	.00676	.00657	.00639
2.5	.00621	.00604	.00587	.00570	.00554	.00539	.00523	.00508	.00494	.00480
2.6	.00466	.00453	.00440	.00427	.00415	.00402	.00391	.00379	.00368	.00357
2.7	.00347	.00336	.00326	.00317	.00307	.00298	.00289	.00280	.00272	.00264
2.8	.00256	.00248	.00240	.00233	.00226	.00219	.00212	.00205	.00199	.00193
2.9	.00187	.00181	.00175	.00169	.00164	.00159	.00154	.00149	.00144	.00139
3.0	.00135									
3.1	.00097									
3.2	.00069									
3.3	.00048									
3.4	.00034									
3.5	.00023									
3.6	.00016									
3.7	.00011									
3,8	.00007									
3.9	.00005									
4.0	.00003									

5 Percentage points of the normal distribution

The table gives the 100α percentage points, z_α, of a standardized normal distribution. Thus z_α is the value of a standardized normal variable which has probability α of being exceeded.

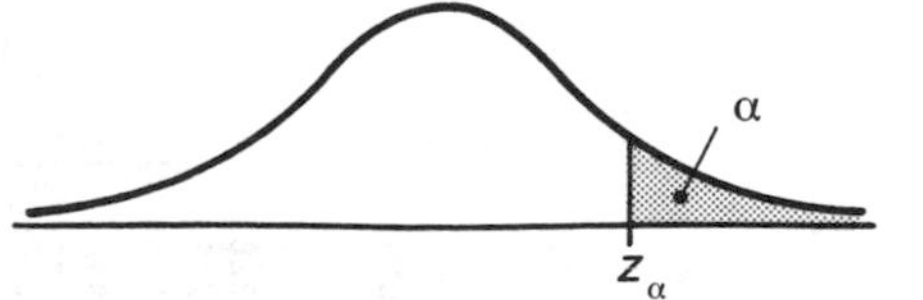

α	z_α	α	z_α	α	z_α	α	z_α	α	z_α	α	z_α
.50	0.0000	.050	1.6449	.030	1.8808	.020	2.0537	.010	2.3263	.050	1.6449
.45	0.1257	.048	1.6646	.029	1.8957	.019	2.0749	.009	2.3656	.010	2.3263
.40	0.2533	.046	1.6849	.028	1.9110	.018	2.0969	.008	2.4089	.001	3.0902
.35	0.3853	.044	1.7060	.027	1.9268	.017	2.1201	.007	2.4573	.0001	3.7190
.30	0.5244	.042	1.7279	.026	1.9431	.016	2.1444	.006	2.5121	.00001	4.2649
.25	0.6745	.040	1.7507	.025	1.9600	.015	2.1701	.005	2.5758	.025	1.9600
.20	0.8416	.038	1.7744	.024	1.9774	.014	2.1973	.004	2.6521	.005	2.5758
.15	1.0364	.036	1.7991	.023	1.9954	.013	2.2262	.003	2.7478	.0005	3.2905
.10	1.2816	.034	1.8250	.022	2.0141	.012	2.2571	.002	2.8782	.00005	3.8906
.05	1.6449	.032	1.8522	.021	2.0335	.011	2.2904	.001	3.0902	.000005	4.4172

6 Percentage points of the t distribution

The table gives the value of $t_{\alpha;\upsilon}$ – the 100α percentage point of the t distribution for υ degrees of freedom.

Note: The tabulation is for one tail only i.e. for positive values of t. For $\pm t$ the column headings for α must be doubled.

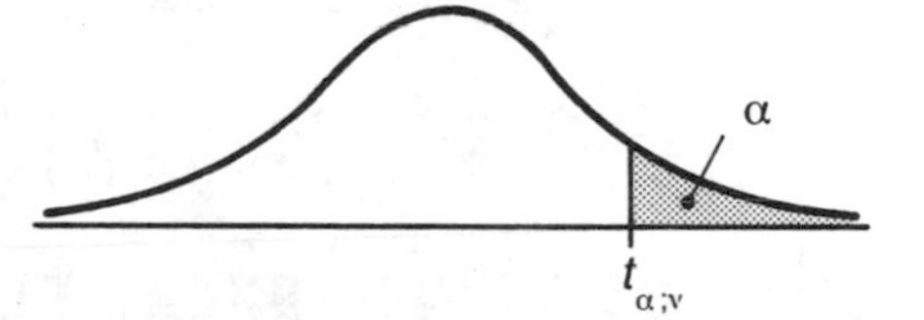

$\alpha =$	0.10	0.05	0.025	0.01	0.005	0.001	0.0005
$\nu = 1$	3.078	6.314	12.706	01.821	63.657	318.31	636.62
2	1.886	2.920	4.303	6.965	9.925	22.326	31.598
3	1.638	2.353	3.182	4.541	5.841	10.213	12.924
4	1.533	2.132	2.776	3.747	4.604	7.173	8.610
5	1.476	2.015	2.571	3.365	4.032	5.893	6.869
6	1.440	1.943	2.447	3.143	3.707	5.208	5.959
7	1.415	1.895	2.365	2.998	3.499	4.785	5.408
8	1.397	1.860	2.306	2.896	3.355	4.501	5.041
9	1.383	1.833	2.262	2.821	3.250	4.297	4.781
10	1.372	1.812	2.228	2.764	3.169	4.144	4.587
11	1.363	1.796	2.201	2.718	3.106	4.025	4.437
12	1.356	1.782	2.179	2.681	3.055	3.930	4.318
13	1.350	1.771	2.160	2.650	3.012	3.852	4.221
14	1.345	1.761	2.145	2.624	2.977	3.787	4.140
15	1.341	1.753	2.131	2.602	2.947	3.733	4.073
16	1.337	1.746	2.120	2.583	2.921	3.686	4.015
17	1.333	1.740	2.110	2.567	2.898	3.646	3.965
18	1.330	1.734	2.101	2.552	2.878	3.610	3.922
19	1.328	1.729	2.093	2.539	2.861	3.579	3.883
20	1.325	1.725	2.086	2.528	2.845	3.552	3.850
21	1.323	1.721	2.080	2.518	2.831	3.527	3.819
22	1.321	1.717	2.074	2.508	2.819	3.505	3.792
23	1.319	1.714	2.069	2.500	2.807	3.485	3.767
24	1.318	1.711	2.064	2.492	2.797	3.467	3.745
25	1.316	1.708	2.060	2.485	2.787	3.450	3.725
26	1.315	1.706	2.056	2.479	2.779	3.435	3.707
27	1.314	1.703	2.052	2.473	2.771	3.421	3.690
28	1.313	1.701	2.048	2.467	2.763	3.408	3.674
29	1.311	1.699	2.045	2.462	2.756	3.396	3.659
30	1.310	1.697	2.042	2.457	2.750	3.385	3.646
40	1.303	1.684	2.021	2.423	2.704	3.307	3.551
60	1.296	1.671	2.000	2.390	2.660	3.232	3.460
120	1.289	1.658	1.980	2.358	2.617	3.160	3.373
∞	1.282	1.645	1.960	2.326	2.576	3.090	3.291

7 Present value factors

The table gives the present value of a single payment received n years in the future discounted at $x\%$ per year. For example, with a discount rate of 7% a single payment of £1 in six years time has a present value of £0.6663 or 66.63p.

Years	1%	2%	3%	4%	5%	6%	7%	8%	9%	10%	11%	12%	13%	14%	15%	16%	17%	18%	19%	20%	Years
1	.9901	.9804	.9709	.9615	.9524	.9434	.9346	.9259	.9174	.9091	.9009	.8929	.8850	.8772	.8696	.8621	.8547	.8475	.8403	.8333	1
2	.9803	.9612	.9426	.9426	.9070	.8900	.8734	.8573	.8417	.8264	.8116	.7972	.7831	.7695	.7561	.7432	.7305	.7182	.7062	.6944	2
3	.9706	.9423	.9151	.8890	.8638	.8396	.8163	.7938	.7722	.7513	.7312	.7118	.6931	.6750	.6575	.6407	.6244	.6086	.5934	.5787	3
4	.9610	.9238	.8885	.8548	.8227	.7921	.7629	.7350	.7084	.6830	.6587	.6355	.6133	.5921	.5718	.5523	.5337	.5158	.4987	.4823	4
5	.9515	.9057	.8626	.8219	.7835	.7473	.7130	.6806	.6499	.6209	.5935	.5674	.5428	.5194	.4972	.4761	.4561	.4371	.4190	.4019	5
6	.9420	.8880	.8375	.7903	.7462	.7050	.6663	.6302	.5963	.5645	.5346	.5066	.4803	.4556	.4323	.4104	.3898	.3704	.3521	.3349	6
7	.9327	.8706	.8131	.7599	.7107	.6651	.6227	.5835	.5470	.5132	.4817	.4523	.4251	.3996	.3759	.3538	.3332	.3139	.2959	.2791	7
8	.9235	.8535	.7894	.7307	.6768	.6274	.5820	.5403	.5019	.4665	.4339	.4039	.3762	.3506	.3269	.3050	.2848	.2660	.2487	.2326	8
9	.9143	.8368	.7664	.7026	.6446	.5919	.5439	.5002	.4604	.4241	.3909	.3606	.3329	.3075	.2843	.2630	.2434	.2255	.2090	.1938	9
10	.9053	.8203	.7441	.6756	.6139	.5584	.5083	.4632	.4224	.3855	.3522	.3220	.2946	.2697	.2472	.2267	.2080	.1911	.1756	.1615	10
11	.8963	.8043	.7224	.6496	.5847	.5268	.4751	.4289	.3875	.3505	.3173	.2875	.2607	.2366	.2149	.1954	.1778	.1619	.1476	.1346	11
12	.8874	.7885	.7014	.6246	.5568	.4970	.4440	.3971	.3555	.3186	.2858	.2567	.2307	.2076	.1869	.1685	.1520	.1372	.1240	.1122	12
13	.8787	.7730	.6810	.6006	.5303	.4688	.4150	.3677	.3262	.2897	.2575	.2292	.2042	.1821	.1625	.1452	.1299	.1163	.1042	.0935	13
14	.8700	.7579	.6611	.5775	.5051	.4423	.3878	.3405	.2992	.2633	.2320	.2046	.1807	.1597	.1413	.1252	.1110	.0985	.0876	.0779	14
15	.8613	.7430	.6419	.5553	.4810	.4173	.3624	.3152	.2745	.2394	.2090	.1827	.1599	.1401	.1229	.1079	.0949	.0835	.0736	.0649	15
16	.8528	.7284	.6232	.5339	.4581	.3936	.3387	.2919	.2519	.2176	.1883	.1631	.1415	.1229	.1069	.0930	.0811	.0708	.0618	.0541	16
17	.8444	.7142	.6050	.5134	.4363	.3714	.3166	.2703	.2311	.1978	.1696	.1456	.1252	.1078	.0929	.0802	.0693	.0600	.0520	.0451	17
18	.8360	.7002	.5874	.4936	.4155	.3503	.2959	.2502	.2120	.1799	.1528	.1300	.1108	.0946	.0808	.0691	.0592	.0508	.0437	.0376	18
19	.8277	.6864	.5703	.4746	.3957	.3305	.2765	.2317	.1945	.1635	.1377	.1161	.0981	.0829	.0703	.0596	.0506	.0431	.0367	.0313	19
20	.8195	.6730	.5537	.4564	.3769	.3118	.2584	.2145	.1784	.1486	.1240	.1037	.0868	.0728	.0611	.0514	.0433	.0365	.0308	.0261	20
21	.8114	.6598	.5375	.4388	.3589	.2942	.2415	.1987	.1637	.1351	.1117	.0926	.0768	.0638	.0531	.0443	.0370	.0309	.0259	.0217	21
22	.8034	.6468	.5219	.4220	.3418	.2775	.2257	.1839	.1502	.1228	.1007	.0826	.0680	.0560	.0462	.0382	.0316	.0262	.0218	.0181	22
23	.7954	.6342	.5067	.4057	.3256	.2618	.2109	.1703	.1378	.1117	.0907	.0738	.0601	.0491	.0402	.0329	.0270	.0222	.0183	.0151	23
24	.7876	.6217	.4919	.3901	.3101	.2470	.1971	.1577	.1264	.1015	.0817	.0659	.0532	.0431	.0349	.0284	.0231	.0188	.0154	.0126	24
25	.7798	.6095	.4776	.3751	.2953	.2330	.1842	.1460	.1160	.0923	.0736	.0588	.0471	.0378	.0304	.0245	.0197	.0160	.0129	.0105	25
26	.7720	.5976	.4637	.3607	.2812	.2198	.1722	.1352	.1064	.0839	.0663	.0525	.0417	.0331	.0264	.0211	.0169	.0135	.0109	.0087	26
27	.7644	.5859	.4502	.3468	.2678	.2074	.1609	.1252	.0976	.0763	.0597	.0469	.0369	.0291	.0230	.0182	.0144	.0115	.0091	.0073	27
28	.7568	.5744	.4371	.3335	.2551	.1956	.1504	.1159	.0895	.0693	.0538	.0419	.0326	.0255	.0200	.0157	.0123	.0097	.0077	.0061	28
29	.7493	.5631	.4243	.3207	.2429	.1846	.1406	.1073	.0822	.0630	.0485	.0374	.0289	.0224	.0174	.0135	.0105	.0082	.0064	.0051	29
30	.7419	.5521	.4120	.3083	.2314	.1741	.1314	.0994	.0754	.0573	.0437	.0334	.0256	.0196	.0151	.0116	.0090	.0070	.0054	.0042	30
31	.7346	.5412	.4000	.2965	.2204	.1643	.1228	.0920	.0691	.0521	.0394	.0298	.0226	.0172	.0131	.0100	.0077	.0059	.0046	.0035	31
32	.7273	.5306	.3883	.2851	.2099	.1550	.1147	.0852	.0634	.0474	.0355	.0266	.0200	.0151	.0114	.0087	.0066	.0050	.0038	.0029	32
33	.7201	.5202	.3770	.2741	.1999	.1462	.1072	.0789	.0582	.0431	.0319	.0238	.0177	.0132	.0099	.0075	.0056	.0042	.0032	.0024	33
34	.7130	.5100	.3660	.2636	.1904	.1379	.1002	.0730	.0534	.0391	.0288	.0212	.0157	.0116	.0086	.0064	.0048	.0036	.0027	.0020	34
35	.7059	.5000	.3554	.2534	.1813	.1301	.0937	.0676	.0490	.0356	.0259	.0189	.0139	.0102	.0075	.0055	.0041	.0030	.0023	.0017	35
36	.6989	.4902	.3450	.2437	.1727	.1227	.0875	.0626	.0449	.0323	.0234	.0169	.0123	.0089	.0065	.0048	.0035	.0026	.0019	.0014	36
37	.6920	.4806	.3350	.2343	.1644	.1158	.0818	.0580	.0412	.0294	.0210	.0151	.0109	.0078	.0057	.0041	.0030	.0022	.0016	.0012	37
38	.6852	.4712	.3252	.2253	.1566	.1092	.0765	.0537	.0378	.0267	.0190	.0135	.0096	.0069	.0049	.0036	.0026	.0019	.0013	.0010	38
39	.6784	.4619	.3158	.2166	.1491	.1031	.0715	.0497	.0347	.0243	.0171	.0120	.0085	.0060	.0043	.0031	.0022	.0016	.0011	.0008	39
40	.6717	.4529	.3066	.2083	.1420	.0972	.0668	.0460	.0318	.0221	.0154	.0107	.0075	.0053	.0037	.0026	.0019	.0013	.0010	.0007	40
41	.6650	.4440	.2976	.2003	.1353	.0917	.0624	.0426	.0292	.0201	.0139	.0096	.0067	.0046	.0032	.0023	.0016	.0011	.0008	.0006	41
42	.6584	.4353	.2890	.1926	.1288	.0865	.0583	.0395	.0268	.0183	.0125	.0086	.0059	.0041	.0028	.0020	.0014	.0010	.0007	.0005	42
43	.6519	.4268	.2805	.1852	.1227	.0816	.0545	.0365	.0246	.0166	.0112	.0076	.0052	.0036	.0025	.0017	.0012	.0008	.0006	.0004	43
44	.6454	.4184	.2724	.1780	.1169	.0770	.0509	.0338	.0226	.0151	.0101	.0068	.0046	.0031	.0021	.0015	.0010	.0007	.0005	.0003	44
45	.6391	.4102	.2644	.1712	.1113	.0727	.0476	.0313	.0207	.0137	.0091	.0061	.0041	.0027	.0019	.0013	.0009	.0006	.0004	.0003	45
46	.6327	.4022	.2567	.1646	.1060	.0685	.0445	.0290	.0190	.0125	.0082	.0054	.0036	.0024	.0016	.0011	.0007	.0005	.0003	.0002	46
47	.6265	.3943	.2493	.1583	.1009	.0647	.0416	.0269	.0174	.0113	.0074	.0049	.0032	.0021	.0014	.0009	.0006	.0004	.0003	.0002	47
48	.6203	.3865	.2420	.1522	.0961	.0610	.0389	.0249	.0160	.0103	.0067	.0043	.0028	.0019	.0012	.0008	.0005	.0004	.0002	.0002	48
49	.6141	.3790	.2350	.1463	.0916	.0575	.0363	.0230	.0147	.0094	.0060	.0039	.0025	.0016	.0011	.0007	.0005	.0003	.0002	.0001	49
50	.6080	.3715	.2281	.1407	.0872	.0543	.0339	.0213	.0134	.0085	.0054	.0035	.0022	.0014	.0009	.0006	.0004	.0003	.0002	.0001	50

(continued)

Present value factors (*continued*)

Years	21%	22%	23%	24%	25%	26%	27%	28%	29%	30%
1	.8264	.8197	.8130	.8065	.8000	.7937	.7874	.7813	.7752	.7692
2	.6830	.6719	.6610	.6504	.6400	.6299	.6200	.6104	.6009	.5917
3	.5645	.5507	.5374	.5245	.5120	.4999	.4882	.4768	.4658	.4552
4	.4665	.4514	.4369	.4230	.4096	.3968	.3844	.3725	.3611	.3501
5	.3855	.3700	.3552	.3411	.3277	.3149	.3027	.2910	.2799	.2693
6	.3186	.3033	.2888	.2751	.2621	.2499	.2383	.2274	.2170	.2072
7	.2633	.2486	.2348	.2218	.2097	.1983	.1877	.1776	.1682	.1594
8	.2176	.2038	.1909	.1789	.1678	.1574	.1478	.1388	.1304	.1226
9	.1799	.1670	.1552	.1443	.1342	.1249	.1164	.1084	.1011	.0943
10	.1486	.1369	.1262	.1164	.1074	.0992	.0916	.0847	.0784	.0725
11	.1228	.1122	.1026	.0938	.0859	.0787	.0721	.0662	.0607	.0558
12	.1015	.0920	.0834	.0757	.0687	.0625	.0568	.0517	.0471	.0429
13	.0839	.0754	.0678	.0610	.0550	.0496	.0447	.0404	.0365	.0330
14	.0693	.0618	.0551	.0492	.0440	.0393	.0352	.0316	.0283	.0254
15	.0573	.0507	.0448	.0397	.0352	.0312	.0277	.0247	.0219	.0195
16	.0474	.0415	.0364	.0320	.0281	.0248	.0218	.0193	.0170	.0150
17	.0391	.0340	.0296	.0258	.0225	.0197	.0172	.0150	.0132	.0116
18	.0323	.0279	.0241	.0208	.0180	.0156	.0135	.0118	.0102	.0089
19	.0267	.0229	.0196	.0168	.0144	.0124	.0107	.0092	.0079	.0068
20	.0221	.0187	.0159	.0135	.0115	.0098	.0084	.0072	.0061	.0053
21	.0183	.0154	.0129	.0109	.0092	.0078	.0066	.0056	.0048	.0040
22	.0151	.0126	.0105	.0088	.0074	.0062	.0052	.0044	.0037	.0031
23	.0125	.0103	.0086	.0071	.0059	.0049	.0041	.0034	.0029	.0024
24	.0103	.0085	.0070	.0057	.0047	.0039	.0032	.0027	.0022	.0018
25	.0085	.0069	.0057	.0046	.0038	.0031	.0025	.0021	.0017	.0014
26	.0070	.0057	.0046	.0037	.0030	.0025	.0020	.0016	.0013	.0011
27	.0058	.0047	.0037	.0030	.0024	.0019	.0016	.0013	.0010	.0008
28	.0048	.0038	.0030	.0024	.0019	.0015	.0012	.0010	.0008	.0006
29	.0040	.0031	.0025	.0020	.0015	.0012	.0010	.0008	.0006	.0005
30	.0033	.0026	.0020	.0016	.0012	.0010	.0008	.0006	.0005	.0004
31	.0027	.0021	.0016	.0013	.0010	.0008	.0006	.0005	.0004	.0003
32	.0022	.0017	.0013	.0010	.0008	.0006	.0005	.0004	.0003	.0002
33	.0019	.0014	.0011	.0008	.0006	.0005	.0004	.0003	.0002	.0002
34	.0015	.0012	.0009	.0007	.0005	.0004	.0003	.0002	.0002	.0001
35	.0013	.0009	.0007	.0005	.0004	.0003	.0002	.0002	.0001	.0001
36	.0010	.0008	.0006	.0004	.0003	.0002	.0002	.0001	.0001	.0001
37	.0009	.0006	.0005	.0003	.0003	.0002	.0001	.0001	.0001	.0001
38	.0007	.0005	.0004	.0003	.0002	.0002	.0001	.0001	.0001	
39	.0006	.0004	.0003	.0002	.0002	.0001	.0001	.0001	.0000	
40	.0005	.0004	.0003	.0002	.0001	.0001	.0001	.0001		
41	.0004	.0003	.0002	.0001	.0001	.0001	.0001			
42	.0003	.0002	.0002	.0001	.0001	.0001				
43	.0003	.0002	.0001	.0001	.0001					
44	.0002	.0002	.0001	.0001	.0001					
45	.0002	.0001	.0001	.0001						
46	.0002	.0001	.0001	.0001						
47	.0001	.0001	.0001							
48	.0001	.0001								
49	.0001	.0001								
50	.0001									

Years	31%	32%	33%	34%	35%	36%	37%	38%	39%	40%
1	.7634	.7576	.7519	.7463	.7407	.7353	.7299	.7246	.7194	.7143
2	.5827	.5739	.5653	.5569	.5487	.5407	.5328	.5251	.5176	.5102
3	.4448	.4348	.4251	.4156	.4064	.3975	.3889	.3805	.3724	.3644
4	.3396	.3294	.3196	.3102	.3011	.2923	.2839	.2757	.2679	.2603
5	.2592	.2495	.2403	.2315	.2230	.2149	.2072	.1998	.1927	.1859
6	.1979	.1890	.1807	.1727	.1652	.1580	.1512	.1448	.1386	.1328
7	.1510	.1432	.1358	.1289	.1224	.1162	.1104	.1049	.0997	.0949
8	.1153	.1085	.1021	.0962	.0906	.0854	.0806	.0760	.0718	.0678
9	.0880	.0822	.0768	.0718	.0671	.0628	.0588	.0551	.0516	.0484
10	.0672	.0623	.0577	.0536	.0497	.0462	.0429	.0399	.0371	.0346
11	.0513	.0472	.0434	.0400	.0368	.0340	.0313	.0289	.0267	.0247
12	.0392	.0357	.0326	.0298	.0273	.0250	.0229	.0210	.0192	.0176
13	.0299	.0271	.0245	.0223	.0202	.0184	.0167	.0152	.0138	.0126
14	.0228	.0205	.0185	.0166	.0150	.0135	.0122	.0110	.0099	.0090
15	.0174	.0155	.0139	.0124	.0111	.0099	.0089	.0080	.0072	.0064
16	.0133	.0118	.0104	.0093	.0082	.0073	.0065	.0058	.0051	.0046
17	.0101	.0089	.0078	.0069	.0061	.0054	.0047	.0042	.0037	.0033
18	.0077	.0068	.0059	.0052	.0045	.0039	.0035	.0030	.0027	.0023
19	.0059	.0051	.0044	.0038	.0033	.0029	.0025	.0022	.0019	.0017
20	.0045	.0039	.0033	.0029	.0025	.0021	.0018	.0016	.0014	.0012
21	.0034	.0029	.0025	.0021	.0018	.0016	.0013	.0012	.0010	.0009
22	.0026	.0022	.0019	.0016	.0014	.0012	.0010	.0008	.0007	.0006
23	.0020	.0017	.0014	.0012	.0010	.0008	.0007	.0006	.0005	.0004
24	.0015	.0013	.0011	.0009	.0007	.0006	.0005	.0004	.0004	.0003
25	.0012	.0010	.0008	.0007	.0006	.0005	.0004	.0003	.0003	.0002
26	.0009	.0007	.0006	.0005	.0004	.0003	.0003	.0002	.0002	.0002
27	.0007	.0006	.0005	.0004	.0003	.0002	.0002	.0002	.0001	.0001
28	.0005	.0004	.0003	.0003	.0002	.0002	.0001	.0001	.0001	.0001
29	.0004	.0003	.0003	.0002	.0002	.0001	.0001	.0001	.0001	.0001
30	.0003	.0002	.0002	.0002	.0001	.0001	.0001	.0001	.0001	
31	.0002	.0002	.0001	.0001	.0001	.0001	.0001			
32	.0002	.0001	.0001	.0001	.0001	.0001				
33	.0001	.0001	.0001	.0001	.0001					
34	.0001	.0001	.0001							
35	.0001	.0001								
36	.0001									

Glossary

Absolute error – *See* Error.

Annual percentage rate (APR) – Rate of interest, calculated annually, equivalent to given rate of interest.

Annuity – Series of constant cash flows received over a period of years.

Arithmetic progression (AP) – A series of numbers such that consecutive terms have a common difference: $a + (a + d) + (a + 2d) + \dots$.

Bias – Systematic misrepresentation of a statistical result, as distinct from chance misrepresentations which one would expect to cancel each other out.

Capital costs – Cost to a company of committing capital to a project.

Census – Complete enumeration of a population.

Central Limit Theorem – The theorem that states 'as long as a sample drawn from a population is reasonably large, the sample means will have a normal distribution'.

Class interval – A subdivision within a frequency distribution.

Cluster sampling – *See* Sample design.

Conditional probability – The probability of an event occurring, conditional on another event having occurred.

Confidence interval – An interval between which a predetermined percentage of sample statistics fall.

Constant – A value that is fixed.

Constraint – Limitation (of resources), e.g. a limit of 200 hours of labour is a constraint on how much labour can be utilized; *see* Linear programming.

Continuous variable – *See* Variable.

Correlation – Interdependence between variables. The degree to which change in one variable is related to change in another.

Cyclical variation – Variation inherent in data that follows a cyclical (seasonal) pattern; *see* Time series analysis.

Data – Information available or collected, usually for some form of statistical analysis.

1. **Primary data** – Information collected at first hand for some prescribed purpose.
2. **Secondary data** – Information already in existence that is utilized, adopted or adapted for current purposes.
3. **Qualitative data** – Data that reflects qualitative aspects of a population which cannot be expressed numerically, qualities such as 'eye colour' and 'political allegiance'. This is in contrast to quantitative data.
4. **Quantitative data** – Data that can be measured; e.g. height, age, annual salary, diameter of a component, etc.

Decision criterion – Rule for deciding between competing projects.

Decision rule – Rule for deciding on acceptance or rejection of a hypothesis.

Dependent variable – *See* Variable.

Discounted cash flow (DCF) – Cash flow represented in present value terms; *see* Net present value.

Discrete variable – *See* Variable.

Economic batch quantity (EBQ) or **Economic order quantity (EOQ)** – The batch (or order) quantity which minimizes the total variable costs in a stock-control problem.

Error – Mistake in recording/observing data, which may be **unbiased** (over a large sample the effects of such an error would be cancelled out) or **biased** (over a large sample the error would be magnified). Error is referred to in a number of ways:

1. **Absolute error** – The difference between the value of the recorded information and the value of the actual information.
2. **Relative error** – The absolute error represented as a percentage of the actual value.

Event – An outcome of interest in probability problems.

Expected monetary value (EMV) – The average earnings attached to a project, when the project has a number of uncertain outcomes, each with a monetary value.

Extrapolation – Process of deducing a value outside of the range of known data.

Feasible region – That region (often represented graphically) representing the values of a mix of decision variables in linear programming which does not contravene any of a number of constraints; *see* Linear programming.

Frequency distribution – The distribution of the number of times a variable value occurs (e.g. the distribution of the number of occurrences of an event or measurement, or the number of members of a population in specified classes).

Frequency table – The tabulated summary of values taken from a frequency distribution.

Function – A quantity (dependent variable) whose value is determined by the value of another quantity (independent variable); *see* Variable.

Geometric progression (GP) – A series of values in which the difference between consecutive values is a common ratio: $a + ar + ar^2 + ar^3 + \dots$

Holding cost – The annual cost of keeping a unit of merchandise (or units of merchandise, depending on how defined) in stock.

Hypothesis test – In trying to reach decisions relating to a population, it is sometimes necessary to make assumptions or guesses about the population. Such assumptions are called statistical hypotheses. By observing the characteristics of a sample taken from the population in order to test the validity of our hypothesis, we are carrying out a hypothesis test.

Immediate replenishment – The assumption that the time lag between ordering stock and receiving stock is zero.

Independent events – Events such that the outcome of one does not affect the outcome of the other.

Independent variable – *See* Variable.

Index numbers – Numbers which measure the magnitude of economic change over a period of time. They can be calculated relative to the **base period**, an initial point in time, or the **current period.**

Internal rate of return (IRR) – The discount rate that equates the present value of future cash flows with the capital cost of investment.

Interpolation – Process of finding a value within the range of known data.

Lead time – The time lag between placing an order and receiving the goods.

Least-squares method–Method of finding the **line-of-best-fit**, such that the sum of squares of the differences between actual data values and corresponding estimates is minimized.

Linear programming–The method whereby a decision is reduced to an objective function which must be optimized subject to a number of linear constraints (or inequalities).

Measure of centre–A statistic representative of the centre or middle of a data set; an **average**; e.g. the mean, the mode, the median, the geometric mean, etc.

Measure of dispersion–A statistic representative of the way a data set is spread or dispersed; e.g. the range, the mean deviation, the inter-quartile range, the standard deviation, etc.

Moving average–A technique involving the calculation of consecutive averages over time to establish the trend of a time series.

Multi-stage sampling–*See* Sample design.

Mutually exclusive events–Events such that the occurrence of one excludes the possibility of the other occurring.

Net present value (**NPV**)–A cash sum to be received or paid at some time in the future discounted at a given rate (of interest or inflation, etc), giving the present value of the money.

Payback period–The time it takes an investment to return the original capital outlay.

Pilot survey–A 'test' survey used to establish the validity of questions and/or questioning techniques to be used in a planned (wider) survey.

Point estimate–An estimate of a population parameter (sample mean, sample standard deviation, etc) as opposed to an interval estimate; *see* Confidence interval.

Population–Any finite or infinite collection of individuals, measurements, etc (not necessarily human), defined by some characteristic common to these individuals, e.g. measurements, etc.

Present value–*See* Net present value.

Primary data–*See* Data.

Product mix–Term used in linear programming (and elsewhere) to describe the balance of production between competing products relative to certain constraints.

Qualitative data–*See* Data.

Quantitative data–*See* Data.

Quota sampling–*See* Sample design.

Regression–A method of investigating the relationship between variables, by (approximately) expressing that relationship as a (linear) function.

Sample–Collection of information (ideally) representative of the population from which it was taken.

Sample design–Most statistical information is obtained by picking a representative part of the population–a sample. How we select a sample is the concern of sample design. Some examples follow:

1 **Simple random sampling**–A sample whereby every member of the population from which the sample is chosen has an equal chance of being chosen.
2 **Systematic sampling**–In practice it is often difficult or impossible to obtain a true random sample as this requires a good sampling frame. If the population is homogeneous a systematic sample can be taken (accountants checking invoices for errors is such a case). The method is to decide on the sample size (e.g. $2\frac{1}{2}\%$ of the population) and to choose at random a starting point in the population (the

population must be open to numbering). For example, if the population is 100 000, a sample of 2½% would be 2500, every 40th item. A randomly chosen starting point could be 29. The sample chosen would be the items numbered 29, 69, 109, 149, 189, 229. . .

3 **Stratified sampling**–If we wish to guarantee that a sample from a population contains elements from a number of subsets of that population, we need to identify the subsets (called *strata*) before sampling. For example, an enquiry concerning the voting intentions of a group of people might require the sample to contain male respondents, female respondents, respondents of a particular age, etc. The chosen population strata are identified and the sample must reflect the population strata (e.g. if the population contains 20% of people in the age range 20-35, so must the sample).

4 **Multi-stage sampling**–This is similar to stratified sampling but the division is usually geographical. For instance, if 20% of people live in small villages, 30% live in towns and 50% live in cities, then choosing two villages, three towns and five cities will reflect the whole population. A random method is decided upon to choose the people from the 10 chosen areas, possibly using a stratified sample.

5 **Cluster sampling**–This method was devised in the USA to cut down on cost and to operate where there is no satisfactory sampling frame. It involves the random selection of a number of sub-areas from a large area and a 'blanket' survey of these sub-areas (rather than a random sample spread across the whole area). Problems include the possible 'sameness' of chosen sub-areas, resulting in sample bias.

6 **Quota sampling**–Quota sampling is commonly used by market researchers. It usually involves stratifying samples and restricting the sample to a fixed number in each strata. The interviewer controls the choice of sample respondents.

Sampling frame–The list of the population from which one wishes to take a sample. (*Note*: It is not always possible to list a population.)

Secondary data–*See* Data.

Simple random sampling–*See* Sample design.

Stratified sampling–*See* Sample design.

Systematic sampling–*See* Sample design.

Time series–Values of a single variable presented at equidistant points in time (e.g. each month, quarter, etc).

Trend–A long-term movement in a time series.

Variable–Any observable phenomenon that can be measured and that takes on different values (e.g. temperature, shoe size, size of family, etc).

1 **Continuous variable**–A variable that (theoretically) can assume any value between two given values (e.g. temperature).

2 **Dependent variable**–A variable whose variations depend upon (or can be shown to be dependent upon) another specified variable or variables.

3 **Discrete variable**–A variable that can only take discrete, or whole number values (e.g. family size).

4 **Independent variable**–A variable that varies independently of another specified variable or variables.

Index

absolute errors 172
accounting rate of return (ARR) 189
accuracy of data 11, 21-2
  see also errors
acquisition of data *see* collection; questionnaires; sampling
addition law for probabilities
  general 115-16, 135
  simple 113-14
addition of matrices 161
additive model of time series 80, 92
alternative hypotheses 145
annual percentage rate (APR) 182
annuities 187
appraisal, investment 187-97
arithmetic mean 49-50, 71, 72, 73
arithmetic progression (AP) 154-5
  summation 155-6
average errors 173
averages, moving 81-4, 93

bar charts 30-4
base periods 62
Bayes' Theorem 120-1, 136, 139
Bessel's correction factor 60, 72, 144
bias
  rounding of data 172, 173
  sampling 14-15, 25
bimodal data 50
binomial distribution 124-6, 135
  normal approximation to 134-5
  Poisson approximation to 127-8
binomial probability table 228-31
bivariate data 101

calculators 6
calculus *see* differentiation; integration; maximum points; minimum points; points of inflection; stationary points
capital costs 219-20
capital recovery factors 185
census sampling 14
Central Limit Theorem 140-2
centering trends 83-4
centre, measures of 49-54
chain based indices 68-9
chance nodes 121-3
classification of data 26-7
cluster sampling 17
collection of data 18-19
  see also questionnaires; sampling
column vectors 161
combinations 124
complementary events 113
component bar charts 30-1, 41-2, 44
compound interest 182, 196, 197
conditional probability 117, 138-9
confidence intervals 142-3, 149, 151
constants of integration 171
constraints (in linear programming) 198, 199, 200-2, 203
  mutually exclusive 205-6
continuous replenishment (stock control with) 220-1
continuous variables 13
correlation 96, 107, 108, 109
  coefficient 100-5, 106, 107
cost minimization 170-1
  linear programming 199-200, 204-5, 207-8, 213-14
critical ratio 147, 152
cumulative frequencies 29
  curves 37
cyclical variation 78, 79, 80

data
  bivariate 101
  classification 26-7
  collection 18-19
  qualitative/quantitative 13
    see also questionnaires; sampling
  deseasonalization 90-2
  errors in 21-2, 172-3, 179-80
  presentation *see* presentation
  primary 11, 13-14
  rounding 86, 172-3
  secondary 11, 12-13
  sources 11-13
decision making 117-23
  hypothesis testing 147-8
decision trees 121-3
deflating a series 69-70
degrees of freedom 148
dependent variables 96-7
depreciation 183-5
deseasonalization 90-2
determination, coefficient of 103, 107
deviation
  mean 58-9
  standard 59-61, 72, 73, 126
differentiation 167-71, 177-8
discounting 186, 194-5
discounts, stock control with 218-19
discrete variables 13
dispersion 54-61

economic order/batch quantity (EOQ/EBQ) 217, 219, 221, 223
model 215-18
equations
  linear 157, 174
    line of best fit 98, 106-7
  quadratic 158-60, 174-5
  simultaneous 157-8, 164-5, 178
errors 21-2, 172-3, 179-80
  sampling 147-8
exhaustive sampling 14
expected values 118-19

expenditure index 64-5
extrapolation 86-7, 89, 99, 107

factorial notation 124
factorization (of quadratic equations) 159
feasible region (in linear programming) 202, 203
Fisher's index 67-8
fixed interest stock 187
forecasting
by extrapolation 86-7, 89, 99, 107
from line of best fit 98-9, 101, 107, 109
random variation of residuals 89-90
freedom, degrees of 148
frequency distributions 28-9, 46-7
see also binomial distribution; normal distribution; Poisson distribution
frequency polygons 34

geometric mean 53-4
geometric progression (GP) 156, 174
summation 156-7
gradients 165-6
see also differentiation
graphs
linear programmes 200-5
mode and median 52-3
representation of data 36-40
grouped data 28-9
arithmetic mean 50
median 51-2
grouped frequencies 28-9

harmonic mean 54
histograms 32-4, 45
holding costs 216
errors 222
variable 219-20
hypothesis testing 144-7, 149-50, 152
identity matrix 162
independent events 114-15
independent variables 96-7
index numbers 62-70, 75-6
chain based 68-9
index points 63
indexation 73
integration 171, 177
interest 181-2, 196, 197
internal rate of return (IRR) 191-3
interpolation 99
interquartile range 57-8
interval estimates 142-3, 149, 151
interviews 18-19
inverse matrices 162-4
investment 181-97
appraisal 187-97
increasing/decreasing sum 183
see also reducing balances
sinking fund method 184-5

Laspeyres' price index 65-6
Laspeyres' volume index 67
least squares 105-7, 109, 110
level of significance 146
line charts 34
linear equations 157, 174
linear programming 198-214
lines of best fit 98, 101, 105-7, 109
loans 185
Lorenz curves 37-9

market price calculations 187
matrices 160-5, 177
inverse 162-4
multiplication 161-4, 178
maximax decision criteria 118
maximin decision criteria 118
maximization 168-70, 175-6
linear programming 199, 200-3, 208-10
maximum points 168-70
maximum stock level 220-1
mean
arithmetic 49-50, 71, 72, 73
geometric 53-4
harmonic 54
population 49
probability distributions 126
sample 49, 144-7
mean deviation 58-9
measures of centre 49-54
measures of dispersion 54-61
measures of skewness 61-2, 74-5
median 51-3, 71
minimax decision criteria 118
minimization 170-1
linear programming 199-200, 204-5, 207-8, 213-14
minimum points 168-70
mode 50, 52-3, 71
money, time value of 181-97
mortgages 185
moving averages 81-4, 93
multi-item index numbers 63-4
multiple bar charts 32
multiplication law for probabilities
general 116-17
simple 114-15
multiplication of matrices 161-4, 178
multiplicative model of time series 88-9, 91
multi-stage sampling 17
mutual exclusiveness (of events) 114
mutually exclusive constraints (in linear programming) 205-6

net present value (NPV) 189-91, 192, 193-4, 197
see also present value
normal distribution 128-35, 149, 150
standard 130-2, 138, 235-6
standardization 132-3
tables 236, 237
null hypotheses 145-8, 149
objective functions (in linear programming) 198
optimization 202-3, 203-4
observation 18
ogives 37
one-tailed tests 145, 146, 147
ordering costs 216

Paasche price index 66

Paasche volume index 67
payback period 188-9, 196
Pearson's coefficient of skewness 62, 74
Pearson's product moment correlation coefficient 101-2, 107
percentage bar charts 30
permutations 123-4
pictograms 35, 43-4
pictorial representation of data 29-35, 41-5
pie charts 34, 42, 43-4
pilot surveys 25
point estimates 142
points of inflection 168-70
Poisson distribution 126-8
 normal approximation to 134-5
Poisson probability table 232-5
population 13-14, 60, 141-2
population mean 49
population proportion 144, 152
population variance 144
postal questionnaires 19
posterior probabilities 120-1
present value factors 186
 table 238
present values 185-7
 net 189-91, 192, 193-4, 197
presentation of data 27-47
 graphs 36-40
 pictorial 29-35, 41-5
 tables 27-9
price indices 62-3, 65-6, 74
 deflating series using 69-70
price relatives 65-6
primary data 11, 13-14
prior probabilities 120
probability 112-17
 distributions *see* binomial distribution; normal distribution; Poisson distribution
profit maximization 168-70, 175-6
 linear programming 199, 200-3, 209-10
profitability index 191
progression 154-7, 174

quadratic equations 158-60, 174-5
qualitative data 13
quantitative data 13
quantity indices 62-3, 67-8, 74
questionnaires 19-21
quota sampling 17-18

random number tables 15, 227
random sampling 15
random variation 78, 79, 80
 forecasting and 89-90
range 57
 interquartile 57-8
rates of change 165
 see also differentiation
reducing balances 183-4
regression 80, 98, 105-7
relative errors 172
residuals *see* random variation
Retail Price Index (RPI) 62, 63, 69
rounding of data 86, 172-3
row reduction 163-4

sample mean 49, 140-1
 testing 144-7
sample statistics 60-1
sampling 14-18, 22-5, 140-51
 bias 14-15, 25
sampling distribution 140, 141
sampling frame 14, 23
scalar multiplication of matrices 161
scatter diagrams (scatter graphs) 36-7
 correlation 96-7, 107
seasonal variation 78, 79, 80
 calculation of 84-6, 93-4
 future 87
 removal of 90-2
second derivatives 169-70
secondary data 11, 12-13
sensitivity analysis 206, 222
sequences 154
 see also progressions
series 154
 see also progressions
set-up cost error 222
significance level 146
simple addition law for probabilities 114
simple aggregate index 63-4
simple events 113
simple interest 181-2, 196
simple multiplication law for probabilities 114-15
simultaneous equations 157-8, 178
 matrix solutions to 164-5
single item index numbers 62-3
singular matrices 164
sinking fund method 184-5
size of sample 18, 143-4, 148-9
skewness 61-2, 74-5
small samples 148-9
source of data 11-13
Spearman's rank correlation coefficient 104-5, 107
standard deviation 59-61, 72, 73
 probability distributions 126
standard error 140, 142
standard normal distribution 130-2, 138
 tables 236-7
standardization of normal distribution 132-3
stationary points 166, 168-70
statistical maps 35
statistical probability 112
stock control 215-25
straight line depreciation 183
strata graphs 39
stratified sample 16-17
subjective probability 113
sum of digits method 183
summation of progressions 155-6, 156-7
systematic sample 15

t distributions 148-9, 237
tables (use of in data presentation) 27-9
theoretical probability 112
time, data changes over 39-40
time-constrained tests 8

time period, moving averages and 81, 82-3
time series 39, 77-95
additive model 80, 92
multiplicative model 88-9, 91
time value of money 181-97
total variable costs 217-18, 222, 223, 224, 225
tree diagrams 116, 119-20, 136, 137
decision trees 121-3
trends 77-8, 79, 80
calculation of 80-4, 92, 93
future 86-7
trust funds 185
two-tailed tests 145, 146, 147

unbiased errors 172, 173
ungrouped frequency distribution 28
variables
continuous 13
dependent and independent 96-7
discreet 13
variable holding costs 219-20
variance 59-60, 72, 73
variation, coefficient of 72-3
volume indices 67-8

weighting 64

Z charts 40
'zero-intercept' 110